Hazards and Monitoring of Volcanic Activity 3

SCIENCES

Geoscience, Field Director – Yves Lagabrielle

Lithosphere–Asthenosphere Interactions,
Subject Head – René Maury

Hazards and Monitoring of Volcanic Activity 3

Gravimetric, Electric and Magnetic Fluids, Products and Methods

Coordinated by
Jean-François Lénat

iSTE

WILEY

First published 2022 in Great Britain and the United States by ISTE Ltd and John Wiley & Sons, Inc.

ISTE Ltd
27-37 St George's Road
London SW19 4EU
UK

www.iste.co.uk

John Wiley & Sons, Inc.
111 River Street
Hoboken, NJ 07030
USA

www.wiley.com

Library of Congress Control Number: 2022938792

British Library Cataloguing-in-Publication Data
A CIP record for this book is available from the British Library
ISBN 978-1-78945-046-0

ERC code:
PE10 Earth System Science
PE10_5 Geology, tectonics, volcanology
PE10_7 Physics of earth's interior, seismology, volcanology

Contents

Foreword

Claude JAUPART
Institut de physique du globe de Paris, Université de Paris,
Académie des sciences, Paris, France

Volcanoes are fascinating because of the beautiful landscapes they form and their superb eruptions. Many books have documented the most catastrophic eruptions, the different eruptive regimes and the physical mechanisms involved. Others have popularized tales of adventurers getting close to volcanic explosions and flows at the risk of their lives. In comparison, the discreet and dedicated work of volcanologists monitoring volcanoes has remained little known. The methods and techniques they use have improved greatly in recent decades and now allow them to predict eruptions with very few limitations. These advances are to be presented and explained in the three volumes of this book.

Volcanoes are built on top of a vast network of underground plumbing, and it is within this network that eruptions occur. Near the surface, the permeable rocks are gorged with water that circulates, heats up at depth and vaporizes under widely varying conditions. The result is a myriad of manifestations, including fumarole fields with changing flows, small earthquakes and ground deformation. This high background noise makes monitoring difficult. An eruption is preceded by the setting in motion of magma from one or more reservoirs located several kilometers underground. This onset is often very discrete and the associated signals are not easily distinguished from the background noise. Once the magma is near the surface, the signs are numerous and leave no room for doubt, but things can move very quickly and it is often too late to evacuate the area. Simply recognizing that an eruption is imminent is not enough: one must also assess its intensity and regime. Sometimes an eruption can even occur without magma and take the form of phreatic explosions, where the water contained in the surface rocks of the volcano vaporizes explosively. The volcanologist's work does not stop when the eruption begins; they must follow it over time and be able to distinguish between a temporary stop and its true end.

Faced with these multiple challenges, volcanologists have adopted methods that can be divided into two broad categories. The first is the historical study that reconstructs the past eruptions of a volcano and the time intervals between them. Knowing that an eruption covers the deposits of those that preceded it and destroys most of them, establishing a reliable chronology and estimating the volumes ejected rely on particular sampling strategies and frequent round trips between the field and the laboratory. New dating methods had to be developed to determine the ages of deposits older than a few tens of thousands of years. The second category covers all the physical and chemical methods used to determine the deep structure of a volcano and to locate the perturbations that it hosts. Rocks are difficult to penetrate and do not allow us to observe the reservoirs and conduits that feed eruptions. The information we obtain is indirect and often ambiguous. For example, small earthquakes are recorded, but they can be caused by the opening of cracks in a hydrothermal system or by magma that moves or by isolated landslides in places that are not easily accessible. An area of abnormal electrical conductivity can be detected, but it may be weathered rock or rock with water-filled fractures. Volcanologists have improved the uncertainty by combining several methods and have added to their toolbox over the years.

Remarkable progress has been made in the last four decades. Previously, the equipment available was limited to a few heavy and unwieldy devices designed for larger scale studies. Measurements have become much more precise, the number of sensors has increased enormously and the mathematical techniques of analysis have been refined. Well-instrumented observatories have been installed on many active volcanoes and in particular on the three volcanoes of the French national territory in Guadeloupe, Martinique and Reunion Island. Nowadays, a "typical" observatory maintains more than a hundred sensors of all kinds. The last two decades have seen the advent of very efficient satellite tools. However, it should not be inferred that volcanology has become the business of pure measurement experts or laboratory researchers. Knowledge of the special features of each volcano is needed to advance. The active volcano of Santorini in the Cyclades, which grows in the middle of a large caldera, is not monitored in the same way as Piton de la Fournaise on Reunion Island, which rises nearly 7 km above the sea floor and grows on the flanks of the ancient Piton des Neiges. It would be absurd to sample the deposits and install sensors randomly or only in easily accessible locations. Every volcanologist, whether geologist, geophysicist or geochemist, has studied their volcanoes for several years. This patient work has rarely been described. In this book, specialists from all the major disciplines of volcanology share their work and their discoveries. They explain how they decipher and interpret their measurements. One is likely to be surprised by the weakness of the signals detected, which can only be measured with sophisticated instruments, in comparison with the enormity of the eruptive phenomena. But it is thanks to these signals that we are able to travel to the very heart of volcanoes.

Preface

Jean-François LÉNAT
Laboratoire Magmas et Volcans, CNRS, IRD, OPGC,
Université Clermont Auvergne, Clermont-Ferrand, France

The impact of natural disasters has become a major concern of our modern societies. Volcanic eruptions, although statistically less deadly and causing less damage than earthquakes or certain atmospheric phenomena, can have devastating local or global effects.

The methods used to determine hazards related to volcanic activity and to monitor the latter are part of many Earth science curricula, both at master and thesis levels.

There are many publications in these areas, but the information is fragmented, requiring teachers to consult a large number of documents to develop their teaching. The aim of this book is to provide them with a single resource, written by specialists, on the methods of monitoring and determining hazards.

The subject is vast, which has led us to present it in three volumes. The first is devoted to geological and historical approaches. The next two are devoted to monitoring methods. The aim of each chapter is not to be encyclopedic. Rather, the intention is to provide the reader with the basic fundamentals of each of the topics covered. On the other hand, each author has taken care to provide bibliographic references that will allow readers to find the detailed information they may need.

This book deals with a scientific field that is constantly evolving. The progress in scientific concepts, approaches, observations and techniques has been spectacular during the last decades. There is no reason why this dynamic should slow down in the future. A logical consequence is that updates should be made periodically to avoid obsolescence of such a book. We therefore hope that it will be useful in the present period and that future editions will enable it to retain its value over time.

March 2022

List of Abbreviations

ACTIVE	Array of Controlled Transient-electromagnetics for Imaging a Volcanic Edifice
AMS	Anisotropic Magnetic Susceptibility
AMT	Audio-magnetotelluric
ASHER	Disdrometer dedicated to the detection of ash
CSD	Crystal Size Distribution
DOAS	Differential Optical Absorption Spectroscopy
FDEM	Frequency Domain Electromagnetics
FOAMS	Fast Objects Analysis and Measurement System
FTIR	Fourier Transform Infrared Spectroscopy
LOTEM	Long Offset Transient Electromagnetics
MC-ICPMS	Plasma source mass spectrometer
MDR	Mass Discharge Rate
MEMS	Microelectromechanical systems
MT	Magnetotelluric

MultiGas	MultiGas analysis system
PDC	Pyroclastic Density Currents
PSD	Particle Size Distribution
SEM	Scanning Electron Microscopy
SP	Spontaneous Polarization
TDEM	Time Domain Electromagnetic
TFV	Terminal Fall Velocity
TGSD	Total Grain Size Distribution
TIMS	Solid source mass spectrometer
VEI	Volcanic Explosivity Index
VLF	Very Low Frequency
VSD	Vesicle Size Distribution

1

Monitoring of Volcanic Fluids

Séverine MOUNE[1,2,3] and Émilie ROULLEAU[1,4]
[1] *Laboratoire Magmas et Volcans, CNRS, IRD, OPGC, Université Clermont Auvergne, Clermont-Ferrand, France*
[2] *OVSG, Gourbeyre, Guadeloupe*
[3] *Institut de physique du globe de Paris, Université de Paris, France*
[4] *Laboratoire GéoSciences Réunion, Université de La Réunion, Stratagem, Saint-Denis, France*

1.1. Introduction

The study of degassing is important because magmatic gases play a key role in the behavior of volcanoes. Magmatic degassing is one of the main manifestations of volcanic activity: most active volcanoes experience persistent degassing activity (whether open vent, fumarolean field or diffuse through the ground). Moreover, large quantities of gas are also released during more violent eruptive episodes. This phenomenon is central to many scientific questions in modern volcanology through its major implications for eruption dynamics (Wilson and Head 1981; Jaupart 1996, 1998), global geochemical cycles of volatile elements (Fischer 2008) and the environmental impact of volcanism (Oppenheimer et al. 2003, 2014; Mather 2015).

Since volcanic gases are generally more mobile than the silicate fluid from which they originate, they carry information to the surface that can tell us about the state of the magma system at depth, changes in physicochemical conditions in intermediate magma reservoirs and the path to the surface. These changes may be due to magma re-injections into the reservoir or conduit (Symonds et al. 1994; Goff et al. 1998; Allard et al. 2005), deep injections of CO_2-rich fluids (Spilliaert et al. 2006; Metrich and Wallace 2008; Moretti et al. 2018) or changes in internal gas

Hazards and Monitoring of Volcanic Activity 3,
coordinated by Jean-François LÉNAT.

pressure (Giggenbach 1996). The chemical composition of the gas phase allows us to trace the sources and processes of degassing and other phenomena (gas-rock interaction, boiling, evaporation, etc.) taking place at varying depths within the volcanic system.

However, studying volatile components in a wide range of emission conditions is always a real challenge. Their analysis can be done by sampling the gas phase itself (fumarole, plume), diffuse gases (gases that pass through the volcanic edifice and react with hydrothermal and/or meteoric waters), as well as by analyzing melt inclusions, since it is accepted that these magma droplets trapped in phenocrysts do not undergo degassing during magma ascent (Anderson 1974; Lowenstern 2003). In the last two decades or so, new perspectives have been opened with the development of remote sensing instruments for the gas phase (DOAS/FLYSPEC (Galle et al. 2003; Horton et al. 2006); OP-FTIR spectroscopy (Oppenheimer et al. 1998; Allard et al. 2005; Bernard et al. 2006); UV camera (Bluth et al. 2007); see Volume 2, Chapter 4), in situ measurements (MultiGas (Aiuppa et al. 2005a; Shinohara 2005)) as well as for the analysis of dissolved volatile components in magmas by electron microprobe, ion microprobe and vibrational micro-spectrometry (Métrich et al. 2004; Di Muro et al. 2006; Venugopal et al. 2020). This evolution of observation and analysis techniques as well as the understanding of degassing processes has led to significant advances in the monitoring of volcanoes via volcanic fluids.

This chapter will provide an overview of the usefulness of in situ measurements of volcanic gases as a tool for volcanological monitoring. Ground and satellite remote sensing techniques for gas monitoring are described and illustrated in Chapters 3 and 4 of Volume 2. After a brief introduction to basic concepts, the in situ techniques used to monitor the composition and fluxes of the main types of volcanic fluids will be described. Then, we will present an example of a volcanic fluid study that has contributed to a better understanding of degassing processes and volcano behavior. We will conclude by summarizing what we know about volcanic gases and by suggesting ways forward in terms of monitoring.

1.2. Composition, solubility and exsolution of magmatic gases

1.2.1. *Composition*

The chemical composition of volcanic gases shows significant variability. The factors controlling this variability are the composition of the source magmas, the chemical reactions in the gas phase itself, the geological and tectonic context, the solubility of the gases dissolved in the magma and the interaction, at shallow depths, with a hydrothermal system, groundwater or rainwater (see Figure 1.1). Despite this variability, a general trend shows that the gas phase consists of major compounds, which are in order of importance: H_2O (50–90% of total emissions), CO_2 (1–40%), SO_2 (1–40%), H_2S (1–10%), HCl (1–10%) (Textor et al. 2004), followed by other minor constituents like HF, HBr, N_2, He, Ar, Ne, CO, CH_4, H_2, etc. The volcanic gas phase contributes to the natural atmospheric cycles of many volatile elements (Aiuppa et al. 2003) which are injected in large quantities into the atmosphere during eruptions, thus contributing to atmospheric pollution. During paroxysmal eruptions, this can lead to environmental and climatic disturbances with significant socioeconomic consequences, such as during the Laki (Iceland) eruption in 1783 (Robock 2000; von Glasow et al. 2009; Oppenheimer et al. 2014; Jones et al. 2016), or during the Eyjafjallajökull eruption in Iceland in 2010, for example (Gudmundsson et al. 2010).

1.2.2. *Solubility and exsolution*

At great depths, the gases are almost entirely dissolved in silicate magma. In general, it is, therefore, preferable to speak of volatile components in magmas. Several processes such as decompression, changes in temperature and composition of the magma induce the partial exsolution of these magmatic volatiles, that is, their passage in gas form. The formation of a fluid phase of low density can generate an overpressure in the reservoirs and considerably reduce the average density of the magmas and thus appears to be one of the main driving forces for their ascent to the surface. In particular, the amount of volatiles initially dissolved in the magma and the gas content of the magma that rises to the surface will influence the fragmentation process and thus the explosive potential of magmas (Sparks and Wilson 1976; Wilson 1976; Wilson et al. 1978; Jaupart 1996, 1998; Di Muro et al. 2004). Two main eruptive dynamisms are generally recognized (Jaupart 1996, 1998): when degassing takes place in a closed system, when the exsolved gas phase remains confined in the magma, the pressure of the gas increases and the regime is explosive (strombolian and Hawaiian dynamisms for basalts; pyroclastic flows and plinian regimes for differentiated magmas, i.e. those richer in silica). In contrast, in open systems, the gas content tends to be low as gases escape from the magma and the eruptive regime is rather effusive (lava flows, dome formation depending on the acidity of the magmas), unless there is a continuous inflow of deep CO_2-rich gases that can lead to high explosiveness (Spilliaert et al. 2006; Metrich and Wallace 2008; Moretti et al. 2018).

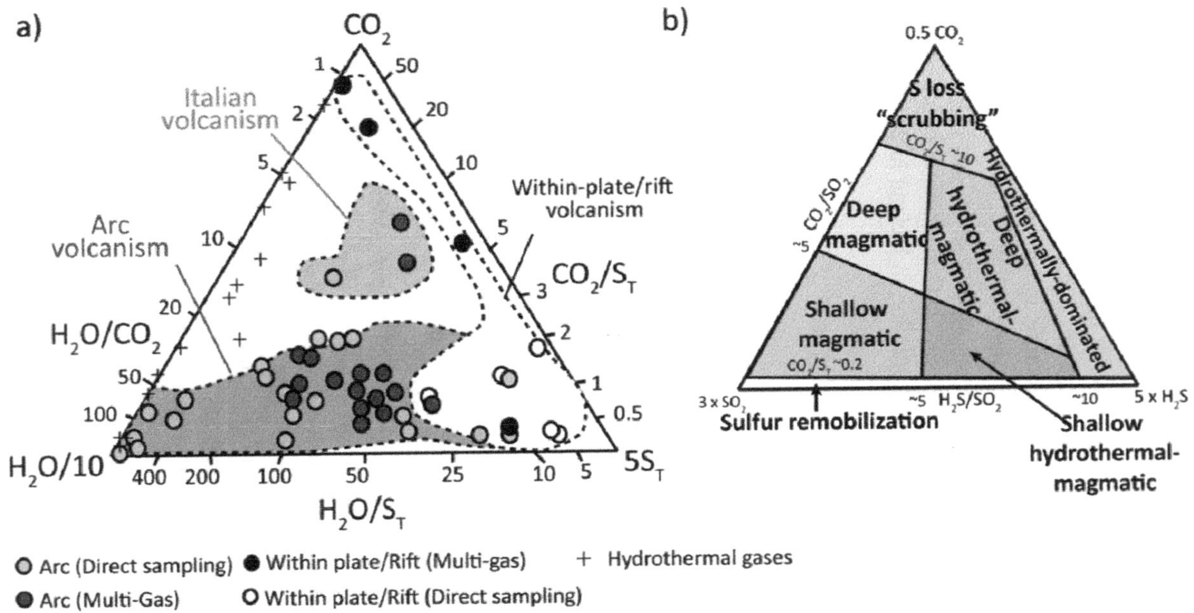

Figure 1.1. *a) Variability in the chemical composition of volcanic fluids (fumaroles and plumes) in H_2O, CO_2 and St (SO_2 + H_2S) for different volcanoes in various geodynamic settings (modified from Aiuppa et al. (2015)). b) This ternary diagram qualitatively defines general areas of chemical composition (SO_2, H_2S, CO_2) of gaseous emissions, from magmatics to hydrothermal*

COMMENT ON FIGURE 1.1.– *It is based on recent studies of Central American volcanoes (Aiuppa et al. 2014; de Moor et al. 2016a, 2016b, 2017). The exact boundaries of these zones vary with tectonic settings and other factors (Aiuppa et al. 2017). Diagram modified from Stix and de Moor (2018). Direct sampling of fumaroles by Giggenbach bottles data are from Symonds et al. (1994), Fischer (2008) and Hammouya et al. (1998). MultiGas measurements are from the UniPa-INGV database and Shinohara and Witter (2005).*

As the gases approach the surface, they may interact with surficial aquifers and/or may be drained along with major volcano-tectonic structures. Thus, depending on these conditions, geochemical monitoring is conducted at different strategic sites representing the main types of volcanic fluids (see Figure 1.2) (Inguaggiato et al. 2011): fumarole gases, plume gases, soil gases and gases dissolved in hydrothermal waters (springs, lake).

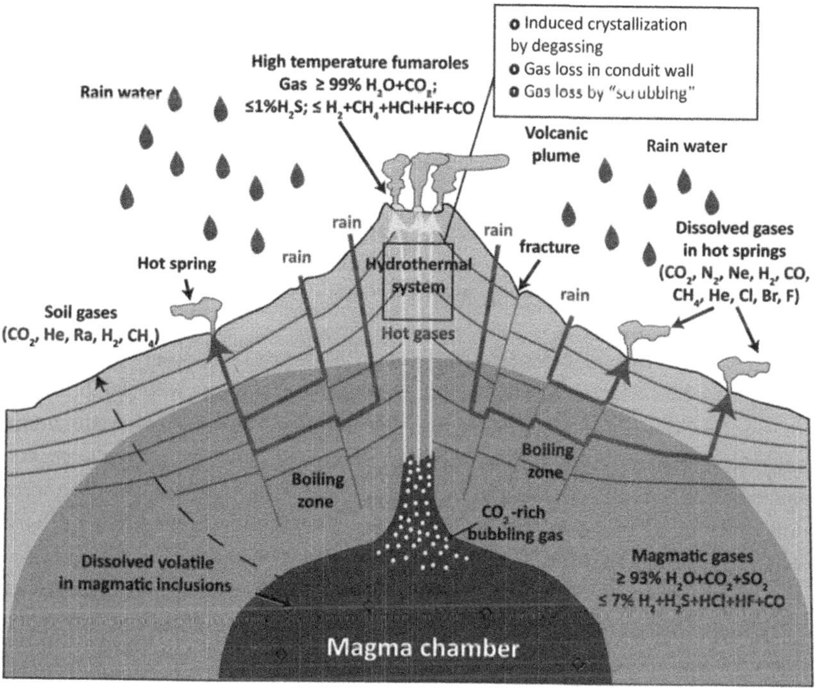

Figure 1.2. Diagram summarizing the different types of volcanic fluids in a magmatic and hydrothermal system. Processes that can alter the composition of volcanic fluids are also shown. Modified from Symonds et al. (2001) and Edmonds (2008)

Many processes can promote interaction between different gas-phase compounds from depth to surface (see Figure 1.2) (King et al. 2018), with examples including water-gas-rock interaction (Symonds et al. 1992), gas-aerosol-ash interaction (Óskarsson 1980; Delmelle et al. 2005; Moune et al. 2006; Delmelle et al. 2018), sulfuric acid-aerosol droplet interaction (Toutain et al. 1995; Mather et al. 2003), sulfuric acid droplet-acid gas interaction (mainly halogens) (Martin et al. 2012), gas-atmosphere interaction (cooling, dilution and oxidation of volcanic gases) (Symonds et al. 1992), trapping of ash and some volatiles by magmatic water and/or by meteoric water (Rose et al. 1995; Textor et al. 2003; Moune et al. 2006; Delmelle et al. 2018). These different phenomena can change the chemical composition of the volcanic plume. Furthermore, if it is a hydrothermal-magmatic system, it is extremely difficult to distinguish between gases from deep or shallow magmatic degassing, as they may be affected by hydrothermal processes such as scrubbing (trapping of reactive gases by the hydrothermal lens at shallow depths), oxidation or re-mobilization of sulfur species, and/or mixtures of gas sources (see Figure 1.1(b)) (Stix and de Moor 2018). These processes are more important for volcanoes in humid climates and affect strongly acidic gases (SO_2, HCl and HF) more than CO_2 and H_2S. Remobilization of elemental sulfur upon temperature increase and/or water table rise (Giggenbach 1996a; Tamburello et al. 2019; Moretti et al. 2020) can also increase the H_2S concentration in gases at the expense of SO_2 (Mizutani and Sugiura 1996). Modeling suggests that CO_2 is the primary species to monitor when scrubbing exists (Symonds et al. 2001). Therefore, these different processes should be taken into account to better assess volcanic activity (see Figure 1.1(b)) and distinguish potential signs of reactivation.

The exsolution of gases depends mainly on pressure (the lower the solubility of an element in the silicate liquid, the deeper the exsolution will be) but also on temperature, oxidation of the system and the composition of the silicate liquid. A general order of exsolution of volatile elements as a function of pressure can be established: CO_2, H_2O, S, Cl and F (Zhang and Cherniak 2010). The relative amounts (ratios) of gaseous species are often relevant to characterize activity. In terms of monitoring, two ratios are paramount: CO_2/S (where $S = SO_2 + H_2S$) and SO_2/H_2S. Because of the contrasting solubilities of CO_2 and sulfur, changes in the CO_2/S ratio are sensitive to changes in magma depth (CO_2 exsolution is deeper than S). These variations reflect the dynamics of degassing. For example, an increase in the CO_2/S ratio may reflect the arrival of deep magma, rich in CO_2, and, consequently, pressurization of the system that may lead to an eruption (Aiuppa et al. 2007, 2009). A decrease in the CO_2/S ratio is an indication of magma ascent (pressure decrease, therefore more S exsolution) (Aiuppa et al. 2007). Following the same principle of contrasting solubilities, ratios such as CO_2/Cl, S/Cl and Cl/F are also sensitive to the change in magma depth and thus vary before and/or during volcanic reactivation, as was the case at Soufrière Hills (Montserrat) (Edmonds et al. 2002). If the emissions result from the interaction of magmatic gases with a hydrothermal system, the scrubbing phenomenon modifies the CO_2/S and CO_2/Cl ratios and confuses the signals. On the other hand, the SO_2/H_2S, H_2/H_2O and CO_2/CH_4 ratios provide information on the thermodynamic conditions of the hydrothermal system (T, P, redox conditions) (Giggenbach 1980), assuming that all these chemical species reach the chemical equilibrium condition at a certain pressure, temperature and redox condition. For example, for sulfur species, the relative abundances of the species, SO_2 and H_2S, are controlled via this equation (Giggenbach 1980):

$$\log\left(X_{SO_2}/X_{H_2S}\right) = \log K_{(1,T)} - 3R_H - \log P.X_{H_2O} \qquad [1.1]$$

where X is the mole fraction of each species, $K_{(1,T)}$ is the equilibrium constant for the reaction $H_2S + 2H_2O = SO_2 + 3H_2$ at a temperature, T, where $R_H = X_{H_2} / X_{H_2O}$ (proxy for redox conditions that replaces the oxygen fugacity typically used in petrology) According to this equation, a strong SO_2/H_2S ratio (>1) characterizes high-temperature gases released from oxidized silicate liquids (low R_H) at atmospheric pressure (Symonds et al. 1994; Giggenbach 1996). In contrast, H_2S dominates (hence SO_2/H_2S is low) in low-temperature fumaroles where emissions originate from reduced deep hydrothermal systems (Giggenbach 1980). Thus, these ratios are indicative of the origin of the gases (magmatic versus hydrothermal) (Chiodini 2009; Moretti et al. 2020) since SO_2, CO_2 and H_2 have a magmatic fingerprint while H_2S, CH_4 and H_2O have a hydrothermal fingerprint (Chiodini 2009). An increase in these ratios can therefore identify periods of magmatic reactivation (Chiodini 2009; Moretti et al. 2020) (see Figure 1.3).

Therefore, the gas phase is a quasi-unique means to understand certain internal mechanisms, such as magma migration or hydrothermal system perturbation. Geochemical signals are therefore an important part of the monitoring of volcanic activity. These signals can be acquired during the quiescence, reactivation or eruption phases of a volcano. They give information on magmatic transfers, thermodynamic conditions of magmatic and hydrothermal systems and on gas sources. Indeed, gases can originate from the mantle, subducted material, crustal sediments or surface fluids (groundwater, seawater and/or meteoric water that infiltrate the system and interact with rising deep gases forming hydrothermal fluids) (Matsuo 1960; Delmelle et al. 2000; Christenson et al. 2017). Thus, a better understanding of the processes and dynamics of magmatic degassing is crucial to advance volcanic monitoring.

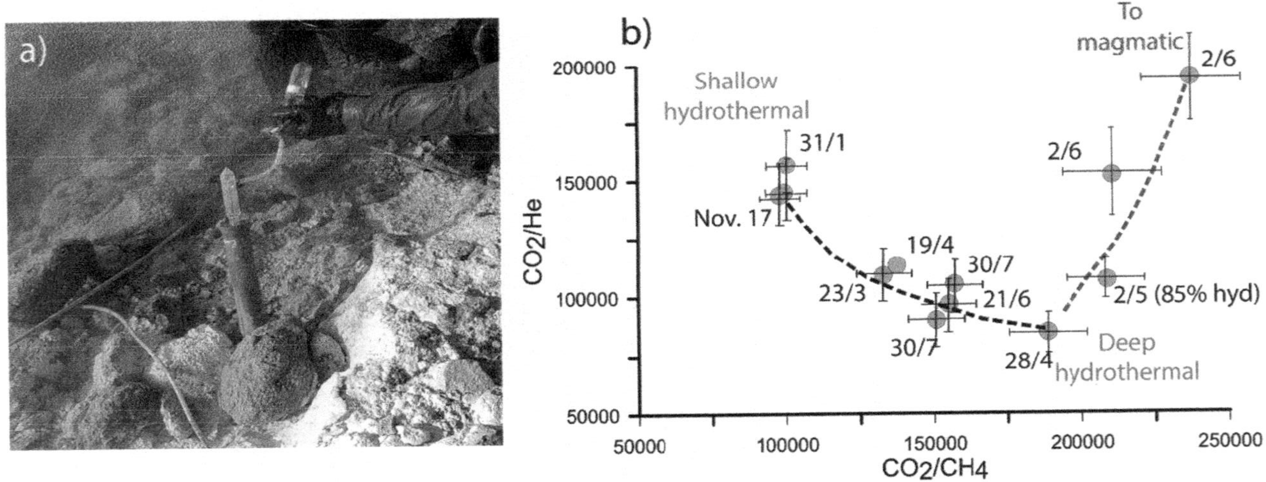

Figure 1.3. *Example of gas sampling and composition at Soufrière, Guadeloupe. a) Direct sampling of fumaroles via Giggenbach bottles (photo. R. Moretti). b) Covariation over time (2018) of CO_2/He and CO_2/CH_4 ratios showing a hydrothermal component (shallow to deep) mixing with a gaseous component from the deep magma. From Moretti et al. (2020). For a color version of this figure, see www.iste.co.uk/lenat/hazards3.zip*

1.3. In situ sampling and analysis techniques for volcanic fluids

Numerous methods for sampling volcanic fluids have been developed. Direct sampling or in situ measurement techniques will be described below. The methods used depend on the types of fluids, the accessibility and the desired frequency of observation.

1.3.1. *Monitoring of fumarole compositions by Giggenbach ampoules*

Direct sampling of fumarole gases is carried out using Giggenbach bottles (Giggenbach and Goguel 1989): the fumarole is channeled through a titanium tube connected to a glass bottle that has previously been partially filled with a basic solution (most often concentrated aqueous sodium hydroxide, NaOH) and the residual volume of which is under vacuum (see Figure 1.3(a)). In the bottle, the gases first pass through the basic solution in which water vapor condenses and acidic gases, such as CO_2, H_2S, SO_2, HCl and HF, dissolve. Uncondensable gases, such as N_2, O_2, H_2, CO, Ar and He, are collected in the upper empty space of the bottle and are then analyzed by gas chromatography or mass spectrometry to determine their concentration. Dissolved species concentrations are analyzed by ion chromatography or conventional wet chemistry techniques.

A disadvantage of this method is that sampling can be dangerous or even impossible, depending on the activity. In addition, the analyses need to be performed in the laboratory and therefore the results are not immediate. However, it is the technique that allows for obtaining the most complete composition of the gases. Their thermodynamic conditions of formation and evolution (T, P, redox conditions) can then be estimated (Giggenbach 1980; Hilton et al. 2002; Liotta et al. 2010; Aguilera et al. 2012; Moretti et al. 2017; Moretti et al. 2020) using the chemical ratios of the H_2O-CO_2-H_2-CO-CH_4, system, as described by Chiodini and Marini (1998). This approach is based on the assumption, which is difficult to test, that all these species are in chemical equilibrium for the pressure, temperature, and redox conditions in the system and that each species has a particular affinity for either the magmatic or the hydrothermal system (see section 1.2). With a time series of analyses, the proportion between magmatic and hydrothermal fluids can be estimated as a function of time (Chiodini et al. 1993; Chiodini 2009; Chiodini et al. 2012). For example, recently, Moretti et al. (2020) showed that in April 2018 at La Soufrière in Guadeloupe, gases from the deep magma mixed with the more superficial hydrothermal system (see Figure 1.3(b)), which could have led to an eruption if the fractured state of the dome had not allowed for a natural release of pressure.

1.3.2. *Monitoring of the compositions and gas fluxes of the diluted plume*

For safety reasons, studying the dilute volcanic plume from a distance is often the only way to acquire information on the chemical composition of the gas phase. Volcanic plumes can be generated by magmatic degassing (passive or eruptive) or by strong fumarolean degassing. They are complex physical mixtures of gaseous solutions (water, gas) and particles (ash, aerosols) that cool and dilute in the atmosphere. Measurements can be made by ground techniques close to the vent or by airborne techniques (drone, airplane/helicopter) and/or remote sensing techniques (ground up to 10 km from the volcano or by satellites; see Volume 2, Chapter 4). The dilution of the plume in the air leads to considering only the concentration ratios and not the absolute concentration of the species.

1.3.2.1. *MultiGas analysis system*

Since a volcanic plume is a mixture of volcanic gases and air, measurement of volcanic H_2O and CO_2 is difficult because of the large amount of these species in the atmosphere (<0.5 ~ 5% H_2O; 410 ppm CO_2; IPCC data). This difficulty can be reduced if it is possible to make in situ measurements of plumes near the crater or active fumaroles. This is possible with the installation of the MultiGas instrument on the ground.

The MultiGas (see Figure 1.4(b)) is a lightweight and compact device consisting of electrochemical detectors (CO, H_2, H_2S and SO_2), non-dispersive infrared (NDIR) sensors (CO_2) as well as air temperature (T), atmospheric pressure (P) and relative humidity (Rh) sensors, which allow for the determination of H_2O concentration via the Buck (1981) equation:

$$H_20 = \frac{(6.1121*(1.007+3.46*P^{-6})*\exp\left(\frac{17.502*T}{240.97+T}\right)*\frac{Rh*10^6}{100}}{P} \quad [1.2]$$

This combination of sensors allows for accurate and simultaneous analysis of H_2O, CO_2, CO, H_2, H_2S and SO_2 from volcanic plumes (Aiuppa et al. 2005b; Shinohara 2005; Roberts et al. 2012). Analysis of the volcanic contribution is obtained after subtracting the atmospheric signal, a signal obtained by measuring atmospheric air uncontaminated by the plume (Aiuppa et al. 2005a; Shinohara 2005). Measurements can be made at a high frequency (0.1 Hz) with an accuracy on molar concentrations typically about 5–10%. This high frequency allows the geochemical data to be correlated with geophysical data also at high frequency (seismology, deformation, etc.). For about 15 years, MultiGas has been widely used, both for point studies of volcanic and hydrothermal emissions, and also for monitoring (Aiuppa et al. 2007, 2009, 2011; Roberts et al. 2012; de Moor et al. 2016b; Aiuppa et al. 2017).

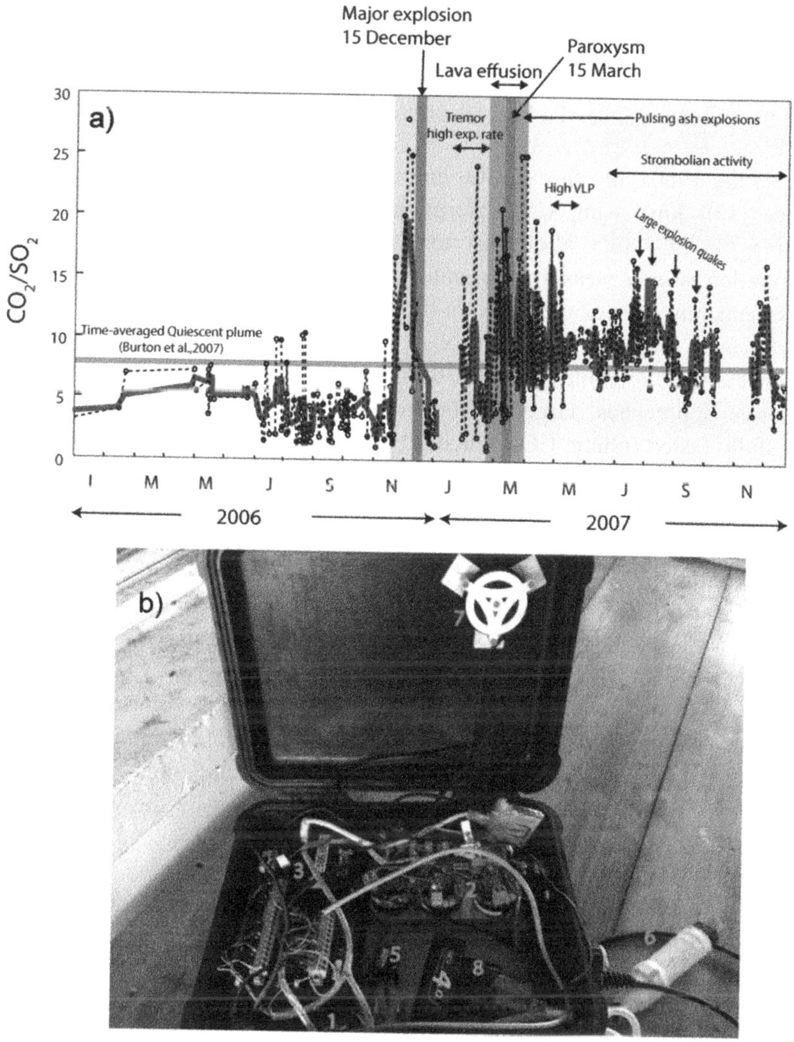

Figure 1.4. *a) Temporal evolution of the CO_2/SO_2 ratio of the Stromboli volcanic plume in 2006–2007. The main volcanic events are also indicated, showing that CO_2 can be considered as a reliable precursor of paroxysmal explosions (modified from Aiuppa et al. (2009)). b) MultiGas configuration and its components. For a color version of this figure, see www.iste.co.uk/lenat/hazards3.zip*

COMMENT ON FIGURE 1.4.– *1: IR sensor; 2: electrochemical sensors; 3: datalogger; 4: Wifi; 5: pump; 6: external sensor for Rh-T; 7: GPS; 8: battery (photo: S. Moune).*

For example, MultiGas data have been used to trace pre-eruptive degassing at Etna (Aiuppa et al. 2007), to better understand deep magma upwelling and detect paroxysms at Stromboli (Sicily; see Figure 1.4(a)) (Aiuppa et al. 2009, 2010), and to record the (re)emergence of a lava lake at Masaya, Nicaragua (Aiuppa et al. 2018).

Since the 1980s, the flux of SO_2 emitted by a volcano has been measured at a distance (<10 km) using UV absorption spectroscopy (Volume 2, Chapter 4). Knowing the SO_2 flux allows us to calculate the flux of other gases whose concentration is only measured: we multiply the concentration ratios X/SO_2, where X represents any species measured in the volcanic plume by MultiGas, by the SO_2 flux. However, when SO_2 is absent or in very low concentration, as is the case for low-temperature fumaroles (<100–300°C), UV spectrometers cannot be used. Alternative approaches, targeting fumarole fluxes of H_2O and CO_2, have been successfully tested (plume CO_2 imaging by diode laser spectroscopy (Pedone et al. 2014); ground gas fluxes (Chiodini et al. 2001) or differential absorption lidar (Aiuppa et al. 2015)). However, some of these methods are relatively cumbersome and difficult to implement in the field, requiring low topography and favorable weather conditions. For example, it was recently shown on La Soufrière in Guadeloupe that fluxes from volcanoes with low-temperature fumarole activity can be reliably determined by performing gas plume tomography with one or more portable MultiGas (Allard et al. 2014; Tamburello et al. 2019) (see Figure 1.5). Fluxes are calculated by coupling the measured total gas composition with the vertical (measured at different heights; see Figure 1.5) and horizontal distribution of H_2S (or CO_2 if the measured plume concentration is much higher than ambient air; see Figure 1.5(b)) in the plume cross-sections, multiplied by the wind speed (see Figure 1.5(b)). The advantage of using CO_2 as a volcanic marker to estimate fluxes is that the response kinetics of the CO_2 sensor is 1.5 times higher than that of the H_2S sensor, which allows, during cross-sections, not to miss gas arrivals, and therefore not to underestimate fluxes, as was shown in Tamburello et al. (2019). In terms of monitoring, following the evolution of fluxes in parallel with concentrations is paramount since CO_2 fluxes can serve as a proxy for magmatic input (Elias and Sutton 2012) and gas flux variations can also be indicative of the opening or closing of a volcanic system. For example, at La Soufrière in Guadeloupe, gas fluxes increased by a factor of 3 between 2012 and 2016, which can be explained by an opening of fractures present on the dome, generating an increase in the permeability of the system (Allard et al. 2014).

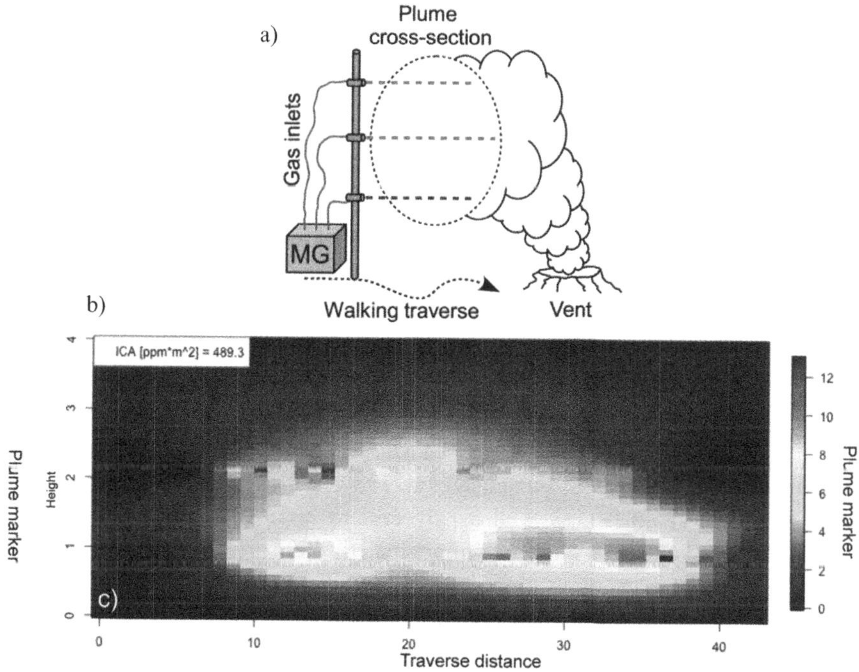

Figure 1.5. *a) Gas concentrations are measured with a MultiGas at different heights (1, 2 and 3 m) with cross-sections perpendicular to the main plume axis (from Tamburello et al. (2019)). b) 2D interpolation of H_2S (here volcanic marker) concentrations for each cross-section. For a color version of this figure, see www.iste.co.uk/lenat/hazards3.zip*

COMMENT ON FIGURE 1.5.– *This 2D surface is integrated to obtain gas column concentrations (ICA: Integrated Column Amount).*

1.3.2.2. *Alkaline traps*

In addition to their environmental impact (Mather 2008), halogens (F, Cl and I) are important because their concentration in magmatic gases is almost unaffected by decompression, gas-rock interactions, or dry cooling (i.e. no interaction with a hydrothermal system) (Symonds and Reed 1993). The F/Cl ratio of magmas is therefore preserved in gases ($T > 120°C$; an example of Etna fumaroles) (Aiuppa et al. 2005a). On the other hand, if the gas interacts with a hydrothermal system, the halogens mainly, but also part of the SO_2, will be dissolved and transported into the low-temperature aqueous systems (a process called "scrubbing" which means:

trapping of soluble reactive gases by the hydrothermal lens at shallow depth) (Symonds et al. 2001; Villemant et al. 2005).

Halogens in volcanic fluids remain nevertheless an effective tool to identify their origin and characterize their transfer. To obtain the halogen composition (HCl, HF) of plumes, one can use either remote sensing methods: Fourier Transform Infrared Spectrometer (FTIR, see Volume 2, Chapter 4), or in situ measurements: alkali traps. Both methods give consistent results (see below). The in situ methods are based on the fact that an alkaline solution reacts quickly with acid gases and traps in the form of dissociated ions.

1.3.2.2.1. Active alkaline traps

Over the years, different types of filters (cellulose, nylon, Teflon, etc.) have been impregnated with different types of caustic solutions (NaOH, LiOH, KOH, $NaHCO_3$) (Stoiber et al. 1986; Finnegan et al. 1989; Witter et al. 2004; Mather et al. 2008; Ménard et al. 2014), as well as glycerol to improve absorption efficiency (Huygen 1963a, 1963b). Two to three impregnated filters are arranged in series (see Figure 1.6), and they are often preceded by particulate filters (PFTE) to minimize the trapping of silicate aerosols and particles on the alkaline filters (Finnegan et al. 1989). These filters are connected to a pump that draws in ambient gases. This active alkaline trap is commonly referred to as a Filter-Pack.

Figure 1.6. *Sampling of the dilute volcanic plume from Lascar volcano, Chile, by filtration (Filter-Pack) (modified from Ménard (2014))*

Depending on the plume concentration and pump flow rate, the sampling time is between 5 and 45 minutes and is chosen so that the majority of the acid gases are absorbed on the first filter. Saturation of the filters should be avoided as this leads to unrepresentative measurements of the sampled gases (Mioduszewski and Kress 2008). For this reason, it has been accepted that for a measurement to be valid, the last filter in the Filter-Pack must contain negligible concentrations (<10%).

This technique is easy to set up (lightweight, non-fragile equipment; see Figure 1.6(a)) and is widely used in volcanology (Faivre-Pierret 1983; Kyle et al. 1990; Allen et al. 2000; Moune et al. 2010; Ménard et al. 2014; Gauthier et al. 2016).

1.3.2.2.2. Passive alkaline traps

Passive alkali traps, also called Japanese boxes, consist of an open tank containing a caustic solution (usually about 500 ml NaOH or KOH) exposed to the volcanic plume. The traps are usually placed on the crater rim in the prevailing wind to promote interactions between the gas and the soda. To protect the solution from major contamination by meteoric water and ash, the trap is placed in a box that allows airflow only through holes in the sides. Sampling time varies from one week to more than one month (Noguchi and Kamiya 1963; Alfaro and Zapata 1997; Goff et al. 1998; Witter et al. 2000). This method has been validated on many volcanoes around the world and under extreme conditions, such as Erebus in Antarctica.

In the laboratory, the impregnated filters are leached in 20 ml of ionized water at room temperature for 2 weeks. A few drops of H_2O_2 are added to the solutions obtained for both methods, to convert all the sulfur species into sulfate. These solutions are then diluted and analyzed by ion chromatography (SO_4^{2-}, Cl^-, F^-).

For all methods, in order to overcome plume dilution and also changes in meteorological conditions that may affect element concentrations, element ratios are used. Even if these techniques are punctual and are not automated, they are complementary to high-frequency measurements (MultiGas) and have proven their relevance for monitoring volcano activity. For example, it has been shown that variations in S/Cl and F/Cl ratios can be important markers of basalt injections at Popocatepetl (Mexico) (Goff et al. 1998) and of major changes in fumarole activity (Vulcano, Sicily) (Italiano 1991).

1.3.3. *Representativeness of samples/measurements in fumaroles (undiluted) and plumes: comparison of techniques*

To compare the consistency of concentration ratios measured from different sampling methods, Aiuppa et al. (2004) studied the F0 fumarole (HT fumarole > 300°C; Fossa crater, Vulcano, Italy) collected in situ by Giggenbach bottle as well as the dilute volcanic plume from the latter (analyzed by FTIR, see Volume 2, Chapter 4, and collected by filtration). This study shows a general consistency between the different ratios of volatile elements measured by FTIR and by filtration, demonstrating the validity of these two approaches. Moreover, this study shows a good agreement between the H_2O/CO_2 and HCl/HF ratios measured by the bottle and those measured by filters and FTIR, thus underlining the validity of the studies carried out on the diluted plumes for the gaseous species. For SO_2/HCl and SO_2/HF ratios, however, one must be careful with the speciation of sulfur. Indeed, by FTIR, only SO_2 is measured, the filters measure SO_2 and the aerosols, which are in the H_2SO_4, form, whereas the bottles measure all the sulfur species (SO_2, H_2S, H_2SO_4 and the proportion of S which sublimes at the fumarole itself). Thus, when estimating the proportion of sublimates, aerosols and H_2S, the authors found close SO_2/HCl and SO_2/HF ratios between the two methods. UV measurements, MultiGas measurements, filter and bottle sampling (Aiuppa et al. 2005b, 2007; Witt et al. 2008) show a strong consistency in gas ratios between these techniques when studying HT fumaroles. These studies showed that H_2S oxidation to SO_2 in the atmosphere is slow, essentially negligible up to 10 km, even in dense plumes. However, Allard et al. (2014) advocate plume measurements closer to the vent (<20 m) for humid volcanic plumes in the tropics, because under these conditions H_2S oxidation is faster than in cold, less humid environments. Impregnated filter and Japanese box measurements near and far from vents have shown comparable results for Cl/S and F/S ratios regardless of distance or pre-sampling method (Wardell et al. 2008), with one limitation being rapid filter saturation (Wittmer et al. 2014). This saturation will be more rapid the more humid the atmosphere, limiting the use of this method to low humidity environments (relative humidity <80%).

These approaches validate the possibility of working at a distance, both for high-temperature hydrothermal systems (fumaroles) and, even more convincingly, for open-vent degassing processes, since the abrupt cooling of the plume as it enters the atmosphere does not allow gaseous species to react. However, for low-temperature fumaroles (<100°C), which emit wet vapors (i.e. compositions richer in H_2O), there is a loss of soluble elements by condensation. Moreover, these low-temperature fumaroles often undergo a more important "scrubbing" process, which induces an even greater loss of halogens and SO_2.

1.3.4. *Monitoring of diffuse degassing*

1.3.4.1. *Ground gases and their signals*

Degassing from the soil releases poorly reactive gases such as CO_2, He, H_2 and Rn through diffusion and/or advection processes (Chiodini et al. 1998). Diffusion implies that a concentration gradient exists in the soil according to Fick's first law of diffusion:

$$FCO_2 = -\upsilon.D.\left(dC/d\lambda\right) \qquad [1.3]$$

with F the CO_2 flux, υ the soil porosity and D the diffusion coefficient. Advection, on the other hand, implies the existence of a pressure and temperature gradient in the soil, governed by Darcy's law:

$$FCO_2 = \left(k/\mu\right) \times \left(dP/d\lambda\right) \qquad [1.4]$$

with k the soil permeability and μ the fluid viscosity. The advective process appears to dominate when gas migration in the soil occurs at fractures and when gas fluxes are large (Lewicki et al. 2003; Hernández et al. 2004).

Diffuse degassing from volcanic edifices is generally higher at fractures where permeability is higher (Toutain and Baubron 1999; Chiodini et al. 2001). Mapping of soil degassing therefore often reveals significant volcano-tectonic structures (Giammanco et al. 2006; Lee et al. 2016) that may correspond to preferential zones for hydrothermal circulations (Finizola et al. 2002; Werner et al. 2008).

Although the majority of volcanic gases are emitted in the main crater region, it has been shown that degassing through the soil accounts for a significant fraction of the degassing from volcanic edifices (up to 15% at Etna and 50% at Merapi for CO_2; (Allard et al. 1991; Burton et al. 2003)). This contribution varies with the eruptive state of the volcano: volcanoes with a recurrent eruptive period of less than 10 years generally show a soil/plume degassing ratio of the order of 0.1–1, whereas volcanoes with longer reactivation periods (>100 years) show an exponential increase in this ratio (Hernández et al. 2015). Therefore, considering soil CO_2 fluxes ensures that the CO_2 flux is not underestimated, which is important for studying the global terrestrial CO_2 cycle. A global study of volcanic CO_2 fluxes, including soil degassing, estimated the flux to be 540 Mt/yr (Burton et al. 2013). Significant fluctuations in soil degassing have been observed prior to eruptions (e.g. Iwojima volcano in Japan and Stromboli volcano in Italy; Brusca et al. 2004; Notsu et al. 2005). This has shown the interest in regular monitoring of diffuse CO_2 emission to constrain the evolution of the gas source and/or the fractured medium traversed.

This regular monitoring has been successfully applied in El Hierro (Pérez et al. 2012) (see Figure 1.7), Etna (Liuzzo et al. 2013), and Piton de la Fournaise (Boudoire et al. 2017), among others. Pérez et al. (2012) showed the potential of continuous measurements of soil CO_2 flux as an early warning sign of an eruption. Soil CO_2 flux increased 1 month before the 2011 El Hierro (Canary Islands) submarine eruption, reaching its maximum 1 week before the onset of the eruption. Similarly, continuous measurements of CO_2 flux made on the Piton de la Fournaise volcano between 2013 and 2016 demonstrated that, despite environmental influences such as extreme precipitation, soil CO_2 time series allow the detection of major deep magmatic events and events difficult to detect by conventional geophysical techniques. These early signs were observed 3 months before the first eruption of Piton de la Fournaise in 2014, occurring after a 3.5-year pause.

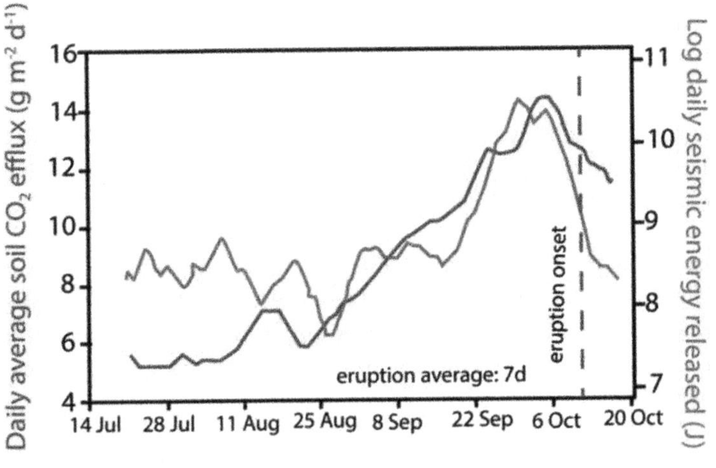

Figure 1.7. *Daily averages of CO_2 fluxes from the ground, as well as daily seismic energy released at El Hierro just before the October 12, 2011 eruption (Pérez et al. 2012). For a color version of this figure, see www.iste.co.uk/lenat/hazards3.zip*

1.3.4.2. *Measurement methods*

There are two methods for measuring soil CO_2 fluxes: accumulation chamber measurement and dynamic concentration measurement (see Figure 1.8). A very complete description of both methods is available in Boudoire (2017). These methods can be automated and many observatories have permanent stations. Weather stations are associated with CO_2 soil stations since rain and temperature can disrupt surface degassing (Harvey and Harvey 2015).

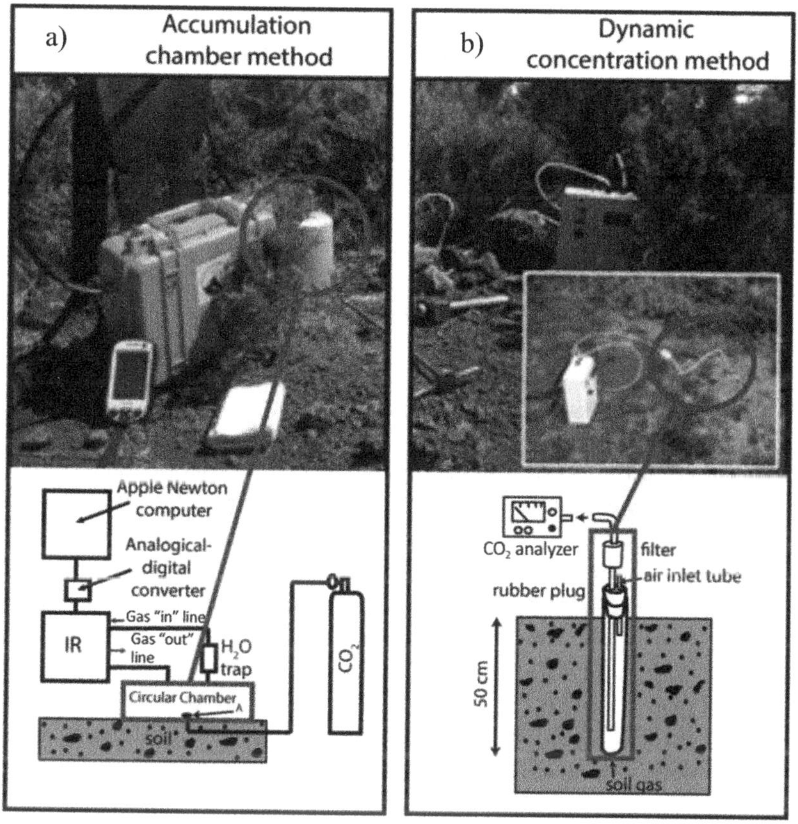

Figure 1.8. *Method for measuring soil CO_2 fluxes. a) Direct method: a hermetic bell jar (Chiodini et al. 1996). b) Indirect method: a system non-hermetic system (Gurrieri and Valenza 1988) (modified from Boudoire (2017)). For a color version of this figure, see www.iste.co.uk/lenat/hazards3.zip*

Direct measurement: For this type of measurement, a sealed bell jar (Chiodini et al. 1998) placed on a flat, unvegetated surface is used to accumulate soil gases. The gases are pumped, the solid particles filtered, and then the gas injected into an infrared spectrometer that measures the CO_2 concentration every second. The analyzed gases are reinjected into the accumulation chamber and homogenized via an internal ventilation system. The concentration of CO_2 in the chamber increases progressively with time, allowing us to calculate a flux (ppm.s^{-1}) from the temporal slope of the concentrations. This slope is linked to the flux via a calibration constant which considers the geometry and volume of the chamber + pipes, the pressure and the temperature of the air.

Indirect measurement: With this method, no accumulation of soil gases is performed, but a dynamic measurement (Camarda et al. 2006) of a mixture of air and soil gases is made. A perforated stainless-steel tube is driven into the soil at a depth of 50 cm. A pipe in the tube is connected to a pump that draws in the mixture at a constant rate. The mixture is then analyzed in an NDIR spectrometer to measure the molar concentration of CO_2. Laboratory experiments have defined an empirical relationship that allows the calculation of the CO_2 flux from its dynamic concentration (i.e. over time), as a function of the system geometry, pump flow rate, probe depth and soil permeability. Unlike the sensor in the chamber, this one automatically corrects for temperature and pressure effects.

1.3.5. *Dissolved gases in hot springs*

The presence of fluids and heat sources in the summit zones of active volcanoes favors the development of hydrothermal systems that are the seat of important transfers of matter and energy. The interaction of rainwater and the hydrothermal-magmatic system is at the origin of hot springs (Figure 1.2). The monitoring of these hot springs has several advantages: located at the base of the volcanoes, they allow the monitoring of the volcanoes at a distance, and they also represent an integrated signal on the whole surface of the edifice.

Since 1980–1990, the interest in monitoring physico-chemical parameters (pH, conductivity and temperature) and the geochemistry of dissolved volatiles in hot springs (CO_2, N_2, H_2, CO, CH_4, He, Ne, I, Br, Cl and F) has been shown for the study of the evolution of internal volcanic activity (Giggenbach and Soto 1992). Recent technical and analytical advances in geochemistry and isotopy open new perspectives in terms of tools for monitoring volcanic activity. The most relevant data are the monitoring of halogens (F, Cl and Br), noble gases (He and Ne), and stable isotopes ($\delta^{13}C$–CO_2, $\delta^{15}N_2$) (Carapezza et al. 2004; Villemant et al. 2005; Hilton et al. 2010; Villemant et al. 2014; Sano et al. 2015; Kagoshima et al. 2016; Tassi et al. 2017).

The halogens (F, Cl, Br and I) are of particular interest because they show simple behaviors during magma degassing and differentiation because they have low mineral/melt partition coefficients and high H_2O vapor/melt partition coefficients (very H_2O dependent). Cl, Br and I are extracted from the magma as halogen acids (HCl, HBr, HI) with similar vapor/melt partition coefficients, while F is not extracted significantly and remains concentrated in the melt. Halogen contents in magmatic gases are only slightly modified by decompression (kinetic fractionation), gas-rock interactions and cooling (to temperatures of 120°C) during the dry transfer path, that is, without interaction with the shallow hydrothermal system or atmospheric gases. In contrast, if gases interact with the hydrothermal system, halogens are dissolved (leaching effect) and transported at low temperatures (Symonds et al. 2001; Villemant et al. 2005). On this basis, the abundance ratios of halogens (Cl, Br and I) in degassed magmas are considered to be conserved in high and medium temperature gases (above 120°C) during their transfer in dry conditions to the surface, as observed at Etna (Aiuppa et al. 2005a). The ratios will also be conserved in hot springs (elemental fractionation is nevertheless possible in case of condensation, boiling or precipitation of halides). Thus, these halogen ratios allow tracing of the geochemical variations of the volcanic system.

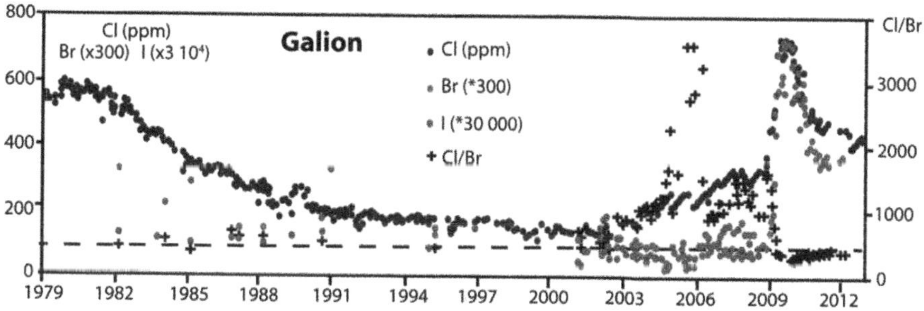

Figure 1.9. *Time series of Cl (ppm), Br (*300), I (*3*104) and Cl/Br composition of the Galion hot spring at the foot of the Soufrière volcano of Guadeloupe. The Cl concentration shows two crises in 1979 and 2012. Between 2001 and 2009, the high Cl/Br ratio indicates halogen fractionation in the Galion spring (Villemant et al. 2014). For a color version of this figure, see www.iste.co.uk/lenat/hazards3.zip*

A recent study on the Soufrière volcano of Guadeloupe (Villemant et al. 2014) shows that regular sampling of hot springs and associated gases over long time series (>30 years) allows us to associate halogen variations with a magmatic signature (see Figure 1.9). During the 1976–1977 seismo-volcanic crisis, a magma intrusion was emplaced at a shallow depth (about 3 km). The volcanic edifice being strongly fractured, the magmatic gases migrated easily, causing an increase in dissolved gases in the hot springs. This increase in dissolved halogens can be observed particularly in the Galion spring, which shows enrichment in Cl, Br and I in 1979 (see Figure 1.9). After the crisis of 1976–1977, the concentration gradually decreased until 2001. Between 2001 and 2009, it increased, as did the Cl/Br ratio. The increase in the latter is characteristic of fractionation between Cl and Br, probably related to interaction in the conduit (such as kinetic fractionation related to decompression, gas-rock interaction or cooling). In 2011, a significant increase in halogen concentration, coupled with an increase in temperature (see Figure 1.6 of Villemant et al. (2014)), suggests a new phase of magmatic intrusion.

Because of their chemical characteristics, noble gases (He, Ne, Kr, Ar and Xe) are good tools for monitoring volcanic activity and tracing magmatic sources. Indeed, they are chemically stable (saturated valence layers) and do not participate in chemical reactions, which allows conservation of the geochemical signature from the magma chamber to the surface. Helium and its isotopes have a strong potential because the isotopic ratios are very contrasted between the different terrestrial reservoirs (see Figure 1.10); there is a primordial isotope (^{3}He) stored in the mantle and a radiogenic isotope (^{4}He) that comes from the decay of U and Th in the crust. The upper mantle has ^{3}He/^{4}He ratios of 8 Ra (where Ra is the ^{3}He/^{4}He ratio of air) while the value for the deep mantle is 16 Ra. In the crust, it is only 0.001 Ra (Ozima and Podosek 2001). Typical values are 16 Ra for hot spot volcanism, 8 Ra for subduction or oceanic rift volcanism and are lower if there is an interaction between the magma and the crust (addition of ^{4}He). The ^{3}He/^{4}He ratio of dissolved gases from hot springs constrains the origin of the magma. Its temporal or spatial variations can reflect magmatic phenomena (e.g. arrival of magma in the system) or hydrothermal phenomena (degassing of the hydrothermal system).

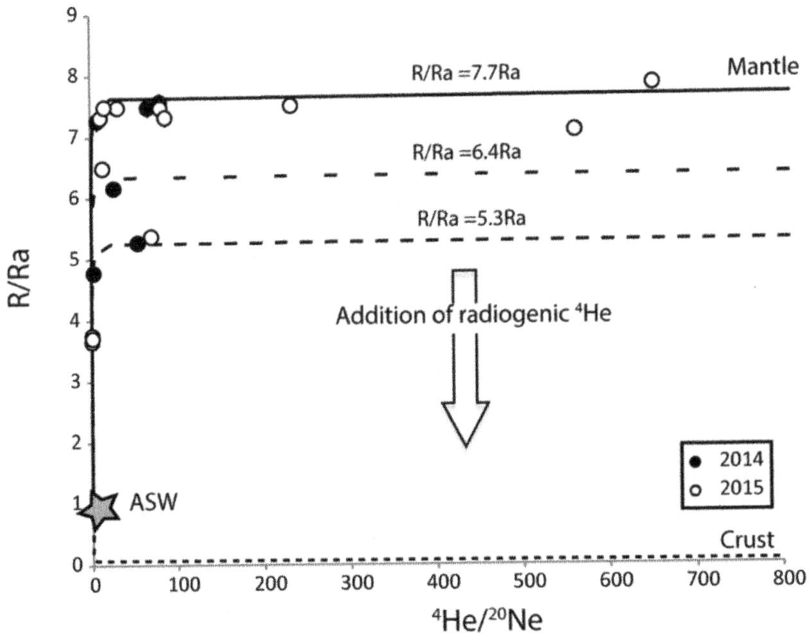

Figure 1.10. *^{3}He/^{4}He Ra (R/Ra) versus ^{4}He/^{20}Ne measured in hot springs from Copahue volcano, Chile (Roulleau et al. 2016). The helium and neon isotopes in the waters are explained by mixing between a mantle and an air source with the dilution of this signature by a crustal component (^{4}He addition)*

The case of the Ontake volcano in Japan helps illustrate the contribution of geochemical hot-spring monitoring. A violent phreatic eruption occurred in 2014 without the usual precursors (seismicity, deformation, etc.) of eruptions (Yamaoka et al. 2016). Monitoring data for a hot spring are available since 1980 (see Figure 1.11) (Sano et al. 2015; Kagoshima et al. 2016). Between 1981 and 2003, the average ratio in $^3He/^4He$ was 6.6 Ra and the ^3He flux was 2.86 × 10^{-6} atom/m^2.sec. Between 2005 and 2009, the mean $^3He/^4He$ ratio increased to 7.3 Ra with a ^3He flux of up to 3.56 × 10^{-6} atom.m^{-2}.sec, while $\delta^{13}C$-CO_2 did not show a significant change prior to the eruption, indicating that no juvenile magma upwelling occurred. The increase in ^3He suggests an increasing contribution of mantle helium related to significant magmatic degassing at depth due to water-magma interaction (increased H_2O fluxes recorded). This led to an overpressure of the hydrothermal system until it exploded (the products emitted are mostly hydrothermalized rocks). This example demonstrates that for certain types of volcanic activity, geochemical and helium isotopic data are crucial to anticipate eruptions like the 2014 Ontake eruption, as geophysical precursors may be absent or too discrete.

The pH and conductivity of hot water are usually measured in the samples. The pH value, hydrogen potential, depends on the acidity of the water. Acidic water has a low pH between 1 and 6, basic water has a high pH between 8 and 14 and neutral water has a pH of 7. The pH is measured with a pH meter probe that is calibrated with control solutions of 4 for acidic environments and 11 for basic environments. The pH of hot springs ranges from 1 to 8 (Villemant et al. 2014), which is consistent with the acidic input from magmatic gases. A decrease in hot spring pH will be symptomatic of magmatic gas input.

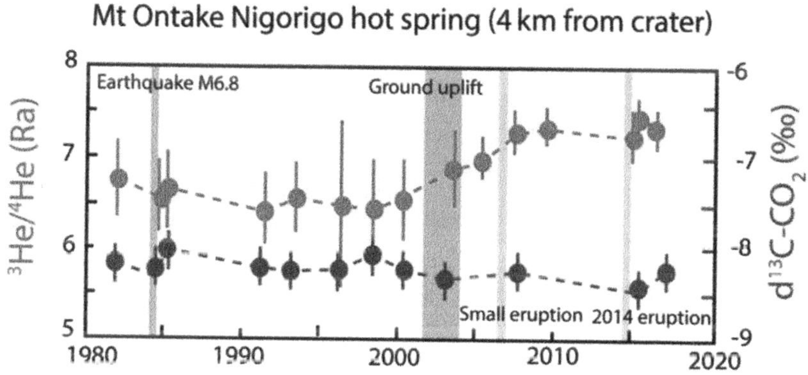

Figure 1.11. *Variations in $^3He/^4He$ (Ra) and $\delta^{13}C$–CO_2 over time in thermal waters of Ontake volcano (Japan). For a color version of this figure, see www.iste.co.uk/lenat/hazards3.zip*

COMMENT ON FIGURE 1.11.– *The blue and pink bars represent geotectonic events and eruptions, respectively (modified from Sano et al. (2015) and Kagoshima et al. (2016)). The observed increase in Helium-3 would be due to an overpressure of the magma system directly related to the increase of water vapor (water-magma interaction), typical of phreatomagmatic eruptions.*

The conductivity of water, measured in $\mu S.cm^{-1}$, is a measure of the concentration of inorganic salt ions and is also temperature-dependent. This measurement is made with a conductivity probe calibrated for a temperature of 25°C. In natural, low mineralized and cold water, it is basic (200 $\mu S.cm^{-1}$) while in hydrothermal, high mineralized water, it is higher (240–>5,000 $\mu S.cm^{-1}$ (Villemant et al. 2014)). An increase in conductivity, accompanied by a decrease in pH and an increase in temperature, indicates an enrichment in compounds of magmatic origin.

1.3.5.1. *Sampling and measurement methods*

Hot water (and associated gases) are collected with double-valved alkaline cylinders (see Figure 1.12(a)), or/and copper tubes closed at each end by steel clamps (see Figure 1.12(b)). This type of cylinders and copper tubes makes it possible to limit as much as possible the diffusion of rare gases (in particular helium) through the container.

Figure 1.12. *a) Example of Sakurajima (Kyushu, Japan) with direct hot spring sampling via a double-valved alkaline cylinder. b) Example of Massif Central (France) with hot spring gas sampling via a copper tube closed with steel clamps. For a color version of this figure, see www.iste.co.uk/lenat/hazards3.zip*

The water and gases are then analyzed in the laboratory using high-precision mass spectrometers (Helix SFT). Sampling can also be done periodically by an automatic sampler. This automated system, named SPARTAH (Barry et al. 2009), consists of collecting water in a series of copper tubes (diameter: 3.18 mm) using a syringe. The sampling rate is adjustable (days, weeks, etc.). Afterward, the samples are analyzed in the laboratory in the same way as the manually collected samples. In both cases, no real-time information is obtained. However, a Swiss team has just addressed this drawback by offering the market a portable mass spectrometer that allows in situ quantification of the composition (He, Ar, Kr, N_2, O_2, CO_2, CH_4, H_2) of dissolved gases in hot springs (Brennwald et al. 2016).

1.4. Contribution of melt inclusions to geochemical monitoring

Melt inclusions (MI; see Figure 1.13) allow a study of the volatile phases associated with silicate fluid. Primary melt inclusions are small-volume (1–300 m in diameter) silicate fluid droplets trapped in minerals at a given time *t* during their crystallization. Primary MIs are an important source of information on magma composition and volatile solubility during its ascent to the surface (Metrich and Clocchiatti 1996; Moune et al. 2007) since it is accepted that these objects behave, if unaffected by fractures, as closed systems protected by their host crystal, and that they do not undergo degassing during magma ascent (Anderson 1974; Lowenstern 2003). MIs provide access to the dissolved major, trace and volatile element chemical composition (H_2O, CO_2, S, Cl, F) of pre-eruptive magma (Anderson 1974). In the presence of glassy inclusions in different minerals, formed at different pressures, the exsolution history of the volatiles can be inferred (Metrich et al. 1993; Moune et al. 2007). The composition of the degassed magma is found in the glass constituting the matrix of the slag or lava.

The concentrations of the five main volatile species (H_2O, CO_2, S, Cl and F) are determined in the laboratory via various analytical techniques (electron microprobe, ion probe, vibrational and infrared spectrometry). Based on the solubility of the two main volatile species (H_2O and CO_2) which are strongly pressure-dependent, trapping pressures can be determined to trace the evolution of the magma between the pre-eruptive stages and the degassed lava. By comparing the concentrations of volatiles in MIs and matrix glass, and knowing the volume of magma emitted, it is possible to estimate the amount of each volatile species degassed (Devine et al. 1984). Conversely, these observations can be used to model the composition and fluxes of the expected gas phase at the surface via multicomponent models based experimental data and thermodynamics (Newman and Lowenstern 2002; Moretti and Papale 2004; Papale et al. 2006; Burgisser and Scaillet 2007).

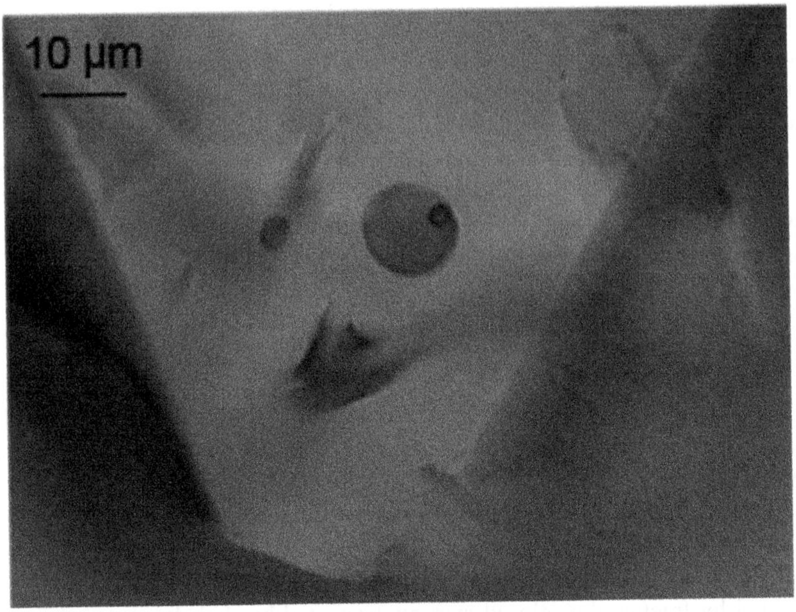

Figure 1.13. *Melt inclusion, with its thermal retraction bubble, trapped in an olivine (Hekla, Iceland, photo: Moune). For a color version of this figure, see www.iste.co.uk/lenat/hazards3.zip*

However, the study of inclusions involves uncertainties. Processes can alter the composition of inclusions: post-trapping modifications by crystallization of new mineral phases, decay, devitrification and loss of volatiles due to the presence of fractures (Lowenstern 1995; Baker 2008). Furthermore, the total volatile content in a magmatic system is not given only by the volatiles dissolved in the inclusion glass, as one must consider the gas that has been exsolved. Estimating the total volatile content, and thus the relative gas and liquid fraction, can be done by computational or even graphical methods on a cogenetic set of inclusions (for $H_2O + CO_2$ (see Papale (2005), Barsanti et al. (2009) and Moretti et al. (2018)). Indeed, inclusions can also contain thermal shrinkage bubbles (see Figure 1.13) and volatiles dissolved in the silicate liquid of the inclusion can diffuse into the bubble (Dunbar et al. 1989). Recently, it has been shown that these shrinkage bubbles can contain up to 60–90% CO_2 (Moore et al. 2015; Wallace et al. 2015) as well as up to 80% sulfur (Venugopal et al. 2020). Thus, volatiles must be measured not only in the silicate melt but also in the bubble to avoid severely underestimating the composition of pre-eruptive magmas. This is important for determining degassing budgets. Indeed, in many studies, an "excess of gas" has been observed in gas measurements made by ground or satellite remote sensing compared to balances obtained from inclusions

(Francis et al. 1993; Shinohara 2008). Although not all observed cases of “excess gas” are identical (Oppenheimer et al. 2003; Wallace 2005), this implies an underestimation of volatiles in inclusions. In the case of the 2007 eruption of Piton de la Fournaise, the excess SO_2 in the satellite measurements could be due to the release of previously stored sulfur, in the hydrothermal system (Gouhier and Coppola 2011). Conversely, in the case of the Grimsvötn sub-Plinian eruption (Iceland, 2011), the satellite-measured SO_2 fluxes were found to be lower than would be predicted from the estimate of available sulfur in the inclusions. This implies a “sink” mechanism trapping some of the sulfur in the surface system, for example in the form of sulfides (Sigmarsson et al. 2013).

Crystals and their inclusions are direct sources of information about the architecture and dynamics of the plumbing system. Inclusion composition and zonations of their host mineral (zonation = variation in chemical composition during mineral growth and/or recrystallization providing information on the diffusion timing of certain elements, primarily Fe, Mg) have been shown to be reliable tools for determining not only time scales regarding storage, mixing, or magma upwelling time (Zellmer et al. 1999, 2003; Kahl et al. 2011, 2013, Flaherty et al. 2018; Ruth et al. 2018), but also the processes that lead to an eruption (Kahl et al. 2011; Albert et al. 2014, 2019).

1.5. Example of monitoring

We present a case study where detailed volcanic fluid research has led to a better understanding of degassing processes, the behavior of the volcano itself, and improved the way we monitor volcanic activity. Other examples are presented in Chapter 4 of Volume 2.

1.5.1. *Gas monitoring of Poas volcano (Costa Rica): identification of phreatic and phreatomagmatic eruption precursors (based on de Moor et al. (2019))*

The processes leading to phreatic and phreatomagmatic eruptions remain poorly understood. Five years of gas monitoring at Poas volcano (Costa Rica), combined with geophysical monitoring data, allows us to characterize the degassing of a volcanic system during phreatic and phreatomagmatic eruptions (see Figure 1.14).

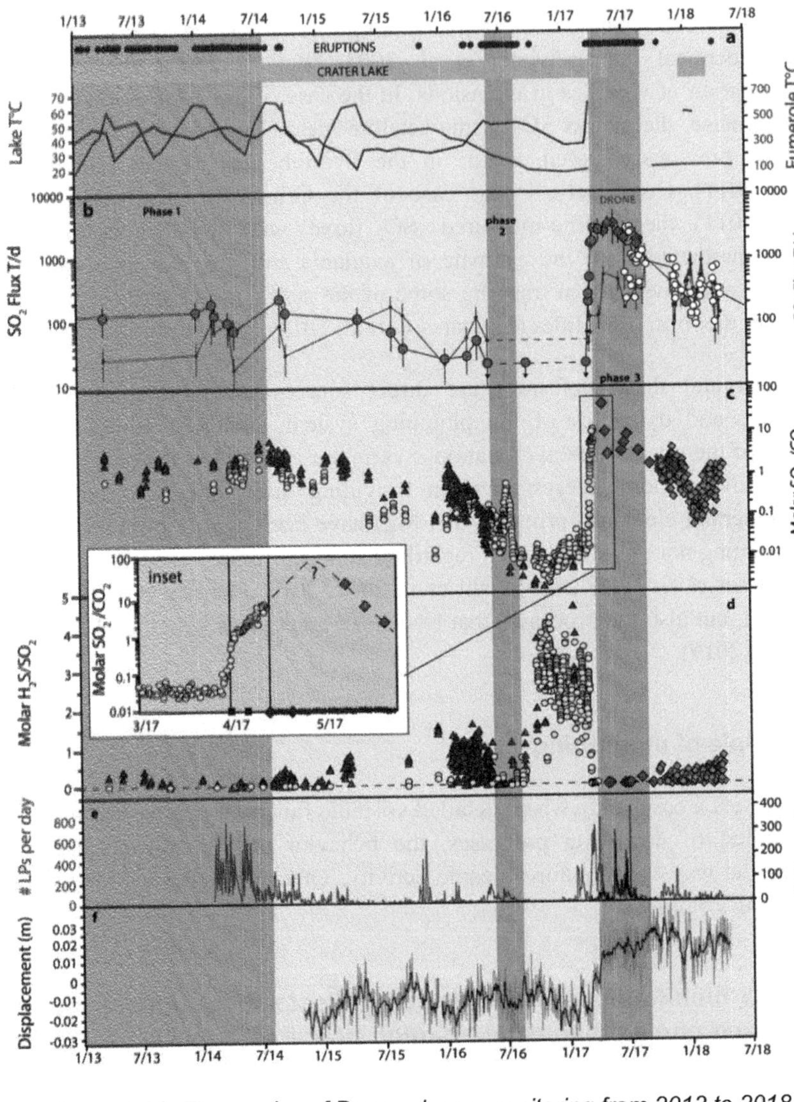

Figure 1.14. *Time series of Poas volcano monitoring from 2013 to 2018. For a color version of this figure, see www.iste.co.uk/lenat/hazards3.zip*

COMMENT ON FIGURE 1.14.– *The green diamonds show the gas data acquired by a drone. a) Chronology of eruption periods (light blue), explosions (dark blue dots) and the presence of a crater lake (gray). Lake and fumarole temperatures are also shown. b) SO_2 fluxes obtained by DOAS (gray circle = traverses, white circles = permanent stations). The blue line shows the CO_2 flux calculated from SO_2 fluxes* *and SO_2/CO_2 ratios). c) and d) MultiGas data (yellow circles = lake-generated plume, red triangles = fumarole-generated plume and orange diamonds = mixed plume after the 2017 eruption. e) Seismic activity: number per day of long-period (LP, red) and volcano-tectonic (VT, black) events. f) Vertical displacement data obtained by GPS. The three blue areas correspond to the three eruptive phases (modified from Moor et al. (2019)).*

SO_2 flux measurements were determined by DOAS (differential optical absorption spectroscopy; Volume 2, Chapter 4), during walking traverses, and also via a mini-DOAS platform on drones during eruptive events and/or via a network of two DOAS permanently installed near the volcano. Key gas ratios, such as SO_2/H_2S and SO_2/CO_2 were measured by a permanent MultiGas station and by a mini-MultiGas system on UAVs as well. The CO_2 fluxes were thus calculated by combining the two techniques (SO_2/CO_2 X fluxSO_2 = fluxCO_2). Figure 1.13 covers 3 phases of eruptive activity. Phase 1 (2012 to 27 August 2014) is characterized by numerous phreatic explosions in the crater lake. Phase 2 (June 5 to September 17, 2016) is similar. In contrast, phase 3 (April 12 to September 24, 2017) is different, both in its notably higher intensity than the first two phases, with the disappearance of the lake and eruptive plumes up to 4 km, and in the emission of magma. The 2017 phreatomagmatic eruption (phase 3) is the largest eruption at Poas since 1953 and was preceded by dramatic changes in geochemical and geophysical parameters (see Figure 1.14). A very large increase in SO_2 flux, SO_2/CO_2 ratio and to a lesser extent CO_2 flux (magmatic signs), with a drastic decrease in H_2S/SO_2 ratio (decrease in hydrothermal contribution), was recorded just before the eruption. These strong geochemical changes were concordant with an increase in lake temperature, a number of long-period earthquakes per day and deformation of the system (see Figure 1.14). The authors interpret this sequence of eruptive activity by a superficial intrusion of magma whose heat and degassing provoke the phreatic activity of the first two phases while exacerbating the hydrothermal activity. The amplification of the latter leads to the deposition of hydrothermal minerals which will decrease the permeability of the system (sealing of the hydrothermal system). This results in an accumulation of volatiles, and thus an increase in pressure which will lead, by fracturing the hydrothermal zone, to the violent phase 3.

1.6. Recommendations for fluid monitoring

Despite considerable advances in gas geochemistry and degassing processes, many questions or uncertainties remain in these areas. Among the most sensitive points are:

– the origin of volatiles in magmas at depth;

– the physico-chemical mechanisms that control major explosive eruptions;

– the time scales involved in the storage and ascent of magma;

– the influence of transfer rates on the behavior of volatiles in the magma and gases at the surface;

– the influence of the interaction of ascending hot magma gases with water, whether of meteoric, hydrothermal, marine origin, etc.

All these processes have a direct impact on the gas concentration measured at the surface. The ability to differentiate between a resting phase, magma upwelling, an increase in the contribution of the hydrothermal system and/or fracture closure (pressurization) is crucial for monitoring volcanic activity. This requires relevant data to have an integrated view of the variations of several parameters. In order to have the best possible monitoring, the following technical means and approaches can be recommended:

– Monitoring over long periods of time to obtain a baseline of concentrations, ratios and fluxes of key chemical species, such as CO_2, SO_2 and H_2S.

– Adaptation of techniques according to the type of activity (presence or absence of SO_2 in measurable quantity, presence of a hydrothermal system, environmental conditions (temperature, humidity, etc.), basaltic or more acidic magmas, etc.).

– Joint analysis of geochemical measurements of gases and geophysical measurements (thermal, seismicity, deformation, etc.) available.

– To take into account the diffuse degassing (soils, springs and lakes) in composition and flux.

– Install continuous measurement stations to record the dynamics of variations. Spot measurements on the same sites and on other sites remain important in order to enrich the monitoring with additional data.

– To take into account the characteristics of the monitored volcano: past behavior, nature of the magmas, volcano-tectonic structures and hydrothermal system.

– The implementation of remote analysis methods for plume gases (see Volume 2, Chapter 4), which are essential in times of crisis.

– Develop techniques and use of gas measurements on airborne platforms (helicopters and drones mainly) to perform measurements in hazardous conditions and locations. MultiGas, Filter-Pack and DOAS are already miniaturized to be adapted to these platforms (Liu et al. 2019; Mandon et al. 2019; James et al. 2020).

– Quantify, in addition to major volatiles, trace elements and isotopes of various volcanic fluids to better identify the sources of volatiles and the different processes that take place between the magma at depth and the surface (de Moor et al. 2013; Ménard et al. 2014; Bonifacie et al. 2019).

– Promote the use of new technologies that are being developed for continuous in situ measurement of fumarolc concentrations and isotopic composition.

1.7. References

Aguilera, F., Tassi, F., Darrah, T., Moune, S., Vaselli, O. (2012). Geochemical model of a magmatic-hydrothermal system at the Lastarria volcano, northern Chile. *Bull. Volcanol.*, 74, 119–134.

Aiuppa, A., Dongarrà, G., Valenza, M., Federico, C., Pecoraino, G. (2003). Degassing of trace volatile metals during the 2001 eruption of Etna. *Geophys. Monogr. Ser.*, 139, 41–54.

Aiuppa, A., Burton, M., Murè, F., Inguaggiato, S. (2004). Intercomparison of volcanic gas monitoring methodologies performed on Vulcano Island, Italy. *Geophysical Research Letters*, 31, L02610. https://doi.org/10.1029/2003GL018651.

Aiuppa, A., Federico, C., Franco, A., Giudice, G., Gurrieri, S., Inguaggiato, S., Liuzzo, M., McGonigle, A.J.S., Valenza, M. (2005a). Emission of bromine and iodine from Mount Etna volcano. *Geochemistry, Geophys. Geosystems*, 6, Q08008. https://doi.org/10.1029/2005GC000965.

Aiuppa, A., Federico, C., Giudice, G., Gurrieri, S. (2005b). Chemical mapping of a fumarolic field: La Fossa Crater, Vulcano Island (Aeolian Islands, Italy). *Geophys. Res. Lett.*, 32, 1–4.

Aiuppa, A., Moretti, R., Federico, C., Giudice, G., Gurrieri, S., Liuzzo, M., Papale, P., Shinohara, H., Valenza, M. (2007). Forecasting Etna eruptions by real-time observation of volcanic gas composition. *Geology*, 35, 1115–1118.

Aiuppa, A., Federico, C., Giudice, G., Giuffrida, G., Guida, R., Gurrieri, S., Liuzzo, M., Moretti, R., Papale, P. (2009). The 2007 eruption of Stromboli volcano: Insights from real-time measurement of the volcanic gas plume CO_2/SO_2 ratio. *J. Volcanol. Geotherm. Res.*, 182, 221–230.

Aiuppa, A., Cannata, A., Cannavò, F., Di Grazia, G., Ferrari, F., Giudice, G., Gurrieri, S., Liuzzo M., Mattia, M., Montalto, P. et al. (2010). Patterns in the recent 2007–2008 activity of Mount Etna volcano investigated by integrated geo-physical and geochemical observations. *Geochemistry, Geophys. Geosystems*, 11, Q09008. https://doi.org/10.1029/2010GC003168.

Aiuppa, A., Burton, M., Allard, P., Caltabiano, T., Giudice, G., Gurrieri, S., Liuzzo, M., Salerno, G. (2011). First observational evidence for the CO_2-driven origin of Stromboli's major explosions. *Solid Earth*, 2(2), 135–142

Aiuppa, A., Robidoux, P., Tamburello, G., Conde, V., Galle, B., Avard, G., Bagnato, E., de Moor, J.M., Martínez, M., Muñóz, A. (2014). Gas measurements from the Costa Rica–Nicaragua volcanic segment suggest possible along-arc variations in volcanic gas chemistry. *Earth Planet. Sci. Lett.*, 407, 134–147.

Aiuppa, A., Fiorani, L., Santoro, S., Parracino, S., Nuvoli, M., Chiodini, G., Minopoli, C., Tamburello, G. (2015). New ground-based lidar enables volcanic CO_2 flux measurements. *Sci. Rep.*, 5, 13614.

Aiuppa, A., Fischer, T.P., Plank, T., Robidoux, P., Di Napoli, R. (2017). Along-arc, inter-arc and arc-to-arc variations in volcanic gas CO_2/ST ratios reveal dual source of carbon in arc volcanism. *Earth-Science Rev.*, 168, 24–47.

Aiuppa, A., de Moor, J.M., Arellano, S., Coppola, D., Francofonte, V., Galle, B., Giudice, G., Liuzzo, M., Mendoza, E., Saballos, A. et al. (2018). Tracking formation of a lava lake from ground and space: Masaya volcano (Nicaragua), 2014–2017. *Geochemistry, Geophys. Geosystems*, 19, 496–515.

Albert, H., Costa, F., Martí, J. (2014). Timing of magmatic processes and unrest associated with mafic historical monogenetic eruptions in Tenerife Island. *J. Petrol.*, 56, 1945–1966.

Albert, H., Costa, F., Di Muro, A., Herrin, J., Métrich, N., Deloule, E. (2019). Magma interactions, crystal mush formation, timescales, and unrest during caldera collapse and lateral eruption at ocean island basaltic volcanoes (Piton de la Fournaise, La Réunion). *Earth Planet. Sci. Lett.*, 515, 187–199.

Alfaro, C.M. and Zapata, G.J. (1997). Acid gas emissions from Galeras Volcano, Colombia, 1989–1994. *J. Volcanol. Geotherm. Res.*, 77, 209–228.

Allard, P., Carbonnelle, J., Dajlevic, D., Bronec, J.L., Morel, P., Robe, M.C., Maurenas, J.M., Faivre-Pierret, R., Martin, D., Sabroux, J.C. et al. (1991). Eruptive and diffuse emissions of CO_2 from Mount Etna. *Nature*, 351, 387–391.

Allard, P., Burton, M., Muré, F. (2005). Spectroscopic evidence for a lava fountain driven by previously accumulated magmatic gas. *Nature*, 433, 407–410.

Allard, P., Aiuppa, A., Beauducel, F., Gaudin, D., di Napoli, R., Calabrese, S., Parello, F., Crispi, O., Hammouya, G., Tamburello, G. (2014). Steam and gas emission rate from La Soufrière volcano, Guadeloupe (Lesser Antilles): Implications for the magmatic supply during degassing unrest. *Chem. Geol.*, 384, 76–93.

Allen, A.G., Baxter, P.J., Ottley, C.J. (2000). Gas and particle emissions from Soufrière Hills Volcano, Montserrat, West Indies: Characterization and health hazard assessment. *Bull. Volcanol.*, 62, 8–19.

Anderson, A.T. (1974). Chlorine, sulfur, and water in magmas and oceans. *Bull. Geol. Soc. Am.*, 85, 1485–1492.

Baker, D.R. (2008). The fidelity of melt inclusions as records of melt composition. *Contrib. Mineral. Petrol.*, 156, 377–395.

Barry, P.H., Hilton, D.R., Tryon, M.D., Brown, K.M., Kulongoski, J.T. (2009). A new syringe pump apparatus for the retrieval and temporal analysis of helium in groundwaters and geothermal fluids. *Geochemistry, Geophys. Geosystems*, 10. https://doi.org/10.1029/2009GC002422.

Barsanti, M., Papale, P., Barbato, D., Moretti, R., Boschi, E., Hauri, E., Longo, A. (2009). Heterogeneous large total CO_2 abundance in the shallow magmatic system of Kilauea volcano, Hawaii. *J. Geophys. Res. Solid Earth*, 114(2009), B12201.

Bernard, M.L., Molinié, J., Petit, R.H., Beauducel, F., Hammouya, G., Marion, G. (2006). Remote and in situ plume measurements of acid gas release from La Soufrière volcano, Guadeloupe. *J. Volcanol. Geotherm. Res.*, 150, 395–409.

Bluth, G.J.S., Shannon, J.M., Watson, I.M., Prata, A.J., Realmuto, V.J. (2007). Development of an ultra-violet digital camera for volcanic SO_2 imaging. *J. Volcanol. Geotherm. Res.*, 161, 47–56.

Bonifacie, M., Bardoux, G., Robert, V., Agrinier, P., Moune, S., Moretti, R., Dessert, C., Komorowski, J.C., Legendre, Y.M. (2019). Chlorine isotopes in volcanology: Example from La Soufrière de Guadeloupe (FWI). https://doi.org/10.1016/j.epsl.2014.12.044.

Boudoire, G. (2017). Architecture et dynamique des systèmes magmatiques associés aux volcans basaltiques : exemple du Piton de la Fournaise. PhD Thesis, Université de la Réunion, Saint-Denis.

Boudoire, G., Di Muro, A., Liuzzo, M., Ferrazzini, V., Peltier, A., Gurrieri, S., Michon, L., Giudice, G., Kowalski, P., Boissier, P. (2017). New perspectives on volcano monitoring in a tropical environment: Continuous measurements of soil CO_2 flux at Piton de la Fournaise (La Réunion Island, France). *Geophys. Res. Lett.*, 44, 8244–8253.

Brennwald, M.S., Schmidt, M., Oser, J., Kipfer, R. (2016). A portable and autonomous mass spectrometric system for on-site environmental gas analysis. *Environ. Sci. Technol.*, 50, 13455–13463.

Brusca, L., Inguaggiato, S., Longo, M., Madonia, P., Maugeri, R. (2004). The 2002–2003 eruption of Stromboli (Italy): Evaluation of the volcanic activity by means of continuous monitoring of soil temperature, CO_2 flux, and meteorological parameters. *Geochemistry, Geophys. Geosystems*, 51. https://doi.org/10.1029/2004GC000732.

Buck, A.L. (1981). New equations for computing vapor pressure and enhancement factor. *J. Appl. Meteorol.*, 20, 1527–1532.

Burgisser, A. and Scaillet, B. (2007). Redox evolution of a degassing magma rising to the surface. *Nature*, 445, 194–197.

Burton, M.R., Allard, P., Murè, F., Oppenheimer, C. (2003). FTIR remote sensing of fractional magma degassing at Mount Etna, Sicily. *Geol. Soc. Spec. Publ.*, 213, 281–293.

Burton, M.R., Sawyer, G.M., Granieri, D. (2013). Deep carbon emissions from volcanoes. *Rev. Mineral. Geochemistry*, 75, 323–354.

Camarda, M., Gurrieri, S., Valenza, M. (2006). CO_2 flux measurements in volcanic areas using the dynamic concentration method: Influence of soil permeability. *Journal of Geophys. Res. Solid Earth*, 111. https://doi.org/10.1029/2005JB003898.

Carapezza, M.L., Inguaggiato, S., Brusca, L., Longo, M. (2004). Geochemical precursors of the activity of an open-conduit volcano: The Stromboli 2002–2003 eruptive events. *Geophys. Res. Lett.*, 31(7), 7620. https://doi.org/10.1029/2004GL019614.

Chiodini, G. (2009). CO_2/CH_4 ratio in fumaroles a powerful tool to detect magma degassing episodes at quiescent volcanoes. *Geophys. Res. Lett.*, 36. https://doi.org/10.1029/2008GL036347.

Chiodini, G. and Marini, L. (1998). Hydrothermal gas equilibria: The H_2O-H_2-CO_2-CO-CH_4 system. *Geochim. Cosmochim. Ac.*, 62, 2673–2687.

Chiodini, G., Cioni, R., Marini, L. (1993). Reactions governing the chemistry of crater fumaroles from Vulcano Island, Italy, and implications for volcanic surveillance. *Appl. Geochemistry*, 8, 357–371.

Chiodini, G., Cioni, R., Frullani, A., Guidi, M., Marini, L., Prati, F., Raco, B. (1996). Fluid geochemistry of Montserrat Island, West Indies. *Bull. Volcanol.*, 58, 380–392.

Chiodini, G., Cioni, R., Guidi, M., Raco, B., Marini, L. (1998). Soil CO_2 flux measurements in volcanic and geothermal areas. *Appl. Geochem.*, 13, 543–552.

Chiodini, G., Frondini, F., Cardellini, C., Granieri, D., Marini, L., Ventura, G. (2001). CO_2 degassing and energy release at Solfatara volcano, Campi Flegrei, Italy. *J. Geophys. Res.*, 106, 16213–16221.

Chiodini, G., Caliro, S., De Martino, P., Avino, R., Gherardi, F. (2012). Early signals of new volcanic unrest at Campi Flegrei caldera? Insights from geochemical data and physical simulations. *Geology*, 40, 943–946.

Christenson, B.W., White, S., Britten, K., Scott, B.J. (2017). Hydrological evolution and chemical structure of a hyper-acidic spring-lake system on Whakaari/White Island, NZ. *J. Volcanol. Geotherm. Res.*, 346, 180–211.

Delmelle, P., Stix, J., Gases, J.V. (2000). *Encyclopedia of Volcanoes*. Elsevier, Oxford.

Delmelle, P., Villiéras, F., Pelletier, M. (2005). Surface area, porosity and water adsorption properties of fine volcanic ash particles. *Bull. Volcanol.*, 67, 160–169.

Delmelle, P., Wadsworth, F.B., Maters, E.C., Ayris, P.M. (2018). High-temperature reactions between gases and ash particles in volcanic eruption plumes. *Rev. Mineral. Geochemistry*, 84, 285–308.

Devine, J.D., Sigurdsson, H., Davis, A.N., Self, S. (1984). Estimates of sulfur and chlorine yield to the atmosphere from volcanic eruptions and potential climatic effects. *J. Geophys. Res.*, 89, 6309–6325.

Di Muro, A., Neri, A., Rosi, M. (2004). Contemporaneous convective and collapsing eruptive dynamics: The transitional regime of explosive eruptions. *Geophys. Res. Lett.*, 31. https://doi.org/10.1029/2004GL019709.

Di Muro, A., Giordano, D., Villemant, B., Montagnac, G., Scaillet, B., Romano, C. (2006). Influence of composition and thermal history of volcanic glasses on water content as determined by micro-Raman spectrometry. *Appl. Geochemistry*, 21, 802–812.

Dunbar, N.W., Hervig, R.L., Kyle, P.R. (1989). Determination of pre-eruptive H_2O, F and Cl contents of silicic magmas using melt inclusions: Examples from Taupo volcanic center, New Zealand. *Bull. Volcanol.*, 51, 177–184.

Edmonds, M. (2008). New geochemical insights into volcanic degassing. *Philos. Trans. R. Soc. A Math. Phys. Eng. Sci.*, 366, 4559–4579.

Edmonds, M., Pyle, D., Oppenheimer, C. (2002). HCl emissions at Sofrière Hills volcano, Montserrat, West Indies, during a second phase of dome building: November 1999 to October 2000. *Bull. Volcanol.*, 64, 21–30.

Elias, T. and Sutton, A.J. (2012). Sulfur dioxide emission rates from Kīlauea Volcano, Hawai'i, 2007–2010. *US Geol. Surv. Open-File Rep.*, 1107, 25.

Faivre-Pierret, R. (1983). SO_2, HCl, and HF detection and dosing in the volcanic gas phase. In *Forecasting Volcanic Events*, Tazieff, H (ed.). Elsevier Science Ltd, Amsterdam.

Finizola, A., Sortino, F., Lénat, J.F., Valenza, M. (2002). Fluid circulation at Stromboli volcano (Aeolian Islands, Italy) from self-potential and CO_2 surveys. *J. Volcanol. Geotherm. Res.*, 116, 1–18.

Finnegan, D.L., Kotra, J.P., Hermann, D.M., Zoller, W.H. (1989). The use of 7LiOH-impregnated filters for the collection of acidic gases and analysis by instrumental neutron activation analysis. *Bull. Volcanol.*, 51, 83–87.

Fischer, T.P. (2008). Fluxes of volatiles (H_2O, CO_2, N_2, Cl, F) from arc volcanoes. *Geochem. J.*, 42, 21–38.

Flaherty, T., Druitt, T.H., Tuffen, H., Higgins, M.D., Costa, F., Cadoux, A. (2018). Multiple timescale constraints for high-flux magma chamber assembly prior to the late bronze age eruption of Santorini (Greece). *Contrib. Mineral. Petrol.*, 173, 75.

Francis, P., Oppenheimer, C., Stevenson, D. (1993). Endogenous growth of persistently active volcanoes. *Nature*, 366, 554–557.

Galle, B., Oppenheimer, C., Geyer, A., McGonigle, A.J.S., Edmonds, M., Horrocks, L. (2003). A miniaturized ultraviolet spectrometer for remote sensing of SO2 fluxes: A new tool for volcano surveillance. *J. Volcanol. Geotherm. Res.*, 119, 241–254.

Gauthier, P.-J., Sigmarsson, O., Gouhier, M., Haddadi, B., Moune, S. (2016). Elevated gas flux and trace metal degassing from the 2014–2015 fissure eruption at the Bárar-Bunga volcanic system, Iceland. *J. Geophys. Res. Solid Earth*, 121(3), 1610–1630.

Giammanco, S., Gurrieri, S., Valenza, M. (2006). Fault-controlled soil CO_2 degassing and shallow magma bodies: Summit and lower East Rift of Kilauea volcano (Hawaii), 1997. *Pure Appl. Geophys.*, 163, 853–867.

Giggenbach, W.F. (1980). Geothermal gas equilibria. *Geochim. Cosmochim. Acta*, 44, 2021–2032.

Giggenbach, W.F. (1996). Chemical composition of volcanic gases. In *Monitoring and Mitigation of Volcano Hazards*, Scarpa, R. and Tilling, R.I. (eds). Springer, Berlin.

Giggenbach, W.F. and Goguel, R.L. (1989). Methods for the collection and analysis of geothermal and volcanic water and gas samples. Report, Chemistry Division, Dept. of Scientific and Industrial Research, Petone, New Zealand.

Giggenbach, W.F. and Soto, R.C. (1992). Isotopic and chemical composition of water and steam discharges from volcanic-magmatic-hydrothermal systems of the Guanacaste Geothermal Province, Costa Rica. *Appl. Geochemistry*, 7, 309–332.

von Glasow, R., Bobrowski, N., Kern, C. (2009). The effects of volcanic eruptions on atmospheric chemistry. *Chem. Geol.*, 263, 131–142.

Goff, F., Janik, C.J., Delgado, H., Werner, C., Counce, D., Stimac, J.A., Siebe, C., Love, S.P., Williams, S.N., Fischer, T. et al. (1998). Geochemical surveillance of magmatic volatiles at Popocatépetl volcano, Mexico. *GSA Bull.*, 110, 695–710.

Gouhier, M. and Coppola, D. (2011). Satellite-based evidence for a large hydrothermal system at Piton de la Fournaise volcano (Reunion Island). *Geophys. Res. Lett.*, 38, L02302. https://doi.org/10.1029/2010GL046183.

Gudmundsson, M.T., Pedersen, R., Vogfjörd, K., Thorbjarnardóttir, B., Jakobsdóttir, S., Roberts, M.J. (2010). Eruptions of Eyjafjallajökull Volcano, Iceland. *Eos, Trans. Am. Geophys. Union*, 91, 190–191.

Gurrieri, S. and Valenza, M. (1988). Gas transport in natural porous mediums: A method for measuring CO_2 flows from the ground in volcanic and geothermal areas. *Rend. Soc. Ital. Miner. Pet.*, 43, 1151–1158.

Hammouya, G., Allard, P., Jean-Baptiste, P., Parello, F., Semet, M.P., Young, S.R. (1998). Pre- and syn-eruptive geochemistry of volcanic gases from Soufriere Hills of Montserrat, West Indies. *Geophys. Res. Lett.*, 25, 3685–3688.

Harvey, M.C. and Harvey, C.C. (2015). Soil CO_2 flux surveys: A review of the technique in geothermal exploration. *Proceedings of the World Geothermal Congress*, Melbourne, April 19–25.

Hernández, P., Pérez, N., Salazar, J., Reimer, M., Notsu, K., Wakita, H. (2004). Radon and helium in soil gases at Cañadas caldera, Tenerife, Canary Islands, Spain. *J. Volcanol. Geotherm. Res.*, 131, 59–76.

Hernández, P., Melián, G., Giammanco, S., Sortino, F., Barrancos, J., Pérez, N.M., Padrón, E., López, M., Donovan, A., Mori, T. et al. (2015). Contribution of CO_2 and H_2S emitted to the atmosphere by plume and diffuse degassing from volcanoes: The Etna volcano case study. *Surv. Geophys.*, 36, 327–349.

Hilton, D.R., Fischer, T.P., Marry, B. (2002). Noble gases and volatile recycling at subduction zones. *Rev. Mineral. Geochemistry*, 47, 319–370.

Hilton, D.R., Ramírez, C.J., Mora-Amador, R., Fischer, T.P., Füri, E., Barry, P.H., Shaw, A.M. (2010). Monitoring of temporal and spatial variations in fumarole helium and carbon dioxide characteristics at Poás and Turrialba Volcanoes, Costa Rica (2001–2009). *Geochem. J.*, 44, 431–440.

Horton, K.A., Williams-Jones, G., Garbeil, H., Elias, T., Sutton, A.J., Mouginis-Mark, P., Porter, J.N., Clegg, S. (2006). Real-time measurement of volcanic SO_2 emissions: Validation of a new UV correlation spectrometer (FLYSPEC). *Bull. Volcanol.*, 68, 323–327.

Huygen, C. (1963a). The sampling of hydrogen fluoride in air with impregnated filter paper. *Anal. Chim. Acta*, 29, 448–452.

Huygen, C. (1963b). The sampling of sulfur dioxide in air with impregnated filter paper. *Anal. Chim. Acta*, 28, 349–360.

Inguaggiato, S., Mazot, A., Ohba, T. (2011). Monitoring active volcanoes: The geochemical approach. *Ann. Geophys.*, 54, 115–119.

Italiano, F. (1991). Monitoring of acid gases at the crater of Vulcano using the method of the alkaline traps. *Acta Vulcanol.*, 1, 249–254.

James, M., Carr, B., D'Arcy, F., Diefenbach, A., Dietterich, H., Fornaciai, A., Lev, E., Liu, E., Pieri, D., Rodgers, M. et al. (2020). Volcanological applications of unoccupied aircraft systems (UAS): Developments, strategies, and future challenges. *Volcanica*, 3(1), 67–114.

Jaupart, C. (1996). Physical models of volcanic eruptions. *Chem. Geol.*, 128, 217–227.

Jaupart, C. (1998). Gas loss from magmas through conduit walls during eruption. *Geological Society*, 145, 73–90.

Jones, M.T., Jerram, D.A., Svensen, H.H., Grove, C. (2016). The effects of large igneous provinces on the global carbon and sulphur cycles. *Palaeogeogr. Palaeoclimatol. Palaeoecol.*, 441, 4–21.

Kagoshima, T., Sano, Y., Takahata, N., Ishida, A., Tomonaga, Y., Roulleau, E., Pinti, D.L., Fischer, T.P., Lan, T., Nishio, Y. et al. (2016). Spatial and temporal variations of gas geochemistry at Mt. Ontake, Japan. *J. Volcanol. Geotherm. Res.*, 325, 179–188.

Kahl, M., Chakraborty, S., Costa, F., Pompilio, M. (2011). Dynamic plumbing system beneath volcanoes revealed by kinetic modeling, and the connection to monitoring data: An example from Mt. Etna. *Earth Planet. Sci. Lett.*, 308, 11–22.

Kahl, M., Chakraborty, S., Costa, F., Pompilio, M., Liuzzo, M., Viccaro, M. (2013). Compositionally zoned crystals and real-time degassing data reveal changes in magma transfer dynamics during the 2006 summit eruptive episodes of Mt. Etna. *Bull. Volcanol.*, 75, 1–14.

King, P.L., Wheeler, V.M., Renggli, C.J., Palm, A.B., Wilson, S.A., Harrison, A.L., Morgan, B., Nekvasil, H., Troitzsch, U., Mernagh, T. et al. (2018). Gas–solid reactions: Theory, experiments and case studies relevant to earth and planetary processes. *Rev. Mineral. Geochemistry*, 84, 1–56.

Kyle, P.R., Meeker, K., Finnegan, D. (1990). Emission rates of sulfur dioxide, trace gases and metals from Mount Erebus, Antarctica. *Geophys. Res. Lett.*, 17, 2125–2128.

Lee, H., Muirhead, J.D., Fischer, T.P., Ebinger, C.J., Kattenhorn, S.A., Sharp, Z.D., Kianji, G. (2016). Massive and prolonged deep carbon emissions associated with continental rifting. *Nat. Geosci.*, 9, 145–149.

Lewicki, J.L., Evans, W.C., Hilley, G.E., Sorey, M.L., Rogie, J.D., Brantley, S.L. (2003). Shallow soil CO_2 flow along the San Andreas and Calaveras Faults, California. *J. Geophys. Res. Solid Earth*, 108. https://doi.org/10.1029/2002JB002141.

Liotta, M., Paonita, A., Caracausi, A., Martelli, M., Rizzo, A., Favara, R. (2010). Hydrothermal processes governing the geochemistry of the crater fumaroles at Mount Etna volcano (Italy). *Chem. Geol.*, 278, 92–104.

Liu, E.J., Wood, K., Mason, E., Edmonds, M., Aiuppa, A., Giudice, G., Bitetto, M., Francofonte, V., Burrow, S., Richardson, T. et al. (2019). Dynamics of degassing and plume transport revealed by proximal Unmanned Aerial System (UAS) measurements at Volcán Villarrica, Chile. *Geochemistry, Geophys. Geosystems*, 20, 730–750.

Liuzzo, M., Gurrieri, S., Giudice, G., Giuffrida, G. (2013). Ten years of soil CO_2 continuous monitoring on Mt. Etna: Exploring the relationship between processes of soil degassing and volcanic activity. *Geochemistry, Geophys. Geosystems*, 14, 2886–2899.

Lowenstern, J.B. (1995). Applications of silicate-melt inclusions to the study of magmatic volatiles. *Magmas, Fluids Ore Depos. Mineral. Assoc. Canada Short Course*, 23, 71–99.

Lowenstern, J.B. (2003). Melt inclusions come of age: Volatiles, volcanoes, and sorby's legacy. In *Developments in Volcanology*, volume 5, De Vivo, B. and Bodnar, V. (eds). https://doi.org/10.1016/S1871-644X(03)80021-9.

Mandon, C.L., Christenson, B.W., Schipper, C.I., Seward, T.M., Garaebiti, E. (2019). Metal transport in volcanic plumes: A case study at White Island and Yasur volcanoes. *J. Volcanol. Geotherm. Res.*, 369, 155–171.

Martin, R.S., Wheeler, J.C., Ilyinskaya, E., Braban, C.F., Oppenheimer, C. (2012). The uptake of halogen (HF, HCl, HBr and HI) and nitric (HNO 3) acids into acidic sulphate particles in quiescent volcanic plumes. *Chem. Geol.*, 296–297, 19–25.

Mather, T.A. (2008). Volcanism and the atmosphere: The potential role of the atmosphere in unlocking the reactivity of volcanic emissions. *Philos. Trans. R. Soc. A Math. Phys. Eng. Sci.*, 366, 4581–4595.

Mather, T.A. (2015). Volcanoes and the environment: Lessons for understanding Earth's past and future from studies of present-day volcanic emissions. *J. Volcanol. Geotherm. Res.*, 304, 160–179.

Mather, T.A., Allen, A.G., Oppenheimer, C., Pyle, D.M., McGonigle, A.J.S. (2003). Size-resolved characterisation of soluble ions in the particles in the tropospheric plume of Masaya Volcano, Nicaragua: Origins and plume processing. *J. Atmos. Chem.*, 46, 207–237.

Mather, T.A., Pyle, D.M., Heaton, T.H.E. (2008). Investigation of the use of filter packs to measure the sulphur isotopic composition of volcanic sulphur dioxide and the sulphur and oxygen isotopic composition of volcanic sulphate aerosol. *Atmos. Environ.*, 42, 4611–4618.

Matsuo, S. (1960). On the origin of volcanic gases. *Jour. Earth Sci. Nagoya Univ.*, 8, 222–245.

Ménard, G. (2014). Comportement des éléments traces au cours des processus de dégazage. Étude des volcans Piton de la Fournaise (Réunion) et Lascar (Chili). PhD Thesis, Université Blaise Pascal – Clermont-Ferrand II, Clermont-Ferrrand.

Ménard, G., Moune, S., Vlastélic, I., Aguilera, F., Valade, S., Bontemps, M., González, R. (2014). Gas and aerosol emissions from Lascar volcano (Northern Chile): Insights into the origin of gases and their links with the volcanic activity. *J. Volcanol. Geotherm. Res.*, 287, 51–67.

Metrich, N. and Clocchiatti, R. (1996). Sulfur abundance and its speciation in oxidized alkaline melts. *Geochim. Cosmochim. Acta*, 60, 4151–4160.

Metrich, N. and Wallace, P.J. (2008). Volatile abundances in basaltic magmas and their degassing paths tracked by melt inclusions. *Rev. Mineral. Geochemistry*, 69, 363–402.

Metrich, N., Clocchiatti, R., Mosbah, M., Chaussidon, M. (1993). The 1989–1990 activity of Etna magma mingling and ascent of H2O-Cl-S-rich basaltic magma. Evidence from melt inclusions. *J. Volcanol. Geotherm. Res.*, 59, 131–144.

Metrich, N., Allard, P., Spilliaert, N., Andronico, D., Burton, M. (2004). 2001 flank eruption of the alkali- and volatile-rich primitive basalt responsible for Mount Etna's evolution in the last three decades. *Earth Planet. Sci. Lett.*, 228, 1–17.

Mioduszewski, L. and Kress, V. (2008). Laboratory calibration of chemical volcanic gas sampling techniques using an artificial fumarole. *J. Volcanol. Geotherm. Res.*, 174, 295–306.

Mizutani, Y. and Sugiura, T. (1996). The chemical equilibrium of the SO2+2H2S$\leftrightarrow$3S+2H2O reaction in solfataras of the Nasudake Volcano. *Bull. Chem. Soc. Jpn.*, 39, 2411–2414.

de Moor, J., Fischer, T.P., Sharp, Z.D., King, P.L., Wilke, M., Botcharnikov, R.E., Cottrell, E., Zelenski, M., Marty, B., Klimm, K. et al. (2013). Sulfur degassing at Erta Ale (Ethiopia) and Masaya (Nicaragua) volcanoes: Implications for degassing processes and oxygen fugacities of basaltic systems. *Geochemistry, Geophys. Geosystems*, 14, 4076–4108.

de Moor, J.M., Aiuppa, A., Avard, G., Wehrmann, H., Dunbar, N., Muller, C., Tamburello, G., Giudice, G., Liuzzo, M., Moretti, R. et al. (2016a). Turmoil at Turrialba volcano (Costa Rica): Degassing and eruptive processes inferred from high-frequency gas monitoring. *J. Geophys. Res. Solid Earth*, 121, 5761–5775.

de Moor, J.M., Aiuppa, A., Pacheco, J., Avard, G., Kern, C., Liuzzo, M., Martínez, M., Giudice, G., Fischer, T.P. (2016b). Short-period volcanic gas precursors to phreatic eruptions: Insights from Poás Volcano, Costa Rica. *Earth Planet. Sci. Lett.*, 442, 218–227.

de Moor, J.M., Kern, C., Avard, G., Muller, C., Aiuppa, A., Saballos, A., Ibarra, M., LaFemina, P., Protti, M., Fischer, T.P. (2017). A new sulfur and carbon degassing inventory for the Southern Central American Volcanic Arc: The importance of accurate time-series data sets and possible tectonic processes responsible for temporal variations in arc-scale volatile emissions. *Geochemistry, Geophys. Geosystems*, 18, 4437–4468.

de Moor, J.M., Stix, J., Avard, G., Muller, C., Corrales, E., Diaz, J.A., Alan, A., Brenes, J., Pacheco, J., Aiuppa, A. et al. (2019). Insights on hydrothermal-magmatic interactions and eruptive processes at Poás Volcano (Costa Rica) from high-frequency gas monitoring and drone measurements. *Geophys. Res. Lett.*, 46, 1293–1302.

Moore, L.R., Gazel, E., Tuohy, R., Lloyd, A.S., Esposito, R., Steele-MacInnis, M., Hauri, E.H., Wallace, P.J., Plank, T., Bodnar, R.J. (2015). Bubbles matter: An assessment of the contribution of vapor bubbles to melt inclusion volatile budgets. *Am. Mineral.*, 100, 806–823.

Moretti, R. and Papale, P. (2004). On the oxidation state and volatile behavior in multicomponent gas–melt equilibria. *Chem. Geol.*, 213, 265–280.

Moretti, R., De Natale, G., Troise, C. (2017). A geochemical and geophysical reappraisal to the significance of the recent unrest at Campi Flegrei caldera (Southern Italy). *Geochemistry, Geophys. Geosystems*, 18, 1244–1269.

Moretti, R., Métrich, N., Arienzo, I., Di Renzo, V., Aiuppa, A., Allard, P. (2018). Degassing vs. eruptive styles at Mt. Etna volcano (Sicily, Italy). Part I: Volatile stocking, gas fluxing, and the shift from low-energy to highly explosive basaltic eruptions. *Chem. Geol.*, 482, 1–17.

Moretti, R., Komorowski, J.C., Ucciani, G., Moune, S., Jessop, D., de Chabalier, J.B., Beauducel, F., Bonifacie, M., Burtin, A., Vallée, M., et al. (2020). The 2018 unrest phase at La Soufrière of Guadeloupe (French West Indies) andesitic volcano: Scrutiny of a failed but prodromal phreatic eruption. *J. Volcanol. Geotherm. Res.*, 393, 106769.

Moune, S., Gauthier, P.-J., Gislason, S.R., Sigmarsson, O. (2006). Trace element degassing and enrichment in the eruptive plume of the 2000 eruption of Hekla volcano, Iceland. *Geochim. Cosmochim. Acta*, 70, 461–479.

Moune, S., Sigmarsson, O., Thordarson, T., Gauthier, P.-J. (2007). Recent volatile evolution in the magmatic system of Hekla volcano, Iceland. *Earth Planet. Sci. Lett.*, 255, 244–253.

Moune, S., Gauthier, P.-J., Delmelle, P. (2010). Trace elements in the particulate phase of the plume of Masaya Volcano, Nicaragua. *J. Volcanol. Geotherm. Res.*, 193, 232–244.

Newman, S. and Lowenstern, J.B. (2002). VOLATILECALC: A silicate melt-H_2O-CO_2 solution model written in Visual Basic for excel. *Comput. Geosci.*, 28, 597–604.

Noguchi, K. and Kamiya, H. (1963). Prediction of volcanic eruption by measuring the chemical composition and amounts of gases. *Bull. Volcanol.*, 26, 367–378.

Notsu, K., Sugiyama, K., Hosoe, M., Uemura, A., Shimoike, Y., Tsunomori, F., Sumino, H., Yamamoto, J., Mori, T., Hernández, P.A. (2005). Diffuse CO_2 efflux from Iwojima volcano, Izu-Ogasawara arc, Japan. *J. Volcanol. Geotherm. Res.*, 139, 147–161.

Oppenheimer, C., Francis, P., Stix, J. (1998). Depletion rates of sulfur dioxide in tropospheric volcanic plumes. *Geophys. Res. Lett.*, 25, 2671–2674.

Oppenheimer, C., Pyle, D.M., Barclay, J. (2003). *Volcanic Degassing*. Geological Society of London, London.

Oppenheimer, C., Fischer, T.P., Scaillet, B. (2014). Volcanic degassing: Process and impact. In *Treatise on Geochemistry*, Holland, H.D. and Turekian, K.K. (eds). Elsevier, Oxford. http://www.sciencedirect.com/science/article/pii/B9780080959757003041.

Óskarsson, N. (1980). The interaction between volcanic gases and tephra: Fluorine adhering to tephra of the 1970 Hekla eruption. *J. Volcanol. Geotherm. Res.*, 8, 251–266.

Ozima, M. and Podosek, F.A. (2001). *Noble Gas Geochemistry*. Cambridge University Press, Cambridge.

Papale, P. (2005). Determination of total H_2O and CO_2 budgets in evolving magmas from melt inclusion data. *J. Geophys. Res. Solid Earth*, 110, 1–13.

Papale, P., Moretti, R., Barbato, D. (2006). The compositional dependence of the saturation surface of H_2O+ CO_2 fluids in silicate melts. *Chem. Geol.*, 229, 78–95.

Pedone, M., Aiuppa, A., Giudice, G., Grassa, F., Cardellini, C., Chiodini, G., Valenza, M. (2014). Volcanic CO_2 flux measurement at Campi Flegrei by tunable diode laser absorption spectroscopy. *Bull. Volcanol.*, 76, 812.

Pérez, N.M., Padilla, G.D., Padrón, E., Hernández, P.A., Melián, G.V., Barrancos, J., Dionis, S., Nolasco, D., Rodríguez, F., Calvo, D. et al. (2012). Precursory diffuse CO_2 and H_2S emission signatures of the 2011–2012 El Hierro submarine eruption, Canary Islands. *Earth Planets Space*, 71, 1 (2019). https://doi.org/10.1186/s40623-018-0980-8.

Roberts, T.J., Braban, C.F., Oppenheimer, C., Martin, R.S., Freshwater, R.A., Dawson, D.H., Griffiths, P.T., Cox, R.A., Saffell, J.R., Jones, R.L. (2012). Electrochemical sensing of volcanic gases. *Chem. Geol.*, 332–333, 74–91.

Robock, A. (2000). Volcanic eruptions and climate. *Rev. Geophys.*, 38, 191–219.

Rose, W.I., Delene, D.J., Schneider, D.J., Bluth, G.J.S., Krueger, A.J., Sprod, I., McKee, C., Davies, H.L., Ernst, G.G.J. (1995). Ice in the 1994 Rabaul eruption cloud: Implications for volcano hazard and atmospheric effects. *Nature*, 375, 477–479.

Roulleau, E., Tardani, D., Sano, Y., Takahata, N., Vinet, N., Bravo, F., Muñoz, C., Sanchez, J. (2016). New insight from noble gas and stable isotopes of geothermal/hydrothermal fluids at Caviahue-Copahue volcanic complex: Boiling steam separation and water-rock interaction at shallow depth. *J. Volcanol. Geotherm. Res.*, 328, 70–83.

Ruth, D.C.S., Costa, F., Bouvet de Maisonneuve, C., Franco, L., Cortés, J.A., Calder, E.S. (2018). Crystal and melt inclusion timescales reveal the evolution of magma migration before eruption. *Nat. Commun.*, 9, 1–9.

Sano, Y., Kagoshima, T., Takahata, N., Nishio, Y., Roulleau, E., Pinti, D.L., Fischer, T.P. (2015). Ten-year helium anomaly prior to the 2014 Mt Ontake eruption. *Sci. Rep.*, 5, 13069.

Shinohara, H. (2005). A new technique to estimate volcanic gas composition: Plume measurements with a portable multi-sensor system. *J. Volcanol. Geotherm. Res.*, 143. Available at: http://www.sciencedirect.com/science/article/pii/S0377027304004238.

Shinohara, H. (2008). Excess degassing from volcanoes and its role on eruptive and intrusive activity. *Rev. Geophys.*, 46, RG4005. https://doi.org/10.1029/2007RG000244.

Shinohara, H. and Witter, J.B. (2005). Volcanic gases emitted during mild Strombolian activity of Villarrica volcano, Chile. *Geophys. Res. Lett.*, 32, 1–5.

Sigmarsson, O., Haddadi, B., Carn, S., Moune, S., Gudnason, J., Yang, K., Clarisse, L. (2013). The sulfur budget of the 2011 Grímsvötn eruption, Iceland. *Geophys. Res. Lett.*, 40.

Sparks, R.S.J. and Wilson, L. (1976). A model for the formation of ignimbrite by gravitational column collapse. *J. Geol. Soc. London.*, 132, 441–451.

Spilliaert, N., Métrich, N., Allard, P. (2006). S-Cl-F degassing pattern of water-rich alkali basalt: Modelling and relationship with eruption styles on Mount Etna volcano. *Earth Planet. Sci. Lett.*, 248, 772–786.

Stix, J. and de Moor, J.M. (2018). Understanding and forecasting phreatic eruptions driven by magmatic degassing. *Earth, Planets Sp.*, 70, 83.

Stoiber, R.E., Williams, S.N., Huebert, B.J. (1986). Sulfur and halogen gases at Masaya Caldera Complex, Nicaragua: Total flux and variations with time. *J. Geophys. Res. Solid Earth*, 91, 12215–12231.

Symonds, R.B. and Reed, M.H. (1993). Calculation of multicomponent chemical equilibria in gas-solid-liquid systems: Calculation methods, thermochemical data, and applications to studies of high-temperature volcanic gases with examples from Mt. St. Helens. *Am. J. Sci. States*, 293, 758–864.

Symonds, R.B., Reed, M.H., Rose, W.I. (1992). Origin, speciation, and fluxes of trace-element gases at Augustine volcano, Alaska: Insights into magma degassing and fumarolic processes. *Geochim. Cosmochim. Acta*, 56, 633–657.

Symonds, R.B., Rose, W.I., Bluth, G.J.S., Gerlach, T.M. (1994). Volcanic gas studies: Methods, results and applications. *Rev. Miner. Geochemistry*, 30, 1–66.

Symonds, R.B., Gerlach, T.M., Reed, M.H. (2001). Magmatic gas scrubbing: Implications for volcano monitoring. *J. Volcanol. Geotherm. Res.*, 108, 303–341.

Tamburello, G., Moune, S., Allard, P., Venugopal, S., Robert, V., Rosas-Carbajal, M., Deroussi, S., Kitou, G.-T., Didier, T., Komorowski, J.-C. et al. (2019). Spatio-temporal relationships between fumarolic activity, hydrothermal fluid circulation and geophysical signals at an arc volcano in degassing unrest: La Soufrière of Guadeloupe (French West Indies). *Geosci.*, 9, 480.

Tassi, F., Agusto, M., Lamberti, C., Caselli, A.T., Pecoraino, G., Caponi, C., Szentiványi, J., Venturi, S., Vaselli, O. (2017). The 2012–2016 eruptive cycle at Copahue volcano (Argentina) versus the peripheral gas manifestations: Hints from the chemical and isotopic features of fumarolic fluids. *Bull. Volcanol.*, 79, 69.

Textor, C., Graf, H.F., Herzog, M., Oberhuber, J.M. (2003). Injection of gases into the stratosphere by explosive volcanic eruptions. *J. Geophys. Res. Atmos.*, 108. https://doi.org/10.1029/2002JD002987.

Textor, C., Graf, H.-F., Timmreck, C., Robock, A. (2004). Emissions from volcanoes BT. In *Emissions of Atmospheric Trace Compounds*, Granier, C., Artaxo, P., Reeves, C.E. (eds). Springer, Dordrecht.

Toutain, J.-P. and Baubron, J.-C. (1999). Gas geochemistry and seismotectonics: A review. *Tectonophysics*, 304, 1–27.

Toutain, J., Quisefit, J., Briole, P., Aloupogiannis, P., Blanc, P., Robaye, G. (1995). Mineralogy and chemistry of solid aerosols emitted from Mount Etna. *Geochem. J.*, 29, 163–173.

Venugopal, S., Schiavi, F., Moune, S., Bolfan-Casanova, N., Druitt, T., Williams-Jones, G. (2020). Melt inclusion vapour bubbles: The hidden reservoir for major and volatile elements. *Sci. Rep.*, 10, 9034.

Villemant, B., Hammouya, G., Michel, A., Semet, M.P., Komorowski, J.C., Boudon, G., Cheminee, J.L. (2005). The memory of volcanic waters: Shallow magma degassing revealed by halogen monitoring in thermal springs of La Soufriére volcano (Guadeloupe, Lesser Antilles). *Earth Planet. Sci. Let.*, 237, 710–728.

Villemant, B., Komorowski, J.C., Dessert, C., Michel, A., Crispi, O., Hammouya, G., Beauducel, F., Chabalier, J.B. (2014). Evidence for a new shallow magma intrusion at La Soufrière of Guadeloupe (Lesser Antilles): Insights from long-term geochemical monitoring of halogen-rich hydrothermal fluids. *J. Volcanol. Geoth. Res.*, 285, 247–277.

Wallace, P.J. (2005). Volatiles in subduction zone magmas: Concentrations and fluxes based on melt inclusion and volcanic gas data. *J. Volcanol. Geotherm. Res.*, 140, 217–240.

Wallace, P.J., Kamenetsky, V.S., Cervantes, P. (2015). Melt inclusion CO_2 contents, pressures of olivine crystallization, and the problem of shrinkage bubbles. *Am. Mineral.*, 100, 787–794.

Wardell, L.J., Kyle, P.R., Counce, D. (2008). Volcanic emissions of metals and halogens from White Island (New Zealand) and Erebus volcano (Antarctica) determined with chemical traps. *J. Volcanol. Geotherm. Res.*, 177, 734–742.

Werner, C., Hurst, T., Scott, B., Sherburn, S., Christenson, B.W., Britten, K., Cole-Baker, J., Mullan, B. (2008). Variability of passive gas emissions, seismicity, and deformation during crater lake growth at White Island Volcano, New Zealand, 2002–2006. *J. Geophys. Res. Solid Earth*, 113. https://doi.org/10.1029/2007JB005094.

Wilson, L. (1976). Explosive volcanic eruptions III. Plinian eruption columns. *Geophys. J. R. Astron. Soc.*, 45, 543–556.

Wilson, L. and Head, J.W. (1981). Morphology and rheology of pyroclastic (of Mount St. Helens) flows and their deposits and guidelines for future observations. *U.S. Geological Survey Professional Paper*, 1250, 513–524.

Wilson, L., Sparks, R.S.J., Huang, T.C., Watkins, N.D. (1978). The control of volcanic column heights by eruption energetics and dynamics. *J. Geophys. Res. Solid Earth*, 83, 1829–1836.

Witt, M.L.I., Fischer, T.P., Pyle, D.M., Yang, T.F., Zellmer, G.F. (2008). Fumarole compositions and mercury emissions from the Tatun Volcanic Field, Taiwan: Results from multi-component gas analyser, portable mercury spectrometer and direct sampling techniques. *J. Volcanol. Geotherm. Res.*, 178, 636–643.

Witter, J.B., Kress, V.C., Calder, E.S., Delmelle, P., Stix, J. (2000). Passive degassing at Volcan Villarrica, south Chile. *EoS, Trans. Am. Geophys. Union*, 81(48).

Witter, J.B., Kress, V.C., Delmelle, P., Stix, J. (2004). Volatile degassing, petrology, and magma dynamics of the Villarrica Lava Lake, Southern Chile. *J. Volcanol. Geotherm. Res.*, 134, 303–337.

Wittmer, J., Bobrowski, N., Liotta, M., Giuffrida, G., Calabrese, S., Platt, U. (2014). Active alkaline traps to determine acidic-gas ratios in volcanic plumes: Sampling techniques and analytical methods. *Geochemistry, Geophys. Geosystems*, 15, 2797–2820.

Yamaoka, K., Geshi, N., Hashimoto, T., Ingebritsen, S.E., Oikawa, T. (2016). The phreatic eruption of Mt. Ontake volcano in 2014. *Earth, Planets Sp.*, Special issue, 68, 175.

Zellmer, G.F., Blake, S., Vance, D., Hawkesworth, C., Turner, S. (1999). Plagioclase residence times at two island arc volcanoes (Kameni Islands, Santorini, and Soufriere, St. Vincent) determined by Sr diffusion systematics. *Contrib. Mineral. Petrol.*, 136, 345–357.

Zellmer, G.F., Sparks, R.S.J., Hawkesworth, C.J., Wiedenbeck, M. (2003). Magma emplacement and remobilization timescales beneath Montserrat: Insights from Sr and Ba zonation in plagioclase phenocrysts. *J. Petrol.*, 44, 1413–1431.

Zhang, Y. and Cherniak, D. (2010). Reviews in mineralogy and geochemistry: Preface. *Rev. Mineral. Geochemistry*, 72, 231–279.

2

Monitoring of Eruptive Products: Deposits Associated with Pyroclastic Fallout

Lucia GURIOLI[1], Alessandro TADINI[1], Simon THIVET[1,2], Étienne MÉDARD[1], Carole BERTHOD[1] and Ivan VLASTELIC[1]

[1] *Laboratoire Magmas et Volcans, CNRS, IRD, OPGC, Université Clermont Auvergne, Clermont-Ferrand, France*

[2] *Department of Earth and Environmental Sciences, Ludwig-Maximilians-Universitat, Munich, Germany*

2.1. Introduction

The word "tephra" (or "pyroclast") refers to all particles ejected during volcanic eruptions. "Pyroclastic fallout" is the process of fallout of these particles, which is one of the most common processes in volcanology and is generally associated with all types of explosive eruptions (from Strombolian to Ultraplinian). This pyroclastic fallout also poses many hazards, potentially affecting aviation, building stability, and also respiratory health as well as agriculture (Blong 1984). For all these reasons, it is important to study carefully the products of pyroclastic fallout, both during (syn-eruptive measurements) and after (post-eruptive studies) an eruption. This chapter aims to show how the study and monitoring of pyroclastic fallout products play a key role in volcanic risk assessment. To do so, we will describe the main characteristics of pyroclastic fallout and associated deposits. For an even more detailed description, we suggest referring to the reviews by Wilson and Houghton (2000) and Houghton and Carey (2015). Next, after discussing the importance of different scales of study when analyzing pyroclastic fallout deposits (from whole

Hazards and Monitoring of Volcanic Activity 3,
coordinated by Jean-François LÉNAT.

deposits to particles), we will provide an overview on syn-eruptive measurements and post-eruptive studies that can be performed taking into account the aforementioned scales of study and measurement conditions.

2.2. General characteristics of pyroclastic fallout deposits

The pyroclastic fallout process is, in its simplest formulation, the sedimentation of pyroclasts through the atmosphere and their deposition on the Earth's surface, either from:

– primary volcanic plumes (for a more detailed description, we suggest referring to the study by Carey and Bursik (2015));

– their associated plumes or clouds (see Figures 2.1(a) and 2.1(b));

– secondary plumes or clouds from pyroclastic flows (see Chapter 3), often called "co-ignimbritic plumes" (see Figure 2.1(c)).

Based on pyroclast size and density, the study by Bursik et al. (1992) distinguishes:

– coarse particles that are ejected from the lower part of the primary plume and follow ballistic trajectories (see Figure 2.1(d));

– fragments that fall from the convective and central part of the primary plume;

– particles that fall from the associated plume;

– fine particles that are transported in the plume but disperse in the atmosphere and are not systematically observed in the fallout deposits.

The last two particle types described above are typical of co-ignimbritic plumes. For a more detailed description of volcanic plume sedimentation, refer to the study by Bonadonna and Phillips (2003).

Ballistic projectiles are hardly affected by their interaction with the atmosphere and follow parabolic trajectories controlled by their initial velocities and ejection angles. For other types of particles, the deposition mechanism is mostly controlled by the vertical settling velocity (often called "terminal velocity") but also by the horizontal wind speed. Deviations from this theory are frequently observed because many secondary mechanisms can occur both within the plume margins, for example, sedimentation waves (Bloomfield and Kerr 2000), or along the plume, for example, ash aggregation (Brown et al. 2012) and filament sedimentation (Freret-Lorgeril et al. 2020).

Figure 2.1. *a) Primary volcanic plume from the Plinian eruption of Pinatubo in 1991. b) Satellite photo of the volcanic plume from the sub-Plinian eruption of Puyehue-Cordòn Caulle in 2011. c) Simplified diagram of the generation of a co-ignimbritic plume from a pyroclastic density current. d) Infrared image of ballistic projectiles from a Strombolian explosion at Stromboli. For a color version of this figure, see www.iste.co.uk/lenat/hazards3.zip*

COMMENT ON FIGURE 2.1.– *a) USGS image. b) NASA image. c) Modified according to Carey and Bursik (2015). d) White lines show the trajectories of identified pyroclasts (modified according to Gurioli et al. (2014)).*

It is very important to be able to correctly recognize pyroclastic fallout deposits from other types of volcanic deposits (e.g. pyroclastic current deposits), as well as to be able to distinguish between different types of fallout deposits, whether they originate from ballistic projections (proximal locations of eruptive vents), primary plumes or co-ignimbritic plumes (Fisher and Schmincke 1984; Cas and Wright 1987; Freundt and Rosi 1998).

Figure 2.2. *a) Juvenile and non-juvenile pyroclasts from the fallout deposit of the Vesuvius eruption, 79 CE, Pompeii. b) Well-sorted fallout deposit of the Huaynaputina eruption, 1600 CE (modified from Prival et al. (2020)). c) Relatively uniform thicknesses of fallout deposits overlain by pyroclastic flow deposits (modified from Gurioli et al. (2007)). For a color version of this figure, see www.iste.co.uk/lenat/hazards3.zip*

COMMENT ON FIGURE 2.2.– *c) The deposits of fallout are the first two levels visible from the bottom. A deposit of pyroclastic flows, thickened at the level of the "paleo-valley" and thinned at the level of the "paleo-ridges", covers them; eruption of Vesuvius, 79 CE, site of Nola, Pompeii.*

Due to selective gravity-controlled sedimentation, pyroclastic fallout deposits from primary volcanic plumes are fairly easily recognizable at all scales in the field (small scale: pyroclasts alone; medium scale: outcrops; large scale: entire deposits) due to three main features:

– The sub-angular to angular shape of the pyroclasts (see Figure 2.2(a)): the absence of mechanical abrasion between each pyroclast (due to their gravitational sedimentation) allows them to keep their original shape, especially for juvenile pyroclasts (pyroclasts emitted during the studied eruption and made of magma emitted during this same eruption), which are generally less resistant to abrasion than non-juvenile fragments (fragments emitted during the studied eruption but made of rocks of different natures, not linked to the magma emitted during this same eruption).

– The generally good size sorting of the outcrops (see Figure 2.2(b)), which is a direct effect of gravity-driven vertical sedimentation of pyroclasts; although size sorting is generally good at the outcrop scale, it is still possible to identify several subpopulations that may reflect the presence of different types of pyroclasts (e.g. juvenile and non-juvenile fragments).

– The tendency to cover the topography uniformly (see Figure 2.2(c)) and to exhibit a decay in deposit thickness from a point of maximum thickness generally near or coincident with the eruptive vent, to more distal locations; this decay is generally exponential, but variations due to secondary depositional mechanisms are possible (see above). A similar decay trend can also be observed on the particle size medians of the deposits.

2.3. Importance of the scale of analysis (from whole deposit to particles)

Considering the above-mentioned characteristics, the study and monitoring of eruptive products are important at all scales, from the whole deposit to the particles that constitute it.

For fallout deposits, the subdivision into proximal, medial or distal deposits depends on the size of the eruption considered. Depending on eruptive styles and eruptive intensities, the fields of fallout deposits can vary greatly, with the size of the dispersal zone directly proportional to the intensity of the eruption (see Figures 2.3(a), 2.3(b), 2.3(c) and 2.3(d)). For a Plinian or sub-Plinian deposit, the proximal extension of a deposit is defined by a deposit thickness that is often highly variable in response to partial plume collapse and rugged topography. Medial and distal deposits are likely to reflect local, subtle, but important phenomena (e.g., secondary deposit thickening and thinning) (Carey and Sigurdsson 1982) and are strongly subject to post-eruptive processes such as remobilization, bioturbation compaction and soil formation (Yang and Bursik 2016).

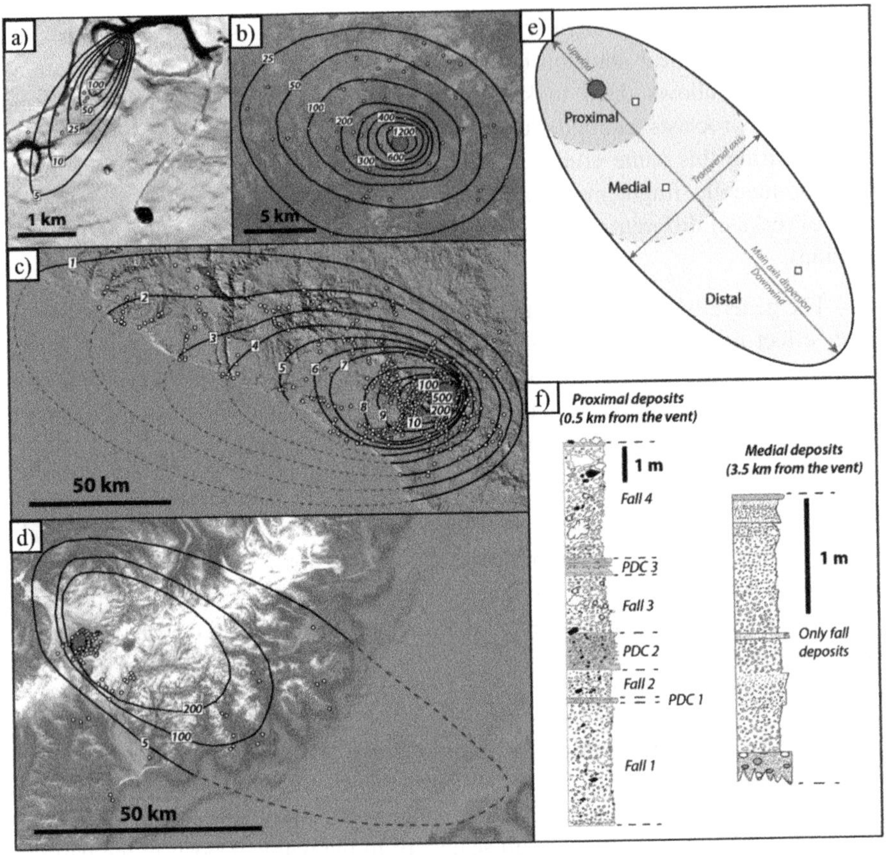

Figure 2.3. *Isopach maps of pyroclastic fallout deposits from different eruptions and spatial variation in deposits. For a color version of this figure, see www.iste.co.uk/lenat/hazards3.zip*

COMMENT ON FIGURE 2.3.– *a), b), c) Isopach maps of pyroclastic fallout deposits from different eruptions. a) Hawaiian activity (lava fountains) of the Kīlauea Iki eruption, 1959 (modified according to Mueller et al. (2018)). b) Violent strombolian activity of Paracutín, 1952 (modified according to Pioli et al. (2008)). c) Plinian eruption of Huaynaputina, 1600* CE *(modified according to Prival et al. (2020)). d) Plinian eruption of Novarupta, 1912 (modified according to Houghton et al. (2004)). e) Simplified schematic of areas that can be identified in a fallout deposit field. f) Complex stratigraphic logs of Plinian deposits from the Novarupta eruption, 1912, showing spatial and temporal variations in depositional characteristics (modified according to Houghton et al. (2004)).*

The geometry of the fallout deposit is strongly related both to the fallout mechanisms (e.g. ballistic or gravity sedimentation corresponding to Strombolian or Plinian dynamism, respectively) and to the wind direction. Because of the wind, fallout deposition fields are generally elliptical in shape (see Figures 2.3(a), 2.3(c) and 2.3(d)), although relatively circular fields may exist due to low wind intensity (see Figure 2.3(b)). In the case of an elliptical dispersion, it is possible to define the main dispersion axis, the windward and leeward sedimentation zones and the transverse zones (see Figure 2.3(e)). The thickness of a fallout deposit at a fixed point is directly proportional to the duration of the fallout event (the thicker the deposit, the longer the fallout event). The variation in particle size, always at a fixed point, can be correlated with the height of the volcanic plume: a normal gradation of particle size within the deposit (decrease in particle size as deposition proceeds) indicates a decrease in plume height, whereas an inverse gradation (increase in particle size as deposition proceeds) indicates an increase in plume height. In conclusion, a stratigraphic log (base to top) reflects syn-eruptive temporal variations in eruptive style, column height, wind direction and intensity. The study and sampling of a fallout deposit should be done with the understanding that the entire depositional area of an eruption reflects the spatiotemporal variations of the eruptive event (see Figure 2.3).

When the aim of the study is to quantify the proportion of different types of pyroclasts present within the deposit, it is important to remember that sedimentation is affected by the density, shape and size of the particles. This is particularly important when different types of pyroclasts, with very different densities, are emitted simultaneously during an eruptive phase: their relative distribution within the deposit may change with distance from the vent (and is therefore not necessarily representative of the population emitted during that eruptive phase), especially in the case of small plumes.

A preliminary study of the entire deposit at different locations is useful in assessing the significance of the identified outcrops. Variations in color, grain size, sorting, sedimentary structures, pyroclast nature and morphology, and any other features associated with post-fragmentation processes, should be evaluated to identify possible lateral and/or vertical variations in deposit facies. Therefore, and in some cases, it may be more appropriate to sample each tephra layer at different locations rather than selecting a single type outcrop.

2.4. Sampling and syn- and post-eruptive parameters

During eruptive crises (e.g. at Etna, Piton de la Fournaise or even Hawaii) a sampling of the eruptive products is generally carried out in the hours following the beginning of each eruption. The objective of this sampling is to identify and classify the different pyroclasts collected in terms of chemical composition (to trace the evolution of the magma composition), density and porosity (to trace the evolution of the degassing processes), rheology (to trace the evolution of the viscosity of the magma which evolves as a function of the chemical composition of the magma liquid, the gas bubble content and the crystal content), as well as the size, abundance, dispersion and morphology of the emitted particles (to identify the fragmentation mechanisms). All these data are integrated in time (from the base to the top of the deposit) and space (from proximal, medial and distal sites to the dispersal axis, upwind and downwind) and provide information on magmatic processes operating from the magma source through the eruptive vent to the volcanic plume.

Figure 2.4. *Different sampling strategies. For a color version of this figure, see www.iste.co.uk/lenat/hazards3.zip*

COMMENT ON FIGURE 2.4.– *a) Samples taken a few hours after an eruption at Villarrica, October 2004 (modified according to Gurioli et al. (2008)). b) Ash samples taken from solar panels after the April 2007 eruption at Piton de la Fournaise, Reunion Island, France (modified according to Michon et al. (2013)). c) Pre-existing surface cleaned before potential sampling at Stromboli (modified according to Pistolesi et al. (2011)). d) Plastic sheeting laid down near the active vent collecting proximal fallout deposits at Stromboli (modified according to Colò et al. (2020)). e) and f) Sampling with various collection vessels (modified according to Lautze et al. (2013) and Bernard (2013), respectively).*

During low-intensity basaltic eruptions, spatter larger than a few centimeters in diameter can be sampled by hand as soon as it falls to the ground under reasonable safety conditions, before being soaked in water. This soaking allows the pyroclasts to solidify rapidly, so that no post-eruptive information unrelated to pre- and syn-eruptive magmatic processes can be recorded. This type of sampling is for example performed at Villarrica and Stromboli (see Figure 2.4(a)) (Gurioli et al. 2008, 2014; Leduc et al. 2015), and also regularly at Piton de la Fournaise (Gurioli et al. 2018; Thivet et al. 2020a, 2020b). Samples can be collected on pre-existing surfaces (e.g. solar panels, see Figure 2.4(b)) cleaned prior to the eruption (see Figure 2.4(c)), on plastic sheets arranged near the vent (see Figure 2.4(d)) or with containers arranged in the field (see Figures 2.4(e) and 2.4(f)). For small pyroclasts (fragments less than about 2 cm in diameter), natural quenching is relatively rapid (Xu and Zhang 2002; Potuzak et al. 2008; Porritt et al. 2012) and is generally sufficient to maintain primary sample conditions.

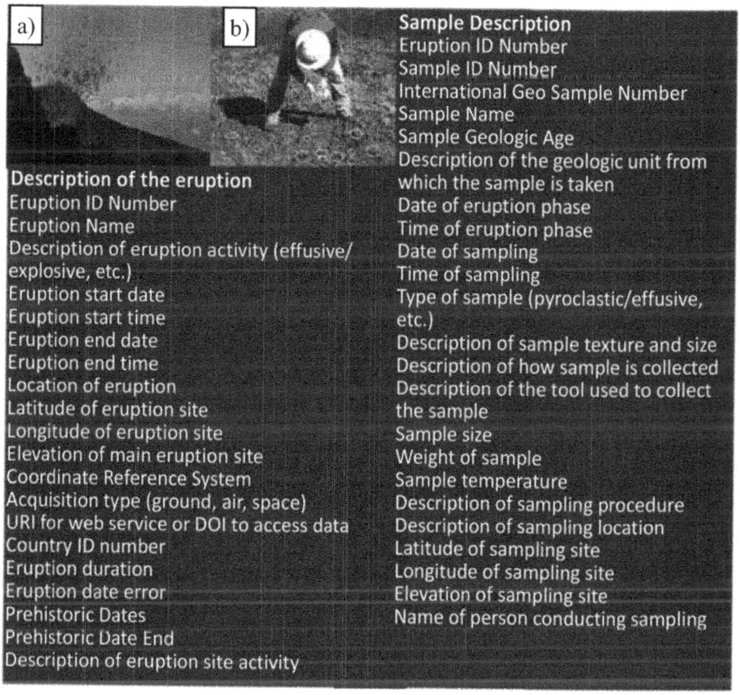

Figure 2.5. *Example of a list of information useful for describing the eruption and samples collected. a) Low-intensity strombolian explosion during the June 2014 eruption at Piton de la Fournaise (modified according to Gurioli et al. (2018)). b) Scoriaceous bombs and lapilli emitted by the eruption shown in (a). For a color version of this figure, see www.iste.co.uk/lenat/hazards3.zip*

When sampling, especially during an eruption, it is necessary to list all the information presented in Figure 2.5 in order to trace the origin of the sample, to locate it precisely at the geographical level and to insert it in the eruptive and stratigraphic sequence. It is also important to define the manner in which the sample was collected to verify whether the primitive signatures (textural and chemical) were preserved, modified or contaminated at the time of sampling.

Scale	Parameter	Syn-eruptive method(s)	Post-eruptive method(s)
Particles	Size (Φ or mm scale) of major axes and particle size distribution	LiDAR, RADAR, optical camera, infrared camera, ASHER disdrometer	Screening Image analysis Coulter counter
	Components (% by mass)	–	Optical and scanning electron microscopy (SEM) Image analysis Manual selection of particles
	Morphology of particles, crystals and vesicles	Infrared camera	Image analysis, SEM, Morphologi G3, X-ray tomography
	Density and porosity (kg/m^3 or g/cm^3)	–	Pycnometry Image analysis
	Vesicle connectivity (%)	–	Helium pycnometry
	Permeability (m^2)	–	Permeameter
	Sedimentation rate (m/s)	ASHER disdrometer	Calculations by numerical models
	Vesicle and crystal content (% by volume) and vesicle and crystal size distribution	–	Optical microscopy, SEM, image analysis, calculations by numerical models
	Total rock composition (major and trace elements)	–	X-ray fluorescence (XRF), optical spectrometry (ICP-AES), mass spectroscopy (ICP-MS)
	Composition of residual materials (major, trace and volatile elements)	–	Electron microprobe (EPMA), laser ablation ICP-MS (LA-ICP-MS), Fourier transform infrared (FTIR) and Raman spectroscopy
	Composition of minerals	–	EPMA LA-ICP-MS

Scale	Parameter	Syn-eruptive method(s)	Post-eruptive method(s)
Particles	Isotopic analysis	–	Mass spectrometry (TIMS)
Outcrop and deposition	Orientation of the dispersion axis of the fallout field (°) and dispersion area (km^2)	Satellite images Aerial images	Field measurements Isopach map
	Mass load (or surface density, kg/m^2)	Installation of tarpaulins ASHER	Isomass map
	Total volume (km^3 or m^3) and total mass (kg) of the deposit	–	Integration of isopach maps
	Total particle size distribution	Infrared and/or RADAR images	Average of single samples
	Height of the eruptive column (m or km)	RADAR LiDAR, infrared images Aerial observation reports	Isopleth map Numerical model calculations
	Mass Discharge Rate (MDR, kg/s)	Infrasound signals RADAR images Satellite images Optical camera	Total mass of the deposit Duration of the eruption Height models of the eruptive column
	Temperature (°C or K)	Infrared images	Petrology Paleomagnetic measurements
	Ballistic trajectories and flight parameters of projectiles	Optical camera Infrared camera	Location of impact craters and sizes of plumes and blocks

Table 2.1. *Important parameters of pyroclastic fallout for hazard assessment*

The characteristics listed in Figure 2.5 and the measurements described in Table 2.1 are important for monitoring the evolution of a volcanic eruption in real time, for identifying the key parameters necessary for a response to a volcanic crisis and for providing fundamental quantitative measurements for various numerical models. In parallel to syn-eruptive measurements, an accurate post-eruptive fallout study represents an important step in the definition of the volcanic hazard, especially in a long-term perspective: fallout deposits can provide information on the general eruptive history of a volcano (with indications on the magnitude, frequency and intensity of expected volcanic events), on the evolution of activity during a single eruption as well as on the occurrence of different specific mechanisms (e.g. particle fragmentation and ash aggregation), and more generally on the phenomenon of fallout and the associated impact area. In the following sections, we will explain the importance of these measurements.

2.5. Pyroclastic scale measurements

2.5.1. *Particle size and distribution*

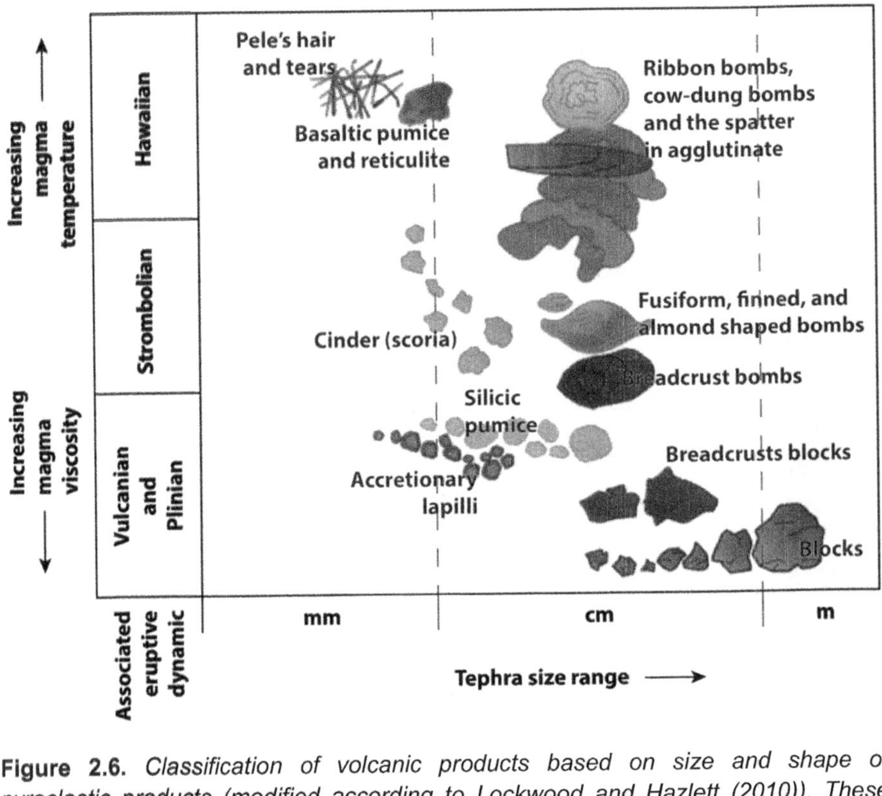

Figure 2.6. *Classification of volcanic products based on size and shape of pyroclastic products (modified according to Lockwood and Hazlett (2010)). These size and shape parameters vary with the chemical composition of the magmas and the type and intensity of the eruptions. For a color version of this figure, see www.iste.co.uk/lenat/hazards3.zip*

Focusing on the size of the pyroclasts, a classification has been established (Fisher 1966; Schmid 1981), describing fragments with an average size greater than 64 mm as bombs (rounded shape, see Figure 2.6) or blocks (angular shape).

Between 64 mm and 2 mm, the fragments are named lapilli. Finally, pyroclasts, with an average size of less than 2 mm, are called ash. The Φ scale is widely used to describe the size of pyroclastic particles (Krumbein 1934, 1938): this logarithmic scale is indeed convenient for visualizing and statistically analyzing the particle size distribution (PSD) of a deposit over a wide range of particle sizes, where $\Phi = -\log_2 d$ (d represents the intermediate particle diameter in mm).

Determination of particle size provides fragmentation characteristics of the associated eruption, genetic interpretation of pyroclastic deposits (fallout or flow) and classification of volcanic events in combination with deposit dispersion (Walker 1971; Cas and Wright 1987; Wohletz et al. 1989; Fagents et al. 2013). Block or ballistic bomb size can provide insights into the kinetic energy of the eruption (Gurioli et al. 2013) and proximity to the eruptive source (Steinberg and Lorenz 1983), while the percentage of ash and fine ash (average size less than 63 μm) can provide insights into fragmentation energy (Thivet et al. 2020a), and also into the hazard associated with our ability to inhale this fraction of fine particles (Horwell and Baxter 2006).

2.5.2. *Syn-eruptive methods for determining pyroclast size and pyroclast size distribution*

In the aftermath of the 2010 Ejafyallajökull (Iceland) eruption, numerous working groups and think tanks (Bonadonna 2014; Brenot et al. 2014) have all identified the importance of real-time monitoring of the PSD emitted by volcanic activity, in synergy with plume modeling, as a fundamental element for better risk assessment and effective volcanic crisis management.

The PSD generated during an explosive eruption is primarily controlled by magmatic fragmentation processes, explosion energy, magma rheology and magma ascent rate within the conduit (Kaminski and Jaupart 1998; Rust and Cashman 2011; Costa et al. 2016). This PSD is also a fundamental parameter for quantifying hazards because it affects volcanic plume dispersal processes and the distribution and impact energy of pyroclastic particles falling to the ground (Hurst and Turner 1999; Costa et al. 2006; Mastin et al. 2009). Thus, an improved, data-driven understanding of magma fragmentation and plume generation processes is needed to inform forecasting models.

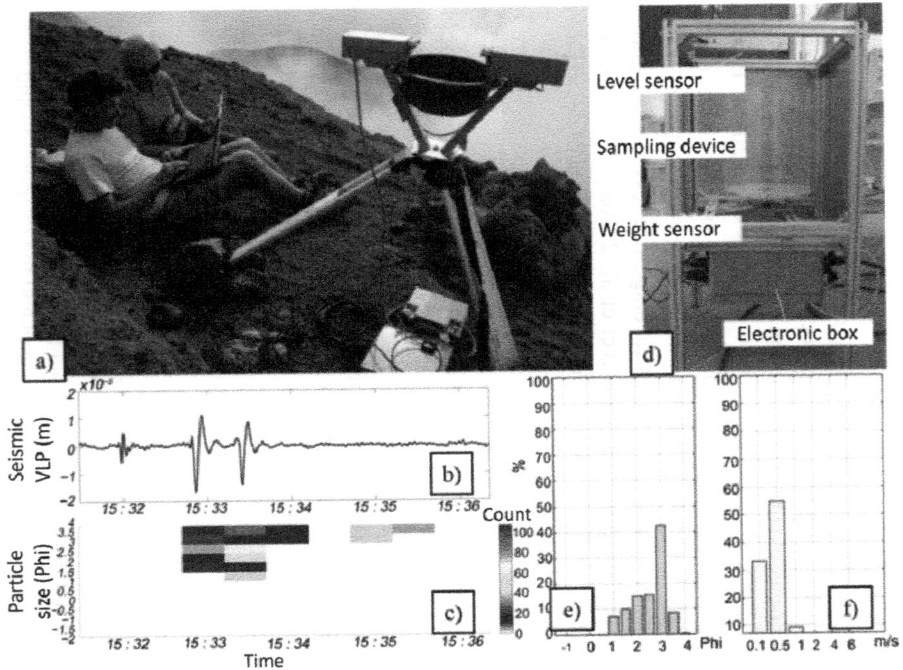

Figure 2.7. *a) Real-time data verification as ash falls within the ASHER light barrier, Stromboli (Italy, July 2016). b) Seismic data (VLP) measured during an explosion. c) In situ counting of ash particles by ASHER, during the same explosion. d) Photo of the ASHER laboratory collector that allows real-time collection of pyroclastic fallout in the field, measuring the accumulation rate of the associated deposit (in terms of mass and thickness). e) In situ particle size distribution. f) Terminal velocity of particles measured during the eruptive event. For a color version of this figure, see www.iste.co.uk/lenat/hazards3.zip*

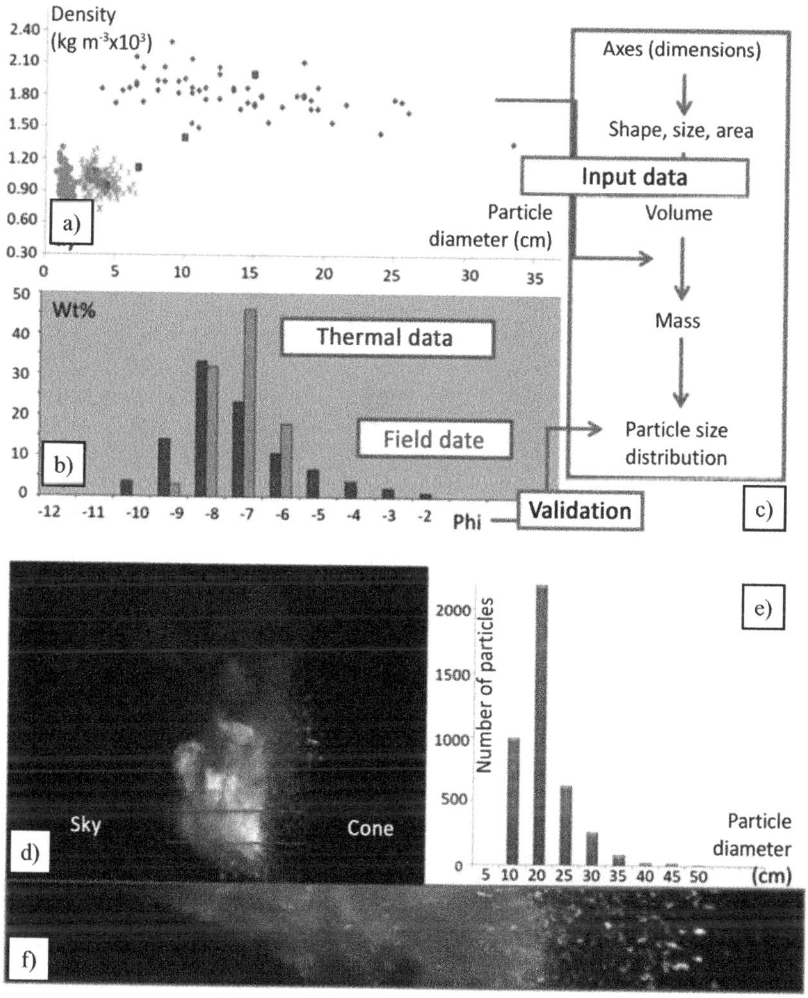

Figure 2.8. *Density and size of bombs and slag from an explosion at Stromboli. For a color version of this figure, see www.iste.co.uk/lenat/hazards3.zip*

COMMENT ON FIGURE 2.8.– *a) Density of scoriaceous lapilli and bombs from Stromboli, as a function of size. b) Particle size distribution from a Strombolian explosion, obtained both by sieving the associated deposit in the laboratory and by thermal camera. c) Schematic showing input and output data for validation of particle size data. d) Plume imaged with the thermal camera. e) Data associated with the image captured in (d). f) Detail of the thermal image shown in (d), modified according to Bombrun et al. (2015).*

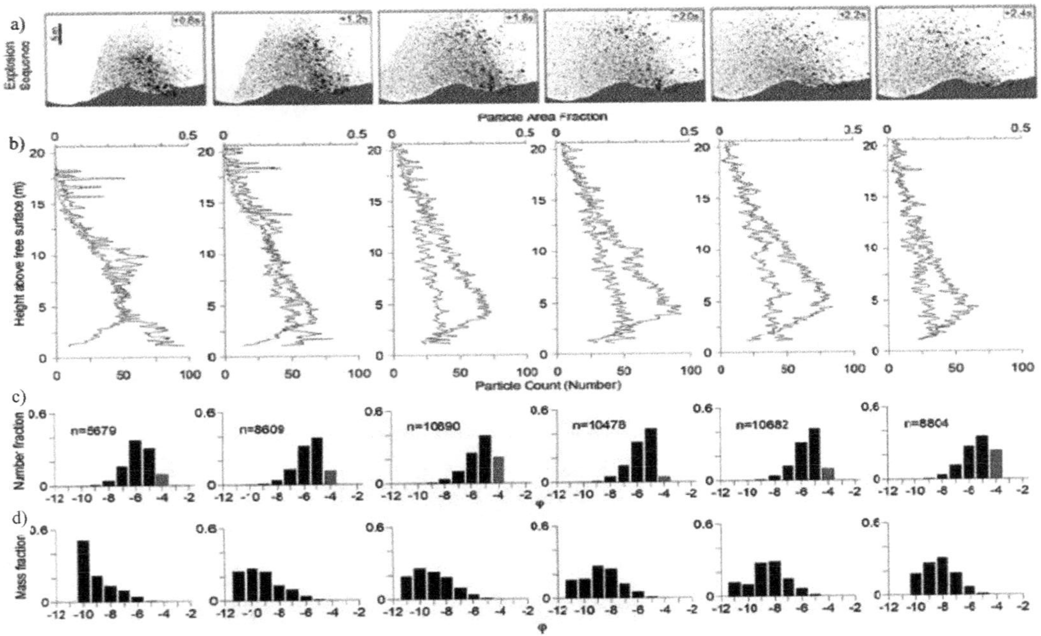

Figure 2.9. *Temporal sequence of a strombolian explosion during the July 2015 eruption at Piton de la Fournaise (modified according to Edwards et al. (2020)). a) Binarized images with pyroclastic particles in black and the volcanic edifice in gray. b) Particle number and cumulative particle area as a function of projection height. Particle size distribution as a function of c) particle number and d) particle mass*

PSDs are currently evaluated based on strategies using satellite data collected in the thermal infrared (Prata and Bernardo 2009; Cerminara et al. 2014; see Volume 2, Chapter 3), and also disdrometers (see Figure 2.7; see also Volume 2, Chapter 4), RADAR systems (Gouhier and Donnadieu 2008; Scollo et al. 2009; see Volume 2, Chapter 4) as well as ground-based thermal cameras (see Figure 2.8) and optical cameras (see Figure 2.9) (Chouet et al. 1974; Ripepe et al. 1993; Bombrun et al. 2015; Pioli and Harris 2019; Edwards et al. 2020; Thivet et al. 2021).

2.5.3. *Post-eruptive methods for determining pyroclast size and pyroclast size distribution*

2.5.3.1. *Pyroclast size*

A manual measurement made with a caliper allows the measurement of the three orthogonal axes of each studied pyroclast. Measurement of the largest pyroclasts deposited around an active vent provides valuable information for determining column height (Carey and Sparks 1986), determining eruptive style (Pyle 1989) and calculating wind speed at the time of pyroclastic fallout (Burden et al. 2011).

2.5.3.2. *Particle size*

On the total samples representative of a deposit or part of a deposit, the measurement of the PSD is carried out by manual sieving. The classical particle size analysis was described by Walker (1971). The deposit to be measured is first dried in an oven at about 60°C and the particle size analyses are then performed with a set of sieves (see Figure 2.10). Each sieve is composed of a gridded mesh (or opening) determining the minimum size of the fragments (intermediate axis) remaining within the sieve. Each of the sieves used has different mesh sizes and is classified according to a size range in Φ scale. The mesh range of the sieves used for this type of measurement is generally spaced ½ or ¼ Φ apart starting from –6 Φ (64 mm) down to 5 Φ (32 μm). Manual vibration is applied to all the sieves for a set time. The mass of the retained samples within each sieve is measured and converted to a percentage and then this percentage is normalized to the total sample mass. The normalized mass percentage of each particle size class can be represented graphically by histograms (see Figure 2.10). This method is fast and sufficiently accurate for most measured deposits.

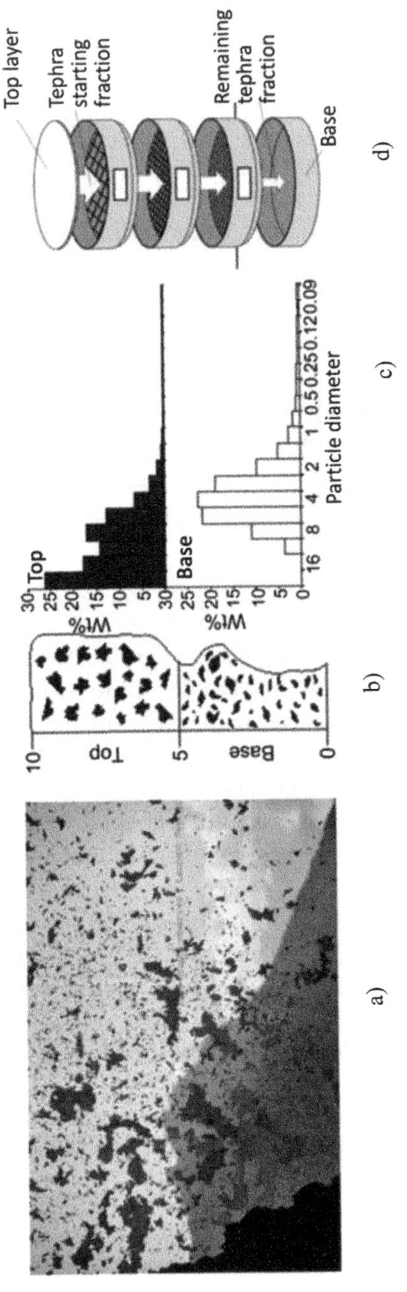

Figure 2.10. *Classical particle size analysis of a pyroclastic deposit*

COMMENT ON FIGURE 2.10.– *a) Strombolian explosion at Piton de la Fournaise during the June 2014 eruption. b) Stratigraphic log of deposits associated with this eruption. c) Particle size distributions for the lower (white histogram) and upper (black histogram) part of the deposit, corresponding to a lava fountain episode and a strombolian explosion episode, respectively (modified from Gurioli et al. (2018)).*

Particle aggregation phenomena (e.g. accreted lapilli, aggregates formed during sample drying) are problematic for particle size measurements. These aggregates mask the properties of the ash particles that constitute them. In order to recover individual ash particles, disaggregation, using ultrasonic baths with distilled water, is possible (Cioni et al. 1992). Sodium hexame-taphosphate, a deflocculant, can also be used (Casalbore et al. 2010). The disaggregated sample is then re-dried and re-sieved.

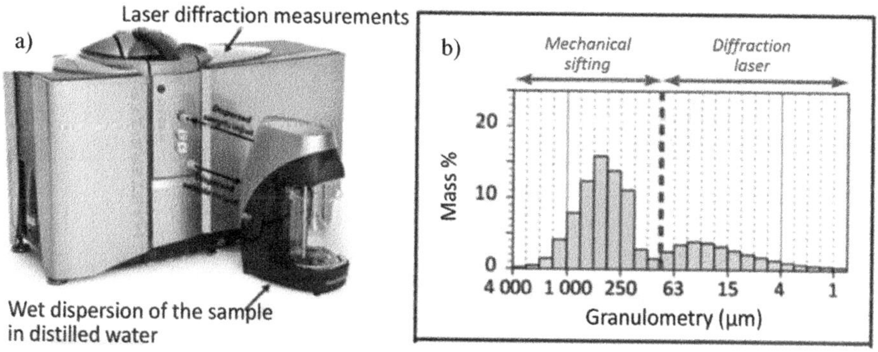

Figure 2.11. *a) Simplified schematic of the Malvern Mastersizer 3000 particle sizer operation using wet sample dispersion. b) Mixed particle size distribution achieved by conventional hand sieving (>63 µm) and laser diffraction (<63 µm)*

For deposits containing fine ash, PSDs are measured using a laser diffraction granulometer (see Figure 2.11). After the instrument has been thoroughly cleaned and the ash sample has been homogeneously dispersed in distilled water, the mixture is introduced into the main unit. The suspended particles diffract the light emitted by a laser. The analysis of the diffraction pattern of this light allows us to find the size distribution of the particles in the sample (Eshel et al. 2004).

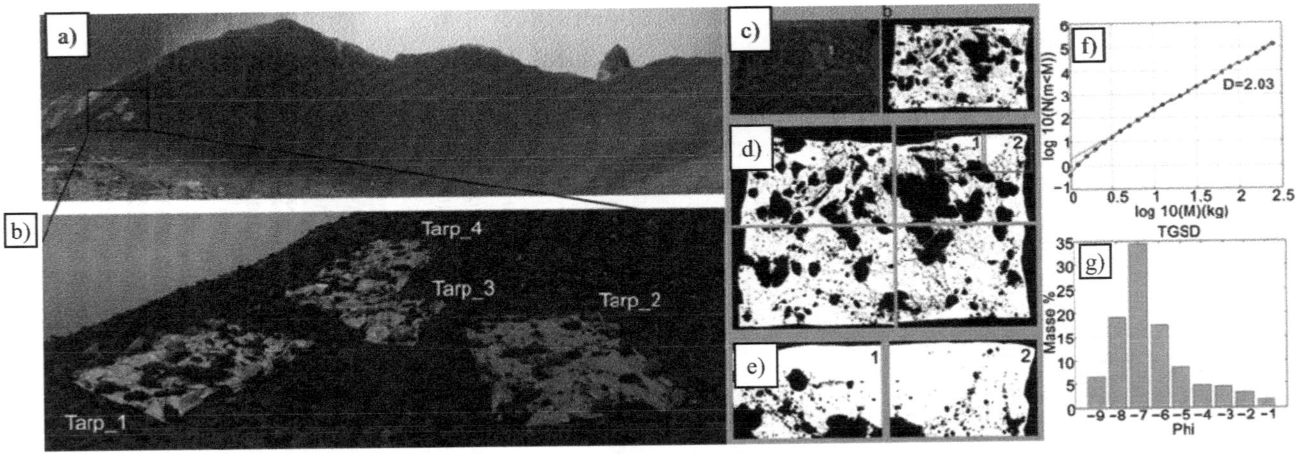

Figure 2.12. *a) View of the Stromboli active area (June 2008) showing the sampling site. b) Plastic sheeting covered with pyroclasts. c), d) and e) Photo and binary image showing the distribution of holes at different magnifications. f) Cumulative pyroclast size distribution. g) Total pyroclast size distribution resulting from mapping the distribution of holes on the sheeting (modified from Colò et al. (2020). For a color version of this figure, see www.iste.co.uk/lenat/hazards3.zip*

Recently, a new sampling strategy using an array of large plastic tarps has been developed to collect pyroclasts during explosions and record the size and mass distributions of pyroclasts landing on them (see Figures 2.4(d), 2.12(a) and 2.12(b)). When conditions are favorable, the tarps can be recovered and pyroclast size can be directly measured (Harris et al. 2013). However, under certain conditions, the high temperature of the pyroclasts falling back on the tarps can burn the plastic, generating holes that can still provide a map with which the pyroclast size distribution can be reconstructed (Colò et al. 2020). Photographs showing the distribution of holes and scattering on each plastic sheet were taken prior to their extraction (see Figure 2.12(c)). Digital images, taken at different magnifications, are created for each tarp. The different magnifications were selected to cover the full-size distribution of the holes. The images are then converted to binary images, black (holes) and white (plastic) (see Figures 2.12(c), 2.12(d) and 2.12(e)). Processing was performed with the Matlab program FOAMS (Fast Object Acquisition and Measurement System) (Shea et al. 2010b) to obtain the cumulative size distribution of the individual pyroclasts (see Figure 2.12(f)). This distribution is expressed in units of volume and must be transformed into mass. Once the tarps were removed from the sample area, the pyroclasts that remained attached to their holes were recovered, measured and weighed. Plotting the area and mass measurements reveals a linear relationship between hole area and pyroclast mass. Equivalent diameters of the pyroclasts were obtained and converted to Φ scale. A pyroclast size distribution can thus be found (see Figure 2.12(g)).

Finally, another approach estimates the volume and mass of a fallout field composed of strombolian bombs (Gurioli et al. 2013). During June 2010 and June 2011 (6 and 18 months after the January 2010 eruption emplaced the strombolian bombs; see Figure 2.13), a mapping of this bomb field was performed (see Figure 2.13(a)). Within this mapped area, the maximum slopes were approximately 30° but the bombs were all flattened on impact (characterized by an ellipsoid shape with relatively constant thickness), but without breaking up, while maintaining their impact positions. The dimensions of each bomb were measured (see Figure 2.13(b)). Photographs were taken to quantify the perimeter and area of some bombs using the Shape Preferred Orientation (SPO) software of Launeau and Robin (1996). A correlation between measured field dimensions, volume and actual bomb mass was established on a set of key samples and extended over all referenced bombs to obtain the total PSD of pyroclastic products associated with the explosion (see Figures 2.13(c) and 2.13(d)) (Gurioli et al. 2013).

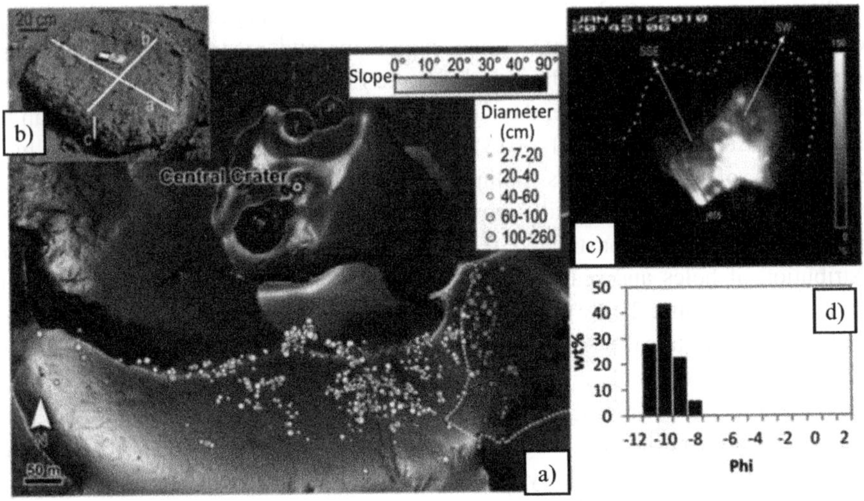

Figure 2.13. *a) Locations of the identified and sized bombs on the summit area of Stromboli. b) Example of the measurement of the three axes of a bomb. c) Thermal image at the beginning of the eruption on January 21, 2010. d) Total size distribution of the bombs (modified from Gurioli et al. (2013)). For a color version of this figure, see www.iste.co.uk/lenat/hazards3.zip*

COMMENT ON FIGURE 2.13.– *a) Yellow circles correspond to bombs emitted during the January 21, 2010 explosion, purple circles correspond to bombs emitted during the November 24, 2009 major explosion and red circles correspond to the sampled bombs. c) Image showing a directed bomb jet producing small bombs (white dotted line), and another bomb jet producing large bombs (red dotted line). The bomb jets are mainly separated into two directions.*

2.5.4. *Particle size parameters*

As already presented in Figures 2.10(c), 2.11(b), 2.12(g) and 2.13(d), the percentage of particles within each particle size class can be represented graphically by histograms where the height of the bars is proportional to the number or mass of particles, with the highest bar representing the main mode of the distribution. When a deposit is well sorted, only one modal class is usually observed. Some distributions, especially for poorly sorted deposits, may have secondary peaks defining subpopulations (see Figure 2.11(b)). The PSD can also be represented by a semi-logarithmic or logarithmic frequency curve (see Figure 2.12(f)) obtained from the cumulative curve of the distribution. For any mass value (y-axis), the cumulative curve maps to a dimension (x-axis) called the "percentile". For example, the first percentile (denoted $\Phi 1$) is the maximum diameter for 1% of the measured particles and the minimum for 99% of the measured particles, and so on. Following this reasoning, and based on the criteria of Folk and Ward (1957) and Inman (1952), statistical parameters are established (see Table 2.2), such as mean, mode, median, sort, skewness and normality.

Particle size parameters			
Parameter	**Meaning**	**Equation according to Inman (1952)**	**Equation according to Folk and Ward (1957)**
Mode	Diameter corresponding to the particles that have the highest frequency in the total distribution	The mathematical determination of this parameter is complex The midpoint of the class with the highest frequency is usually taken as the mode value	
Median (Md)	The median is the value relative to which 50% of the particles have larger dimensions and 50% have smaller dimensions	$Md_\Phi = \Phi 50$	$Md_\Phi = \Phi 50$
Mean (M)	The graphic mean is calculated on two or three percentiles	$M_\Phi = (\Phi 16 + \Phi 84)/2$	$M_\Phi = (\Phi 16 + \Phi 84 + \Phi 50)/3$
Grading (σ)	The grading is the measure of the degree of selection in clast size	$\sigma_\Phi = (\Phi 84 - \Phi 16)/2$	$\sigma_\Phi = [(\Phi 84 - \Phi 16)/4 + [(\Phi 95 - \Phi 5)/6.6]$
Skewness (α)	The skewness value measures the degree and sign of skewness of the particle size distribution curve	$\alpha_\Phi = (\Phi 84 + \Phi 16 - Md_\Phi)/\sigma_\Phi$	$\alpha_\Phi = ½\{[(\Phi 84 - \Phi 16 - 2\,\Phi 50)/(\Phi 84 - \Phi 16)] + [(\Phi 95 - \Phi 5 - 2\,\Phi 50)/(\Phi 95 - \Phi 5)]\}$
Kurtosis (β)	The kurtosis value indicates the relationship between tail selection and the central part of the particle size frequency curve	$\beta_\Phi = [(\Phi 95 - \Phi 5) - \sigma_\Phi]/ \sigma_\Phi$	$\beta_\Phi = [(\Phi 95 - \Phi 5)/2.44\,(\Phi 75 - \Phi 25)]$

Table 2.2. *List of the main particle size parameters*

The data obtained from the particle size analyses are interpreted and compared with each other or used for different types of graphic representations. The particle size mode indicates the transport capacity of the sample, while the maximum diameter indicates its competence. The sorting coefficient expresses the selective capacity of the transport medium. In general, deposits associated with dilute pyroclastic flows and fallout are relatively well sorted compared to deposits associated with dense pyroclastic flows. The asymmetry indicates the degree of homogeneity of the grain size of the deposit and thus provides insight into the sorting mechanisms as well as the sedimentation processes involved. When the asymmetry is negative, the deposit has an excess of coarse particles while when the asymmetry is positive the deposit has an excess of fine particles, compared to a normal distribution. A high asymmetry coefficient coincides with multi-modalities.

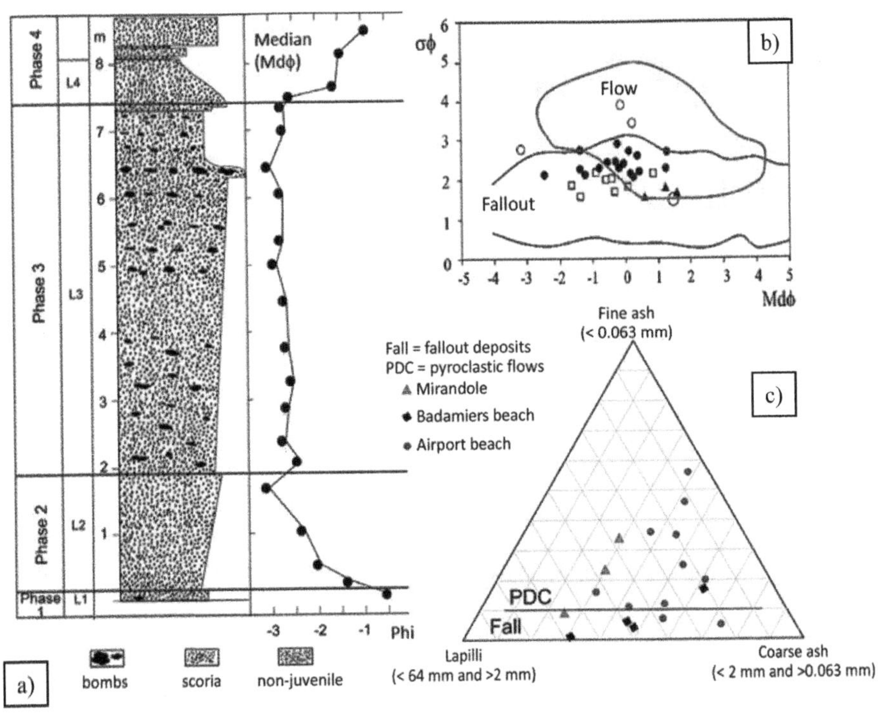

Figure 2.14. *a) Stratigraphic log of fallout deposits from the sub-Plinian basaltic eruption of La Vache and Lassolas (Chaine des Puys) (modified according to Jordan et al. (2016)). b) Deposits from the Breccia di Commenda eruption (Vulcano). c) Plot of lapilli, coarse ash and fine ash content of pyroclastic deposits at Petite Terre (Mayotte). For a color version of this figure, see www.iste.co.uk/lenat/hazards3.zip*

COMMENT ON FIGURE 2.14.– *b) Fallout deposits are characterized by a coarse grain size compared to the more ash-rich flow deposits. The flow depositional field is shown in red and the fallout deposit field is shown in gray. Squares and large circle: fallout facies; small white circles: ballistic facies; black triangles: dilute flow facies; black circles: dense flow facies (modified according to Gurioli et al. (2012)).*

If the parameters are related to the stratigraphic height of the deposits of an eruption (see Figure 2.14(a)), these allow observation of the variations that occurred as an eruption progressed in both energy (particle size) and depositional mechanism (particle sorting). One of the most commonly used comparisons for the particle size study of different deposits is to plot median values against sorting values (Walker 1971, 1973, 1981). In this type of graph, deposits from pyroclastic flows show less variation in particle size and worse sorting than deposits from pyroclastic fallout (see Figure 2.14(b)). Other parameters such as the percentage of particles smaller than 1 mm (F1) and 1/16 mm (F2) in size are used to check for possible depletion or enrichment of fine particles. Triangular diagrams with different clast sizes are also used (see Figure 2.14(c)).

2.5.5. *Identification of pyroclastic components*

Analyses of the different components allow determination of the relative proportions of juvenile (see Figures 2.15(a) and 2.15(b)) and non-juvenile (see Figure 2.15(c)) fragments within a pyroclast sample.

Juvenile fragments may be vesiculated or dense magmatic fragments and free crystals produced by the explosion. Non-juvenile fragments are all products that predate the explosion, such as incidental and/or country rocks (e.g. mantle xenoliths, deep magma chamber rocks, fragments from erosion of eruptive conduits, altered rocks from a hydrothermal reservoir at the same level as the fragmentation and eroded bedrock). The abundance and relative proportions of these components in the different grain size classes provide information on the type of fragmentation (magmatic, phreatomagmatic or phreatic), the magmatic feeder system and the active processes within the eruptive conduit (see Figure 2.16) (Sheridan and Marshall 1983; Wohletz 1983; Barberi et al. 1989; Taddeucci et al. 2002; Eychenne et al. 2015; Thivet et al. 2020a, 2020c).

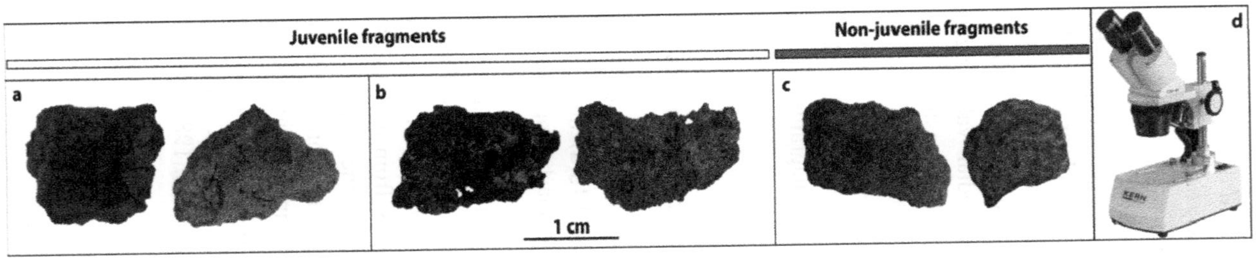

Figure 2.15. *Lithological counting (or component analysis). Images of different types of pyroclasts identified within the deposits of the April 2005 eruption at Karthala (Comoros). a) Dense juvenile fragments. b) Vesiculated juvenile fragments. c) Non-juvenile fragments. d) Binocular microscope used for this type of analysis. For a color version of this figure, see www.iste.co.uk/lenat/hazards3.zip*

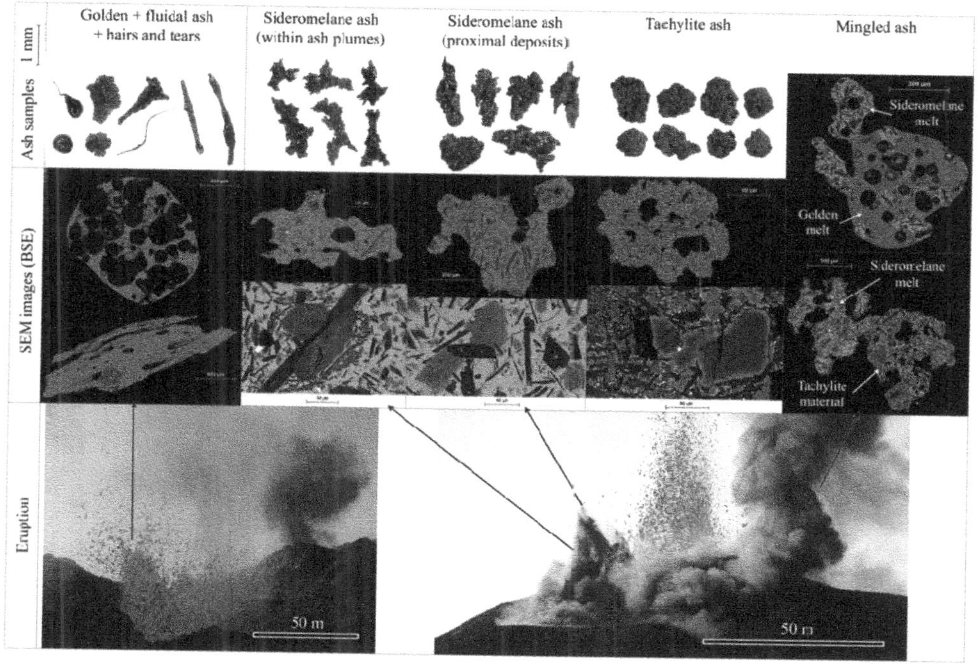

Figure 2.16. *Different pyroclastic components identified within the deposits of the September 2016 eruption at Piton de la Fournaise (modified according to Thivet et al. (2020a)). For a color version of this figure, see www.iste.co.uk/lenat/hazards3.zip*

COMMENT ON FIGURE 2.16.– *The relative proportion of components in each size class, as well as the internal texture of these components, are correlated with the dynamics observed during the eruption. a) Lava fountains producing bombs, lapilli and to a lesser extent ash (with Pele's hair) rich in vesicles and poor in crystals; b) strombolian explosions producing ash and to a lesser extent bombs and lapilli, rich in crystals and poor in vesicles. This variation in granulometry and texture reflects a change in the magmatic processes involved in the fragmentation of the magma.*

2.5.5.1. *Post-eruptive methods for determining pyroclastic components*

Different techniques are used to separate the different components of pyroclastic deposits.

As for grain size, lithological counting can be performed from images of the deposits, by coupling image analysis with stereological conversion (based on image segmentation and stereological conversion techniques).

High-resolution images of selected deposit sections can be taken at different magnifications to characterize the lithological nature and shape of the fragments for a wide range of sizes.

Figure 2.17 shows the methodology adapted by Bernard et al. (2014).

If samples are brought to the laboratory, larger size classes are selected by hand. For size classes from 2 mm to 0.5 mm, manual sampling is performed using a binocular microscope (see Figure 2.15(d)). For the smallest size classes and following the method of Barberi et al. (1989), thin slides (or epoxy resin blobs) containing particles of similar size are made (see Figures 2.18(a) and 2.18(b)). The grains are counted using a scanning electron microscope (SEM) (see Figures 2.18(c), 2.18(d) and 2.18(e)).

After counting enough particles to be representative of the sample (see Figure 2.19) (van der Plas and Tobi 1965), it is necessary to determine a conversion factor in order to obtain an equivalent mass percentage, from the density of the individual grains.

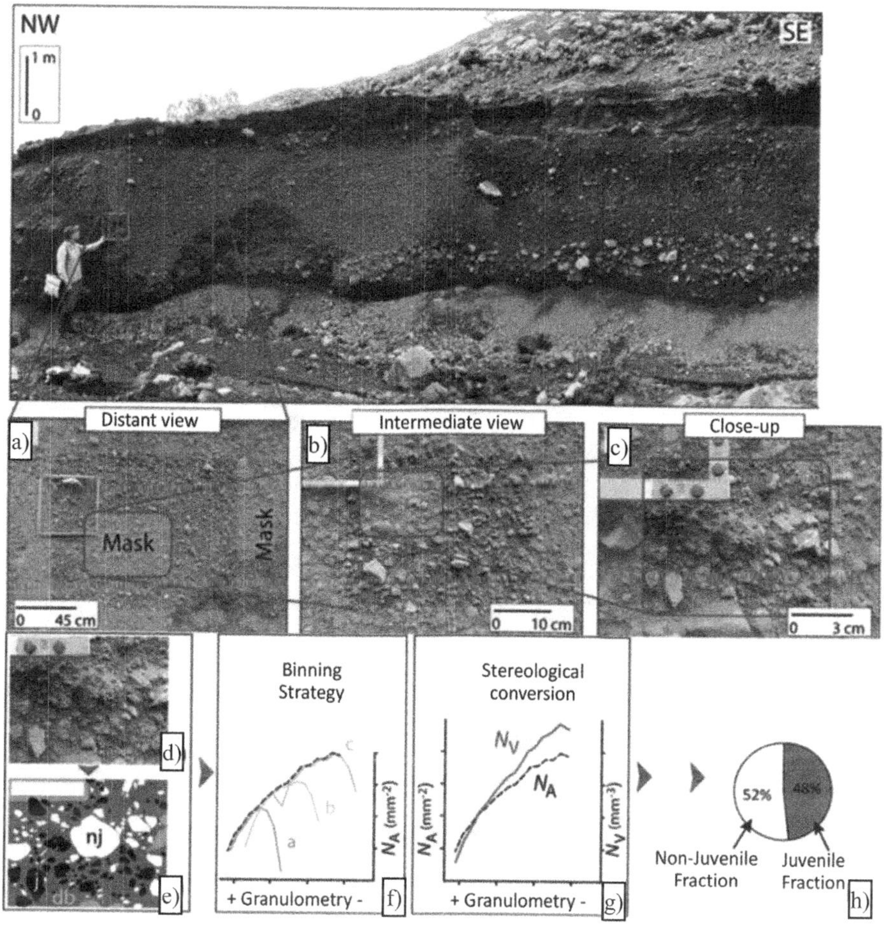

Figure 2.17. *Lithological counting from images of the deposits. For a color version of this figure, see www.iste.co.uk/lenat/hazards3.zip*

COMMENT ON FIGURE 2.17.– *Outcrop of pyroclastic flow deposits (Tungurahua 2006). Images a), b) and c) are taken at different magnifications. d) High-resolution image. e) Filtering and segmentation between juvenile (black) and non-juvenile (white) fragments. f) Numerical density of fragments of a given size per unit area (NA) of the overall image (obtained using the technique in Shea et al. (2010b)). g) NA values converted to unit volume (NV) using the stereological application of Sahagian and Proussevitch (1998). h) NV data converted to mass proportions (Eychenne and Le Pennec 2012; Bernard et al. 2014).*

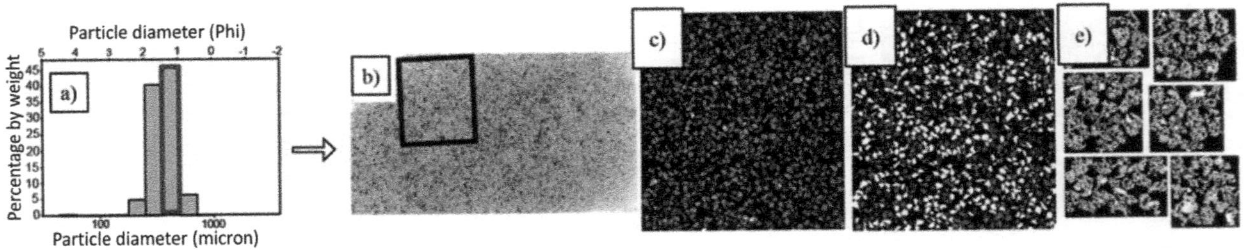

Figure 2.18. *a) Histogram of the particle size distribution of an ash sample from the Fuego eruption, 1974. b) Thin section of ash particles within the particle size mode. c) Back Scattered Electrons (BSE) SEM image acquisition at x 100 magnification. d) Classification of ash particles into five different types. e) Image processing (modified according to Gurioli et al. (2016)). For a color version of this figure, see www.iste.co.uk/lenat/hazards3.zip*

COMMENT ON FIGURE 2.18.– *a) The grain size mode is shown in red. c) Backscattered electron energy analysis allows imaging of the ash interior texture on the thin section. d) White = phenocrysts; green = vesiculated slag; blue = dense slag; red = phenocrysts surrounded by glass; yellow = non-juvenile.*

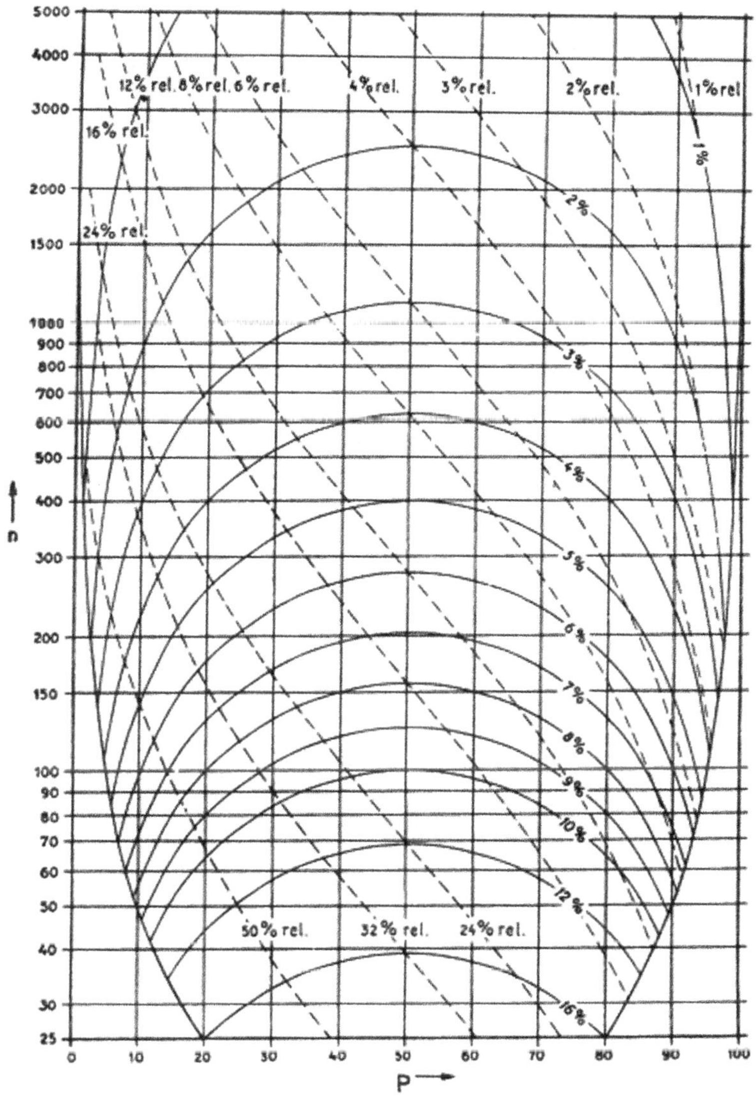

Figure 2.19. *When a large number of modal analyses are required, as many points should be counted as necessary to achieve a predetermined level of reliability*

COMMENT ON FIGURE 2.19.– *This chart is used to judge the reliability of point count results where n = the total number of points counted, p = the estimated percent by volume of items counted and the curves and solid lines represent reliability values at* 2σ *(modified according to van der Plas and Tobi (1965)).*

2.5.5.2. *Particle shape parameters*

Particle shapes record the processes responsible for their genesis and also provide information about the origin and evolution (Dellino and La Volpe 1996; Büttner et al. 1999; Manga et al. 2011; Liu et al. 2015a), properties (Dürig et al. 2012) and behavior of transported particles, such as sedimentation, radiation/diffusion or interaction properties (Riley et al. 2003; Alfano et al. 2011; Kylling et al. 2014; Liu et al. 2015a; Bagheri and Bonadonna 2016). Particle shape parameters provide quantitative and reproducible measurements that minimize the subjectivity associated with descriptive terminology and allow direct comparison between particles. While a perfect sphere can be uniquely described by a single property (its diameter), irregularly shaped particles require measurements of multiple dimensions (Blott and Pye 2008). Volcanic ash, in particular, encompasses a wide range of shapes, each with different physical properties and behavior. Ash particle morphology plays an important role in terms of depositional mechanism because the shape of grains within the ash deposit influences its likelihood of settling or remaining suspended in the atmosphere, which impacts ash hazard analysis and mitigation. The shapes of juvenile ash also reflect the fragmentation mechanisms operating during an eruption, which in turn are controlled by magmatic properties (such as viscosity, temperature and composition) and the conditions under which magma rises to the surface and fragments (Heiken 1972; Liu et al. 2007; Cioni et al. 2008; Liu et al. 2015a; 2015b; Heiken and Wohletz 2016; Thivet et al. 2020c).

2.5.5.3. *Syn-eruptive methods to determine the shape of pyroclasts*

The syn-eruptive methods used for particle size with thermal and visible cameras can be used to obtain the shape parameters (in 2D) of the particles analyzed for low-intensity eruptions (see Figure 2.20).

2.5.6. *Post-eruptive techniques to determine particle morphology*

The morphology or shape of pyroclasts can be measured in 2D or 3D. 2D morphologies are often based on silhouettes (projections) of the measured particles using an SEM (Dellino and La Volpe 1996; Dürig et al. 2012; Jordan et al. 2014;

Avery et al. 2017) or optical methods that automatically measure the size and shape of these particles (Leibrandt and Le Pennec 2015; Schmith et al. 2017; Nurfiani and Bouvet de Maisonneuve 2018; Freret-Lorgeril et al. 2019; Thivet et al. 2020c). Particle morphology can also be determined on sample sections or slices (such as thin sections or epoxy resin mounts containing the particles to be analyzed) imaged with different microscopy techniques (Liu et al. 2015a, 2015b; Jones et al. 2016; Buckland et al. 2018; Verolino et al. 2019). New 3D techniques using micro-tomography allow not only the acquisition of relatively precise data on the shape but also the internal texture of the measured objects (Schipper et al. 2013; Vonlanthen et al. 2015; Dioguardi et al. 2017; Mele and Dioguardi 2018; Mele et al. 2018), such as the presence of vesicles and crystals.

Bombs and lapilli (>2 cm in diameter) can be analyzed by a 3D scanner that produces a 3D digital image of the solid object, based on a simple triangulation system, and can provide a measurement of the main morphological parameters, such as volume and surface area. In this method, the objects are digitized with a coherent light scanner. The digital model is obtained by recording the interference patterns projected onto the object with a dedicated camera (Sugar et al. 2008; Barbero and Ureta 2011). Two images are recorded by the system to cover the entire surface of the object. The images are processed by software that determines the coordinates of each pixel in the 3D image, producing a polygonal mesh of the surface. This mesh is then used to generate a 3D digital model.

Here we focus on the description of the Malvern Morphologi G3 (see Figures 2.21(a), 2.21(b), 2.21(c)) which is a fast, automated optical microscope for particle size and morphology (silhouettes) analysis, potentially useful for monitoring volcanic products. This instrument has the ability to accurately measure particle size and shape parameters ranging from 0.5 μm to a few millimeters using the static image analysis technique, allowing a range of morphological properties to be measured for each particle (see Figure 2.21(d)), such as:

– raw parameters (number, location, size, shape, transparency);

– size parameters (equivalent diameter, diameter, length, width, perimeter, equivalent volume);

– shape parameters (aspect ratio, circularity, convexity, strength, elongation);

– transparency parameters.

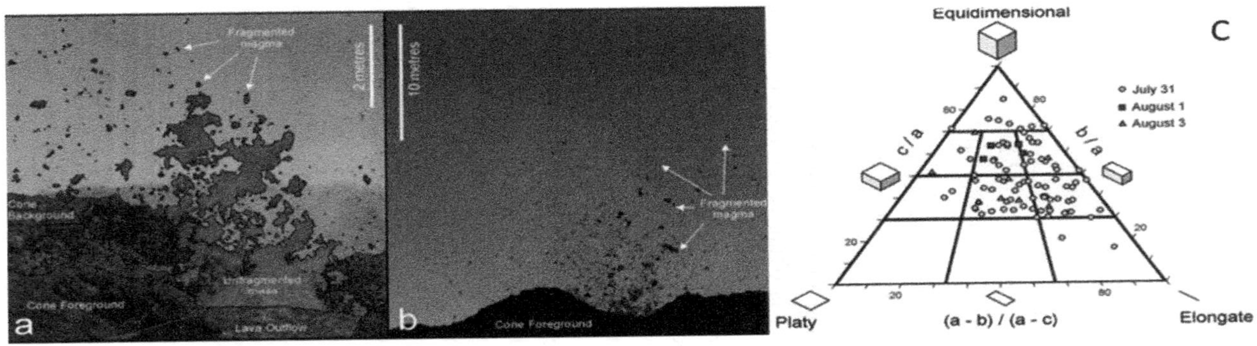

Figure 2.20. *Strombolian explosions during the July 2015 eruption at Piton de la Fournaise. a) Low-intensity explosion. b) Relatively high-intensity explosion. Particles are circled to distinguish them from the background. c) Shape of clasts collected on the ground and measured along the three axes for validation with 2D digital image data (modified from Edwards et al. (2020)). For a color version of this figure, see www.iste.co.uk/lenat/hazards3.zip*

The strength of the instrument is also based on the dispersion unit which allows for dispersing an ash sample on the glass plate of the microscope which can thus record images of individual particles by scanning the whole sample dispersed on the glass plate. The particle size classes analyzed are generally chosen with respect to the representativeness of the total sample, that is, by choosing the most massive and voluminous particle size class of the total deposit (particle size mode). For the sake of compatibility, the results on ash morphology are also compared on similar particle size classes because the morphology of the same ash deposit potentially varies according to grain size. The scope of the glass plate analysis is initially chosen based on an estimate of the number of particles present. Statistically, and thanks to empirical tests, a measurement can be considered robust from about 1,000 particles. Each analysis is focused on a single particle size class determined by manual sieving to be ¼ Φ. Once the particle size class has been selected and isolated from the total deposit, the ash particles are sometimes washed beforehand with water and using ultrasonic baths, in order to remove possible microparticles or secondary deposits that could modify the primary shape of the particles. In the case where the sample is composed of several components, the particles are sorted to potentially analyze the sample, component by component (see Figure 2.22).

The size and shape parameters are automatically calculated by the instrument (see Figure 2.21). In addition to the length (L) and width (l) of the particles, the circle-equivalent diameter of the particle is systematically calculated: this diameter (μm) is equal to the diameter of the disk whose area is equivalent to the measured particle. The direct measurement of the perimeter (P) and the area (A) of each particle, as well as the perimeter and the area of their convex envelope (P_{CH} and A_{CH}, respectively), allow us to define the parameters of convexity (CVX) and solidity (SLD) according to these formulas:

$$\mathrm{CVX} = \frac{P_{CH}}{P} \text{ and } \mathrm{SLD} = \frac{A}{A_{CH}} \quad [2.1]$$

where convexity and solidity are dimensionless values between 0 and 1 (see Figures 2.21(d) and 2.22), and represent textural roughness (small-scale) and morphological roughness (particle-scale), respectively. Leibrandt and Le Pennec (2015) and Liu et al. (2015a) summarize the set of shape parameters existing in the literature.

Free crystals and purely vitreous ashes can also be identified by the amount of light that this type of particle allows to pass through during analysis, unlike opaque microcrystalline particles that do not allow any light radiation to pass through.

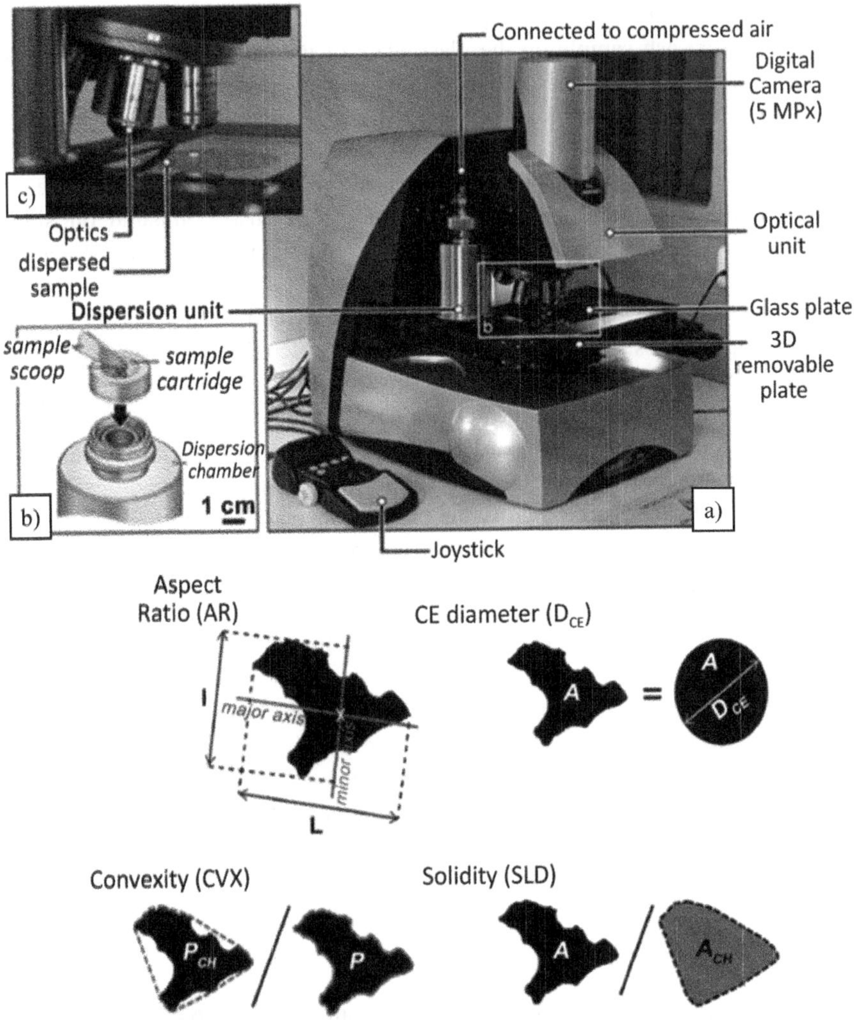

Figure 2.21. *Operation of the Malvern Morphologi G3 (modified according to Leibrandt and Le Pennec (2015))*

COMMENT ON FIGURE 2.21.– *a) Overview of the instrument. b) Optics of the instrument. c) Diagram of the dispersion procedure of the ash sample. d) Main morphometric characteristics of a particle. A certain volume of ash is introduced in the center of the container, itself introduced in the cartridge of the dispersion unit.*

The sample is dispersed on the glass plate of the microscope by injection of compressed air. Automatic particle analysis is performed at fixed magnification. Data are filtered, by removing particles smaller than the grain size range of the sample. Then, particle clusters and poor quality images are removed manually.

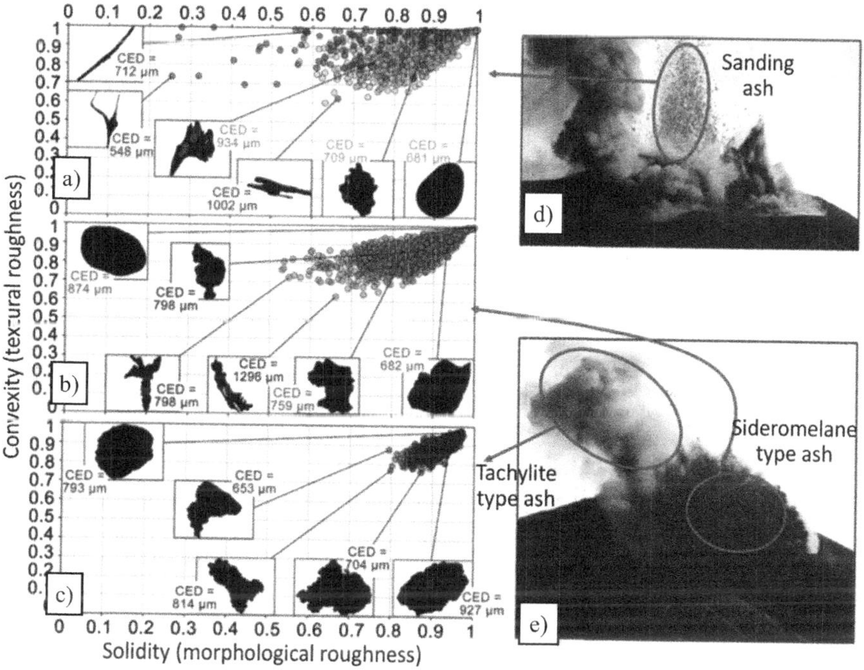

Figure 2.22. *Morphological analysis of a series of ash samples from the September 2016 eruption at Piton de la Fournaise (modified according to Thivet et al. (2020a)). For a color version of this figure, see www.iste.co.uk/lenat/hazards3.zip*

COMMENT ON FIGURE 2.22.– *The morphological parameters chosen here are convexity and solidity (see Figure 2.21(d)), representative of textural and morphological roughness, respectively. Analyses of a) pumice ash (glassy texture) from sustained lava fountain activity; b) sideromelane ash (glassy texture with microlites); and c) tachylite ash (fully crystallized texture) from intermittent strombolithic activity. d) and e) Representative photos of eruptive activity and samples collected in the field, respectively.*

2.5.7. *Terminal fall velocity of the particles*

The transport capacity of pyroclastic particles within the atmosphere depends on their aerodynamic behavior, which is influenced by both the properties of the atmosphere and those of the particles (density, size, shape). The sedimentation of particles injected into the atmosphere depends on their fall velocities or terminal velocities (TFV for Terminal Fall Velocity), which represent the dynamic balance between the gravitational force (which accelerates the fall velocity of the object) and the drag forces (which oppose the fall motion). According to the studies of Wilson and Huang (1979) and Dellino et al. (2005), the fall velocity of a particle is described by the following formula:

$$\mathrm{TFV} = \sqrt{\frac{4gd(\rho_S - \rho)}{3C_D\rho}} \qquad [2.2]$$

where g is the gravitational acceleration (9.8 m/s^2), d is the diameter of the particle, ρ_s and ρ are the densities of the particle and the surrounding fluid and C_D is the drag coefficient (a dimensionless number that depends on the particle shape and the flow regime of the surrounding fluid). The flow regime of the fluid is related to the Reynolds number (Re), which is the ratio of inertial to viscous forces of the fluid. High Reynolds numbers (>500) are associated with a turbulent sedimentation regime, while low Reynolds numbers (<0.4) are associated with a laminar sedimentation regime (Kunii and Levenspiel 1991). For intermediate values, the flow is termed transient (intermediate sedimentation regime). Other more complex formulations take into account particle size (Textor et al. 2006), or variation in Reynolds number and/or variation in drag coefficient (Ganser 1993; Pfeiffer et al. 2005).

Irregularly shaped particles are characterized by higher aerodynamic drag forces than regularly shaped (e.g. spherical) particles, for the same density and size range. The velocity differences with the spherical model increase with particle size up to about 50% and depend on the choice of the fall velocity model and the morphological parameterization used. Therefore, for depositional dispersion modeling purposes, a range of sedimentation velocities is more appropriate than giving an average value (Alfano et al. 2011).

2.5.7.1. *Syn-eruptive methods for determining particle fall velocity*

As shown in Figure 2.7, the ASHER field instrument is capable of counting the number of particles between 90 μm and 4 mm that sediment within the apparatus (see Figure 2.7(c)), measuring their sizes (see Figure 2.7(e)) and also quantifying the

TFV of particle from 0.1 to 6 m/s (see Figure 2.7(f)) every 30 seconds, with access to the data in real time. Other syn-eruptive methods are reported in Chapter 4 of Volume 2.

2.5.7.2. *Post-eruptive methods for determining particle fall velocity*

Accurate determination of aerodynamic drag forces requires detailed quantification of particle morphology (Ganser 1993; Riley et al. 2003). A widely used shape parameter is sphericity (Ψ), defined as the ratio of the area of a sphere of the same volume as the particle (equivalent sphere) to the surface area of the particle. Sphericity can be derived primarily on the basis of simple geometric equivalent shape approximations (Aschenbrenner 1956), either from 2D image analysis of the projected particle surface (Riley et al. 2003), from 3D scanning (Alfano et al. 2011), from micro-tomography analysis (Mele et al. 2018) or from direct surface measurements by gas absorption (Dartevelle et al. 2002; Riley et al. 2003).

The last two methods mentioned above will be presented below.

Recently, the application of X-ray micro-tomography (see Figure 2.23) to measure the shape of particles has led to a significant advance in this field (Ersoy et al. 2010; Bagheri et al. 2015; Rausch et al. 2015). This technique allows for 3D morphological visualization and quantification of the analyzed objects (particles, vesicles, crystals, etc.) with a resolution higher than a micrometer. The study by Dioguardi et al. (2017) proposes new 3D shape parameters, namely, 3D sphericity (Ψ3D) and 3D fractal dimension (D3D), obtained by analyses on volcanic ash and lapilli particles. These two 3D parameters were incorporated into shape-dependent drag force formulas, which resulted in good predictability of sedimentation rates of irregularly shaped pyroclasts.

The gas absorption technique consists of measuring the quantity of gas that can be absorbed by a sample under different pressure conditions. This quantity is in fact related to the total exposed surface of the measured object (specific surface a_s expressed in m^2/g). The Brunauer, Emmett and Teller (BET) model (Brunauer et al. 1938) is widely used to derive it from gas absorption data. Using the determinations of a_s on a known number of particles (i), the average particle area ($\bar{A}$) can be obtained if the total mass of the analyzed particles (m) is known ($\bar{A} = a_s$ m/i).

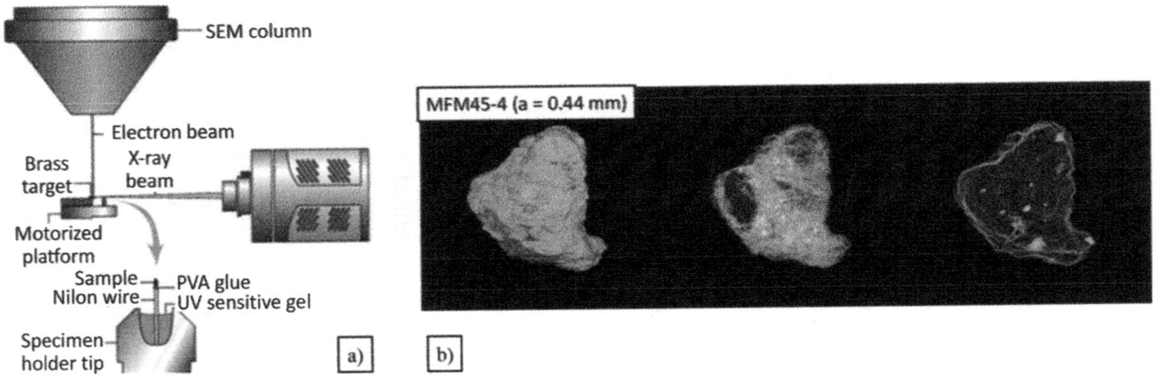

Figure 2.23. *a) Example of how X-ray micro-tomography works. b) Results of imaging a volcanic ash particle (left) with imaging of vesicles (middle) and crystals (right), adapted from Vonlanthen et al. (2015)*

2.5.8. *Pyroclast density and porosity*

Density and porosity measurements of eruptive products, particularly juvenile pyroclasts, are important because they provide a first-order indication of the amount of gas exsolving from a sample, upon fragmentation (Shea et al. 2010a). Indeed, as magma rises to the surface and decompresses, volatile species gradually exsolve from the magma fluid (Shea 2017), forming gas bubbles within the magma. These bubbles materialize as vesicles that are more or less present within magmatic samples collected in the field, lowering the total density (or envelope density) of the samples.

The assumption that the samples are representative of the magma at the fragmentation level relies on three conditions (Gurioli et al. 2016). First, the magma must be quenched immediately after fragmentation in order to avoid post-fragmentation expansion of gases that would alter the porosity and thus the density of the samples (Thomas et al. 1994). Second, porosity varies with sample size. Therefore, it is important to make measurements over small size ranges so that they can then be compared. Finally, the samples must be representative of the eruptive activity in question, in terms of:

– time (the samples must come from narrow stratigraphic intervals and therefore representative of the time considered);

– space (samples should also come from several fallout zones for each eruption event);

– particle size (the sampling method must be chosen appropriately according to the size of the pyroclasts);

– composition (the measured samples must reflect the possible textural heterogeneities of the magma).

Respecting these conditions is essential in order to translate the totality of the initial magmatic information and the properties of the magma within the eruptive conduit and at the fragmentation level.

2.5.8.1. *Post-eruptive methods for determining the envelope density and porosity of pyroclasts*

The determination of the density of the samples is based on various pycnometric methods. For bombs and lapilli (see Figure 2.24(a)), a first method (see Figures 2.24(b), 2.24(c)) can be initially used, repeating the analytical procedure of Houghton and Wilson (1989) and Shea et al. (2010b).

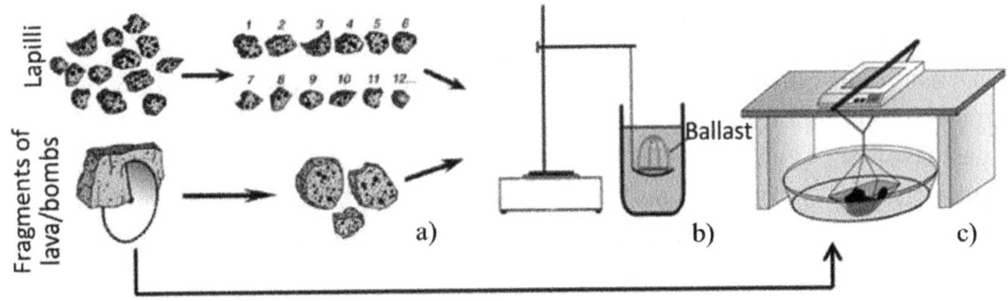

Figure 2.24. *Density measurements*

COMMENT ON FIGURE 2.24.– *a) Illustrative scheme for the preparation of lapilli samples, lava fragments or bombs, adapted from Shea et al. (2010a). b) Density measurements using Archimedes' principle for the fraction of lapilli or cut lava pieces and bombs, including ballasting with a solid cage (ballast) in case the samples are less dense than water. c) Schematic diagram of the weighing device used in the study of Stromboli bombs, adapted from Gurioli et al. (2016).*

This technique, based on buoyancy, calculates the envelope density ($\rho_{envelope}$) of a sample by measuring its mass in air (m_{air}) and in water (m_{water}). In order to waterproof the measured sample, the application of Parafilm ($m_{parafilm}$) around it is often necessary. The envelope density is calculated using this simple relationship:

$$\rho_{envelope} = \frac{m_{air}}{m_{air} - m_{water} + m_{parafilm}} \quad [2.3]$$

Sample densities are expressed in kg/m^3 (see Figure 2.25(a)).

A second possible method uses a granular media pycnometer (see Figure 2.25(b)) to measure the total volume of the samples, and then determine their envelope densities by dividing that volume by their masses.

The density measurements from these two methods are relatively similar. Furthermore, the measurement errors generated by these analytical techniques are negligible compared to the natural density variation of the measured samples.

A third method is applied to measure the density of the ash. This method follows the principle used by water pycnometry for large samples (bombs and lapilli). However, the experimental protocol has been adapted for the measurement of small particles and described by Eychenne and Le Pennec (2012). This protocol consists of measuring the mass of several ash particles in air and in water. The envelope density thus obtained is the average of the density of all the measured particles. For ash particles, the surface tension of water is considered high enough to prevent water from entering the vesicles potentially present in the ash. The pycnometer used is a glass vial with a dipstick on the stopper that gives the precise volume inside the vial. This experimental precision, as well as the use of a precision balance, is necessary for the calculation of water pycnometry because the masses used for these measurements are relatively small.

In order to determine the porosity rate from the envelope density of the samples, the skeletal density or total rock density (ρ_{dense}) of the samples must be known. This density actually corresponds to the density of the eruptive products without taking porosity into account and therefore depends on the composition of the magma and the crystals' content. The density of the magma can be calculated by knowing its chemical composition and using the model of Bottinga and Weill (1970). If the nature and proportion of crystals within the magma liquid are known, the skeletal density can be found. A second method consists of reducing one or more representative samples of a given magma to powder, and then measuring their volume with a helium pycnometer (see Figure 2.25(c)). The use of helium is relatively well adapted for this type of measurement because helium seeps into the smallest asperities, to measure only the solid volume of the powder. The skeletal density is obtained by dividing the mass of each powder by their volume. Finally, thanks to the relation given by Houghton and Wilson (1989), the porosity (P) of the samples is calculated (vol%):

$$P = \frac{100\,(\rho_{dense} - \rho_{envelope})}{\rho_{dense}} \quad [2.4]$$

2.5.8.2. *Vesicle connectivity*

Vesicles within juvenile samples may be more or less interconnected (see Figure 2.25(d)). This rate of connectivity reflects various magmatic processes occurring within the conduit and during magma fragmentation, such as vesicle coalescence, gas percolation (Colombier et al. 2017a, 2017b) and eruptive style (Gurioli et al. 2018; Thivet et al. 2020a, 2020c). Similar to density and porosity measurements, samples should therefore be representative of the magma at the time of fragmentation.

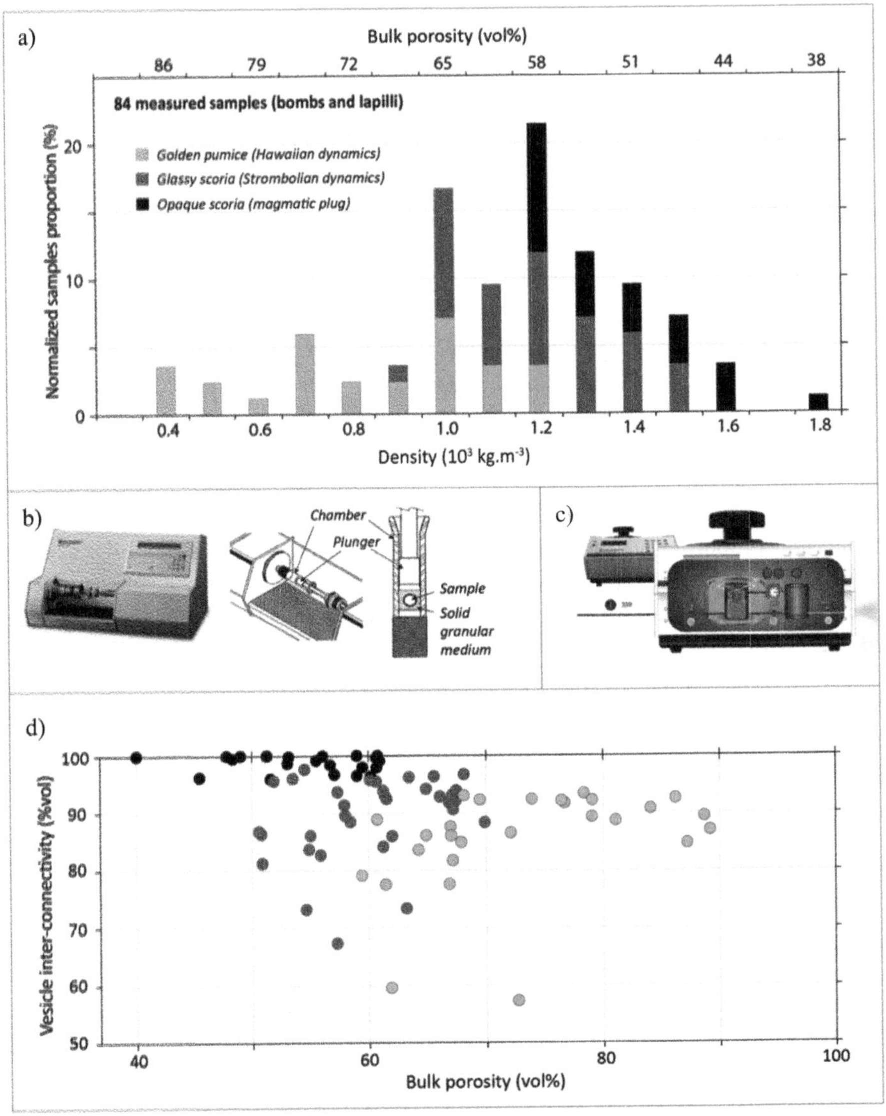

Figure 2.25. *a) Densities and porosities of samples from the September 2016 eruption of Piton de la Fournaise, showing different populations depending on the nature of the pyroclasts. b) Solid granular medium pycnometer, GeoPyc 1365. c) Helium pycnometer, AccuPyc II 1340. d) Porosity of samples as a function of their vesicle connectivity rate (adapted from Thivet et al. (2020a)). For a color version of this figure, see www.iste.co.uk/lenat/hazards3.zip*

COMMENT ON FIGURE 2.25.– *Golden pumice from Hawaiian dynamics is characterized by relatively low vesicle connectivity (55–95 vol%) compared to samples from Strombolian dynamics (65–100 vol%). In the Strombolian activity, both coalescence of vesicles within the eruptive conduit and brittle fragmentation favor fissure formation between vesicles.*

2.5.8.3. *Post-eruptive methods to determine the rate of connectivity between vesicles*

The use of a helium pycnometer, this time with intact ungrinded samples, allows us to measure the connectivity rate between vesicles.

The volume measured here (V_{dense}) corresponds to the skeletal volume of the sample plus the volume of isolated vesicles into which helium cannot penetrate (unlike interconnected vesicles).

A relation given by Formenti et al. (2003) allows us to calculate the percentage of interconnected porosity (P_c) with the envelope volume of the sample, previously determined during the density measurements:

$$P_c = 1 - \frac{V_{dense}}{V_{envelope}} \qquad [2.5]$$

The degree of vesicle connectivity (c) is obtained by normalizing P_c to the total porosity P (vol%):

$$c = \frac{100\, P_C}{P} \qquad [2.6]$$

2.5.9. *Pyroclast permeability*

Permeability represents the ability of a material to allow a fluid to pass under a pressure gradient.

The air permeability of selected magmatic samples can be measured to identify specific processes (see Figure 2.26(a)) such as the pressure increase of a volcanic system due to the accumulation of magmatic gases held by low-permeability materials (Colombier et al. 2017a; Gurioli et al. 2018; Thivet et al. 2020a), especially for Strombolian and Vulcanian dynamisms.

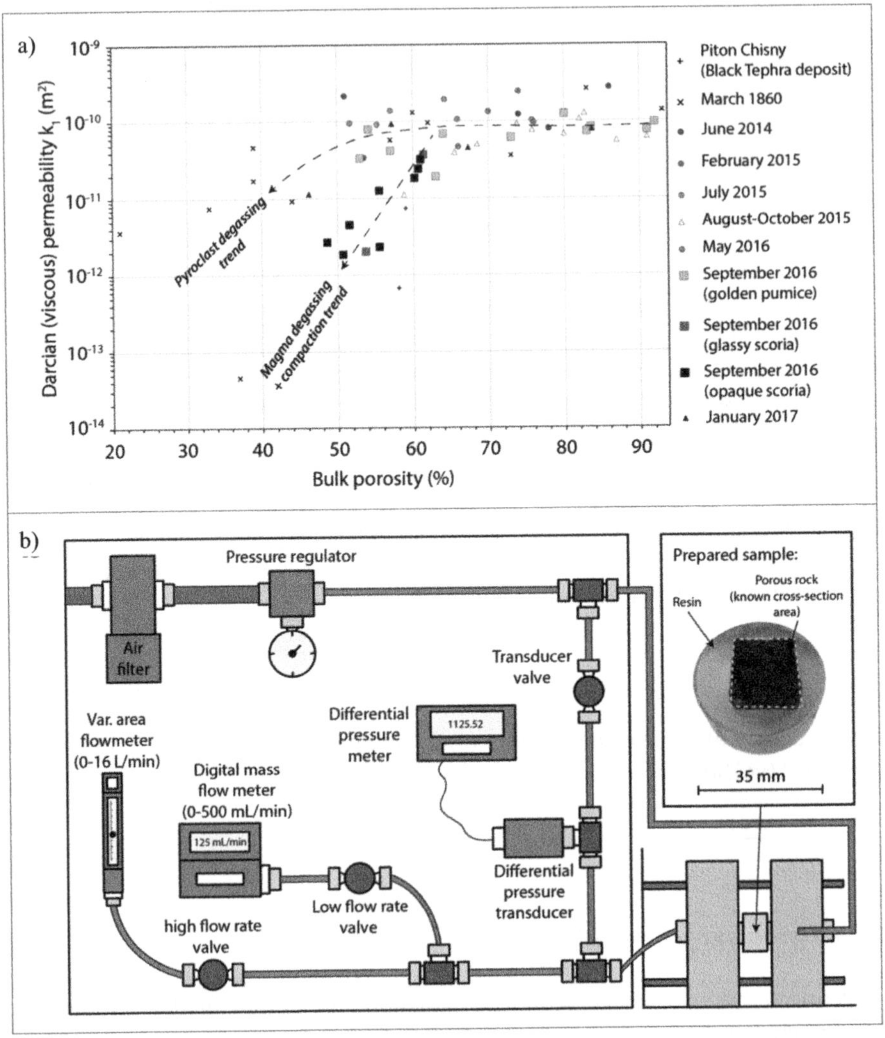

Figure 2.26. *a) Permeability versus porosity of samples from several eruptions of Piton de la Fournaise (adapted from Thivet et al. (2020b)). b) Schematic showing the operation of the permeameter based on the designs of Takeuchi et al. (2008), as well as the final appearance of a sample ready to be analyzed. For a color version of this figure, see www.iste.co.uk/lenat/hazards3.zip*

COMMENT ON FIGURE 2.26.– *For the September 2016 eruption, golden pumice from lava fountain activity (yellow squares) is relatively permeable compared to scoria from strombolithic activity (black squares) caused by the formation of a surface magma plug from which it originated.*

2.5.9.1. *Post-eruptive methods to determine the permeability of pyroclasts*

The use of a permeameter to measure samples previously prepared for this type of measurement (see Figure 2.26(b)) allows the acquisition of permeability data. For the permeameter presented here, the selected fragments are first cut into rectangular prisms, which allow us to know precisely the lengths, areas and volumes of each sample, necessary for the permeability calculation.

These fragments are then coated with resin, while preserving their internal structure and their ends. These prepared samples are then placed in a sample holder directly connected to the permeameter. Once the sample is connected to the air circuit of the instrument, several measurements of air flow rate (L/min) and differential pressure (Pa) are made for different initial air pressures. The correlation between air flow rate and differential pressure allows the solution of the Darcy-Forchheimer equation for a compressible fluid (Rust and Cashman 2004; Yokoyama and Takeuchi 2009; Bai et al. 2010). This equation is applicable to measure the permeability of a porous sample through the inertia of a fluid (in our case, air):

$$\frac{P_2{}^2 - P_1{}^2}{2\,P_0\,L} = \frac{\mu_{gas}}{k_1} q + \frac{\rho_{gas}}{k_2} q^2 \qquad [2.7]$$

where k_1 (m^2) is the viscous (or Darcian) permeability, k_2 (m^2) is the inertial (or non-Darcian) permeability, q is the volumetric air flow rate per sample cross-sectional area (m/s), P_0 (~101,325 Pa = 1 atm) is the pressure at which the airflow is measured, P_2 and P_1 are the inlet (Pa) and outlet pressures (1 atm = P_0) of the sample, respectively, L is the length of the sample (m) and ρ_{gas} and μ_{gas} are the density (1.2041 kg.m^{-3} for an ambient temperature of 20°C) and the viscosity (1.81 10^{-5} Pa.s) of air, respectively. By recalculating q (using the air flow rate and the cross-sectional area of the sample), k_1 and k_2 are obtained from a second-order polynomial. When using air in the range of flow rates used, inertial effects are negligible compared to viscous effects. The viscous permeability (or Darcian) is therefore the parameter chosen for this type of measurement (Bai et al. 2010).

2.5.10. *Vesicle and crystal size distributions*

Quantification of vesicle size distribution (VSD) and crystal size distribution (CSD) as well as their number density (N_V and N_C) within the magma allow the identification of important degassing (Shea et al. 2010a, 2010b) and crystallization

(Higgins 2002) processes occurring both within the magma chamber (pre-eruptive processes) and the conduit (syn-eruptive processes).

2.5.10.1. *Post-eruptive methods to quantify vesicle and crystal size distributions*

Once the measurements of density, porosity, vesicle connectivity and permeability have been carried out, a selection of representative clasts can still be carried out for the microscopic analysis of the vesicles and crystals present within the eruptive products: thin sections (bombs and lapilli) or epoxy resin mounts (ash of the same granulometric size) are made. The objective of this type of analysis is to identify and quantify all the vesicles and crystals present in each type of sample, starting from the largest objects (see Figures 2.27 and 2.28(a)), visible under the optical microscope within the thin section, to the smallest (see Figures 2.27 and 2.28(b)), requiring finer imaging under the SEM, following the methodology established by Shea et al. (2010a).

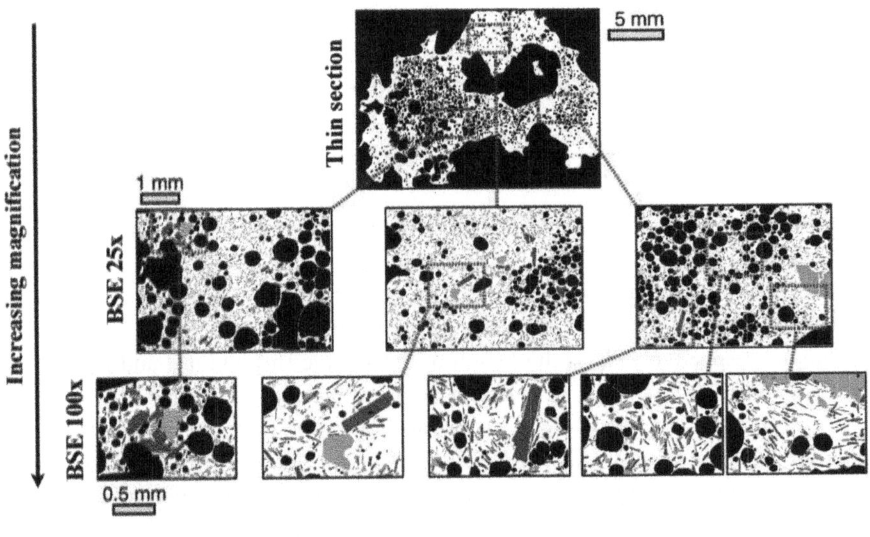

Figure 2.27. *Textural analysis procedure (vesicles and crystals) on samples such as bombs, lapilli or lava fragments, allowing the integrated measurement of the largest (thin section) and also the smallest (SEM images x 25 and x 100) objects (vesicles and crystals). For a color version of this figure, see www.iste.co.uk/lenat/hazards3.zip*

COMMENT ON FIGURE 2.27.– *The images are derived from the binarization of a set of optical images (thin sections) and SEM images (BSE) of a sample (lapilli) from the July 2015 eruption at Piton de la Fournaise.*

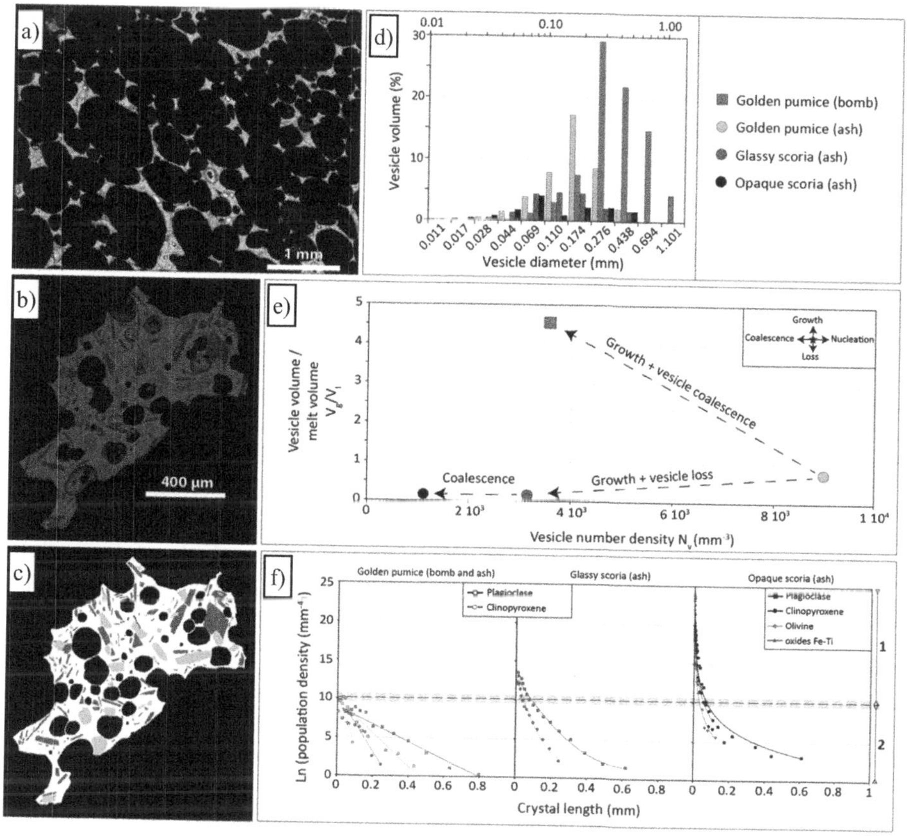

Figure 2.28. *Textural data from the September 2016 eruption at Piton de la Fournaise (adapted from Thivet et al. (2020a)). For a color version of this figure, see www.iste.co.uk/lenat/hazards3.zip*

COMMENT ON FIGURE 2.28.– *a) SEM image (BSE) on a thin section of a golden pumice bomb. b) Image of a sideromelane ash. c) Binarized image of the previous image. d) Vesicle size distribution of four types of clasts. e) V_G/V_L versus N_V (vesicle number density) plot showing the evolution of the outgassing processes. f) Crystal size distribution on the same samples showing the evolution of the crystallization processes: a single population of crystals in terms of size is identified within the golden pumice while a second population of crystals (curved line) appears within sideromelane and opaque scoria.*

For the sake of representativeness, the higher the magnification, the higher the number of images taken into account to have enough vesicles and crystals measured.

Empirically, about 500 measured objects (vesicles or crystals) are sufficient to correctly and statistically represent the porosity or crystallinity of the sample (van der Plas and Tobi 1965) (see Figure 2.19). For textural analysis of ash, SEM images are acquired particle by particle until the number of objects is sufficient for the measurement to be statistically representative of the actual texture of the sample being analyzed.

Once the optical and electronic images of each sample have been acquired, binarization of the images with image processing software is required to facilitate textural analysis and to clearly identify and classify each type of object present in the images (see Figure 2.28(c)). These binarized images are then analyzed by the Matlab program FOAMS developed by Shea et al. (2010a), which measures and characterizes the set of objects identified in these images (vesicles and crystals). One of the main strengths of this program is to apply stereological (3D) conversion schemes to vesicles measured on a 2D section by considering spherical vesicles. The total porosity of each sample, the VSD (see Figure 2.28(d)) as well as the numerical density of vesicles per unit area (see Figure 2.28(e)) are thus calculated. Depending on whether the growth of the bubbles takes place before, during or after the formation of the crystalline phases and in particular of the phenocrysts, a correction on the value of the numerical density of vesicles is made to get rid of the volume of these phenocrysts which potentially form before the bubbles. Considering that the observed vesicles were formed during the magma ascent, the value of the vesicle number density can be indirectly related to the magma decompression rate during its ascent using the model of Toramaru (2006). A final parameter directly related to porosity is the ratio of vesicle volume to magma fluid volume (V_G/V_L) of the samples (Gardner et al. 1996). This ratio, corrected for the rate of phenocrysts that may be present in the samples, can be used to identify certain degassing processes (see Figure 2.28(e)).

The same analytical procedure as for vesicles is applied for the textural analysis of crystals. However, the stereological conversion of crystal surfaces (2D) into volume (3D) is more complex: the precise determination of the three dimensions (small, medium and large axes) of the different types of crystals is necessary using a database, such as CSDslice (Morgan and Jerram 2006), which gathers a set of 2D and 3D shapes for different types of crystals. Second, and similar to the FOAMS program, the CSDcorrections software (Higgins 2000) converts surface units to volume using intersection probability laws, appropriate for non-spherical shapes and known dimensions. The crystallinity, CSD (see Figure 2.28(f)) and crystal number density are calculated in this way. The number and size of microcrystals formed in the conduit also allow the exsolution rate of volatiles to be determined (Toramaru et al. 2008), from which the magma ascent rate can be deduced. Magma ascent rates can also be determined from the diffusion of volatiles into the reentrances (embayments) of some crystals (Liu et al. 2007; Lloyd et al. 2014; Moussallam et al. 2019).

2.6. Petrographic and geochemical analysis of pyroclastic products

Geochemical and petrographic analysis of pyroclasts can constrain the initial conditions from the magma chamber to the surface via the conduit. It can also identify processes that may lead to changes in eruptive activity (recharge, mixing, etc.). Combining textural studies with fieldwork (Mattsson 2010), petrological data (Shea et al. 2009, 2010b) and chemical analyses (Schipper et al. 2011; Shea et al. 2011, 2012, 2014) allow us to deliver comprehensive pictures of explosive eruptions and their dynamics (Gurioli et al. 2014). During an eruptive crisis, the essential contribution of petrographic and geochemical analysis is in tracking variations in magma composition, which can result in variations in eruptive processes and the identification of mixing and recharge processes.

2.6.1. *Petrological observation of pyroclasts*

Observation of juvenile pyroclasts, in thin sections or polished sections, allows the characterization of the different mechanisms of magma genesis (Vernon 2004). Textures are generally visible under optical microscopy, either in reflected or transmitted light. However, the use of SEM or back scattered electrons (BSE) allows a better resolution of the compositional variations. The choice of the smallest juvenile pyroclasts (<2 cm, see above), the glassiest, guarantees a rapid quench in contact with air and avoids the post-eruptive crystallization processes observed in the bombs or in the samples taken from the lava flows. If the emitted magma is perfectly liquid at the beginning of its ascent and rapidly quenched during the eruption, the pyroclasts are glassy, with euhedral microlites formed during ascent in the conduit (see Figure 2.29(a)). This is because the presence of dissolved volatile elements, especially H_2O, decreases the liquidus temperature of the magma, that is, the temperature above which it is completely liquid. Outgassing caused by decompression in the conduit increases the liquidus temperature, resulting in microlite crystallization (decompression-induced crystallization) (Hammer et al. 1999; Hammer and Rutherford 2002; Toramaru et al. 2008; Shea and Hammer 2013, Andrews and Befus 2020).

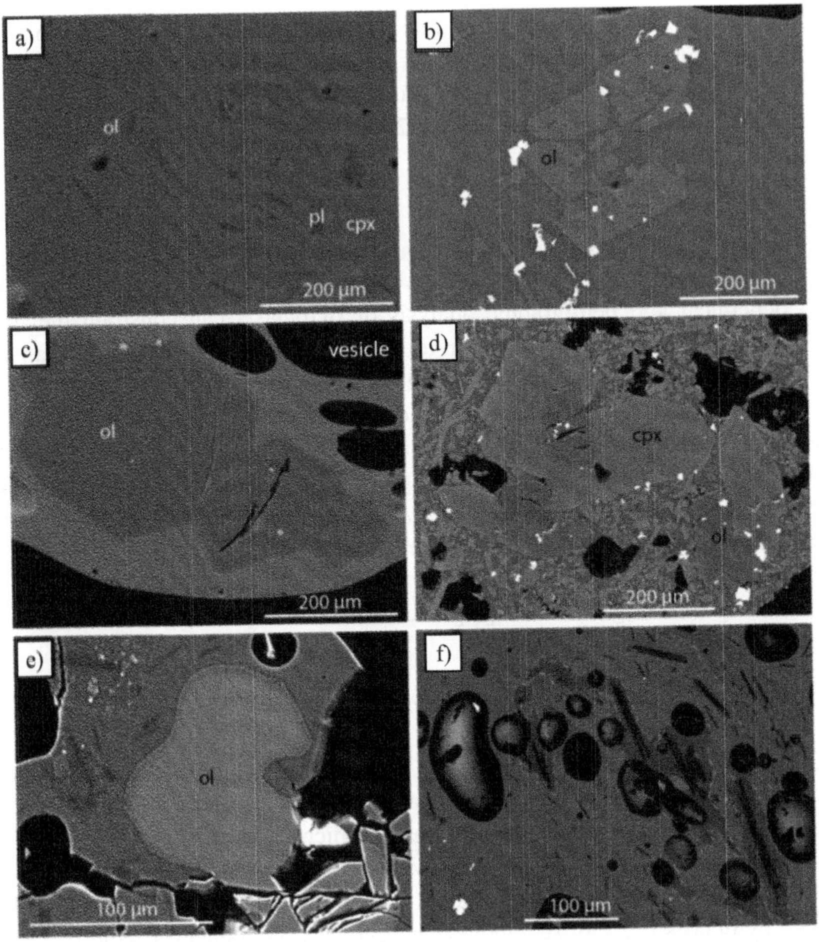

Figure 2.29. *Backscatter electron images of common magmatic textures in juvenile pyroclasts*

COMMENT ON FIGURE 2.29.– *a) Glassy texture with plagioclase feldspar (pl), olivine (ol) and clinopyroxene (cpx) microlites (2017–2018, Ambae). b) Skeletal olivine crystal and magnetite microlites (2018–2021, offshore Mayotte). c) Euhedral olivine phenocrysts (April 2020, Piton de la Fournaise). d) Glomerocryst composed of olivine and clinopyroxene crystals (seamount offshore Mayotte). e) Olivine xenocryst and rounded droplets of more crystallized magma (December 2018, Ambrym). f) Mingling of three magmas. The different compositions of the magmas result in different shades of gray (December 2018, Ambrym).*

The shape of the microlites depends on the cooling rate and degree of undercooling, and can tell us about crystallization kinetics (Faure et al. 2003; Welsch et al. 2013). Euhedral crystals, with well-defined faces, correspond to slower cooling rates, skeletal (see Figure 2.29(b)) or dendritic crystals to fast cooling rates. The presence of larger euhedral crystals, the phenocrysts (usually >200 μm, see Figure 2.29(c)), often reflects storage of magma in a reservoir, allowing slow growth of large crystals. Phenocrysts are particularly interesting for volcanological monitoring because they record the conditions under which they formed, corresponding to the storage conditions of magmas and the variations of these conditions in time. These phenocrysts can sometimes aggregate into glomerocrysts ("microcumulates", see Figure 2.29(d)), small clusters of crystals that are evidence of mineral assemblages crystallizing in magma chambers. As for the anhedral crystals, often rounded (see Figure 2.29(e)), they do not belong to the juvenile magma. They may correspond to fragments of the country rock around the chamber or conduit, but they may also be indicators of magma mixing. In backscattered electron images, it is sometimes possible to directly identify a magma mixing in progress, frozen during the eruption (physical mixing process or "mingling"), either in the form of ribbons (see Figure 2.29(f)) or rounded enclaves corresponding to less differentiated magmas (see Figure 2.29(e)). The multiple textural observations possible with a simple optical or electron microscope, and particularly their evolution with time during the eruption, make it possible to highlight processes of magmatic recharge and/or mixing, as well as stability or evolution of the magma composition. Although these petrological characterizations are post-eruptive, their relative ease makes it possible to quickly obtain (within a few days) information on the nature of the magma emitted and the syn-eruptive variations. The parameters related to the evolution of the magma composition are directly linked to the eruptive parameters and allow anticipation of an evolution of the eruptive scenarios during the syn-eruptive follow-up of long-lasting eruptions (e.g. the eruptions of Kilauea in Hawaii or those of Piton de la Fournaise on Reunion Island).

2.6.2. *Pyroclast composition*

The composition of the pyroclasts provides information on the nature of the magmas emitted. The whole-rock major composition is obtained from a powder, which is either transformed into glass (XRF) or put into solution (ICP-AES). X-ray fluorescence spectrometry (XRF) is used to determine the composition of a glass formed by melting the powder. In the case of optical spectrometry coupled to a plasma torch (ICP-AES), the powder is vitrified using a flux ($LiBO_2$) and then put into solution which will then be introduced into the plasma for analysis.

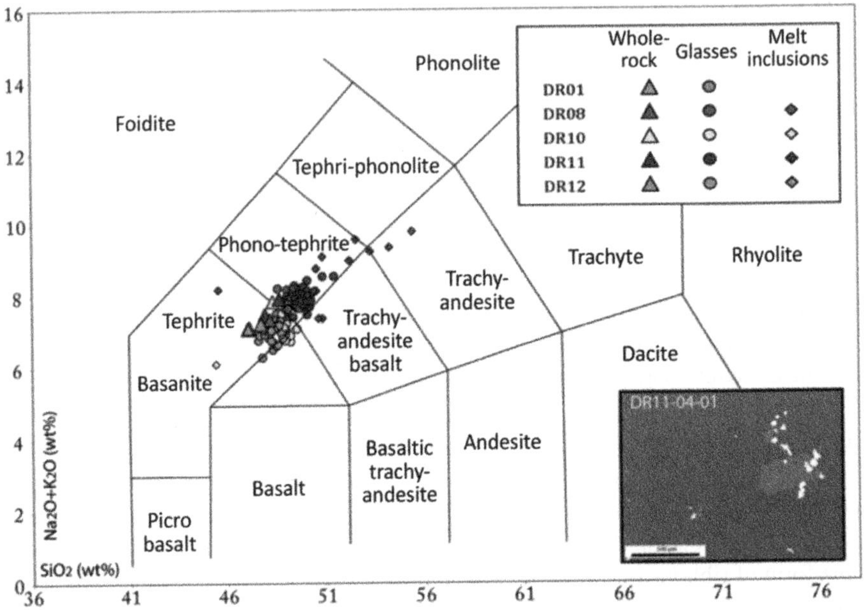

Figure 2.30. *Composition of lavas produced by the submarine eruption (2018–2021) offshore the island of Mayotte in a TAS diagram (Na2O + K2O vs. SiO_2 (Le Maitre et al. 2002)). For a color version of this figure, see www.iste.co.uk/lenat/hazards3.zip*

COMMENT ON FIGURE 2.30.– *The compositions are in the field of tephrites and basanites, and the normative olivine content allows us to specify that they are basanites. The compositions of the glasses are enriched in Na_2O, K_2O and SiO_2 due to the presence of olivine and magnetite crystals – thus the composition of the glasses is never identical to that of the pyroclasts, unless they are entirely glassy. Magmatic inclusions (see Chapter 1) sometimes record early evolutions of the magmatic system or mixtures and may have very different compositions.*

As pyroclasts are small, it can be useful to take several fragments to obtain a representative analysis, but in this case, it is necessary to ensure that they belong to the same class of components. The major elements content of pyroclasts is used to give a name to the rock (see Figure 2.30, for volcanic rocks, we use the TAS classification diagram – $Na_2O + K_2O$ vs. SiO_2 (Le Maitre et al. 2002)). To be able to compare different compositions, it is important to always follow the same procedure, in particular renormalize all analyses to a total of 100% and use the same oxidation state for iron (all FeO, or all Fe_2O_3, or a defined FeO/Fe_2O_3 ratio; Irvine and Baragar 1971). Variations in major element composition allow the identification of changes in magma composition that may reflect the emptying of a zoned magma chamber (see, e.g., the case of Laacher See: (Wörner and Schmincke 1984; Tomlinson et al. 2020)), magma mixing (Thornber et al. 2003); intersection of a differentiated subsurface reservoir during the 2018 eruption of Kilauea in Hawaii (Gansecki et al. 2019), magma re-injections. It is also useful for calculating physical parameters of magmas, such as viscosity, or density of the non-vesiculated fraction (DRE = Dense Rock Equivalent) (Bottinga and Weill 1970).

From rock powders, it is also possible to determine trace element compositions or isotopic compositions that allow a more accurate monitoring of magma evolution during long-lasting eruptions or eruptive cycles (see section 2.6.6).

2.6.3. *Magma temperature*

The measurement of pre-eruptive temperatures is a key parameter for detecting changes in the nature of the emitted magma. The temperature can be measured during the eruption, either by infrared remote sensing (see Volume 2, Chapters 3 and 4) or by direct measurement by dipping a thermocouple in the lava. However, these methods are only applicable to lava flows. Precise temperature values can however be obtained from post-eruptive analysis of pyroclasts. For monitoring purposes, and thus for rapid acquisition of data for monitoring activity, the most useful thermometers are those based on the composition of the glass, analyzed with an electron microprobe. Helz and Thornber (1987) have shown experimentally, in the case of the Kilauea lavas in Hawaii, the existence of an excellent correlation between the MgO content of residual melt in equilibrium with olivine and temperature:

$$T = 1{,}014°C + 20.1\ \text{MgO (mass\%)} \quad [2.8]$$

This easy-to-use empirical geothermometer provides the emission temperature of pyroclasts. More complicated versions of this equation have been proposed to generalize it to other volcanic systems (Sugawara 2000; Putirka 2008). However, the gain in accuracy remains limited and the Helz and Thornber equation is easily applicable in routine for many basaltic volcanic systems. It is possible to gain slightly in accuracy by recalibrating an equation for a specific volcano (see, e.g., Rhéty et al. (2017), Figure 2.31 for the case of Piton de la Fournaise). This approach is particularly useful in the case of a frequently active volcano (Kilauea, Piton de la Fournaise) for which the monitoring team will have to make measurements on many different eruptions.

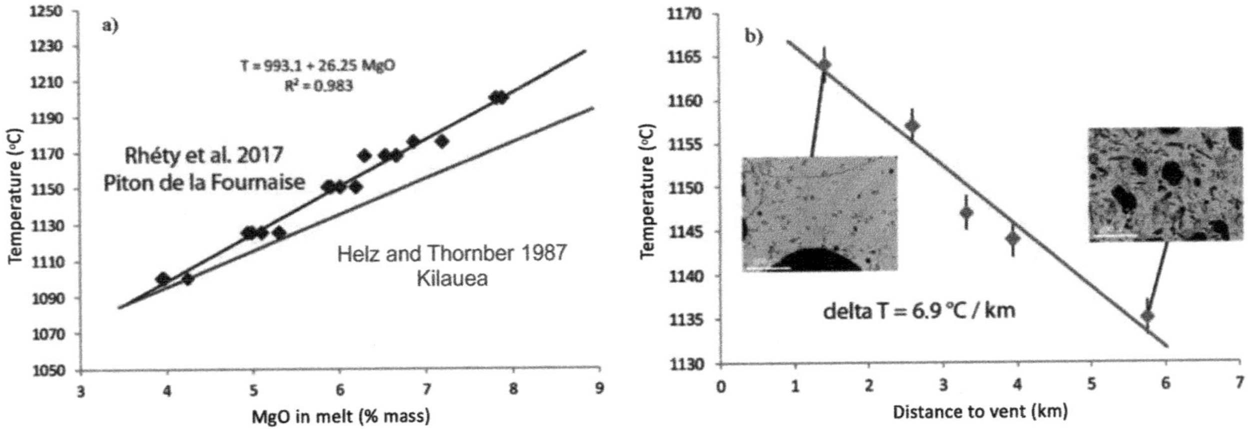

Figure 2.31. *a) Evolution of MgO content with temperature for glasses in equilibrium with olivine at Piton de la Fournaise. Calibration by Rhéty et al. (2017) based on experimental data from Villeneuve et al. (2008) compared to the calibration of Helz and Thornber (1987) for Kilauea lavas. b) Use of thermometry to quantify lava flow cooling as a function of distance from the vent (Mauna Ulu, Kilauea) (Robert et al. 2014)*

COMMENT ON FIGURE 2.31.– *The figure also shows the evolution of textures as a function of temperature, from a completely glassy texture with a few dendritic olivine crystals formed during quenching at 1,164°C to a texture rich in olivine, clinopyroxene and plagioclase microlites at 1,135°C.*

The procedure used for the determination of pre-eruptive temperature requires an important work of quality control. Indeed, if the pyroclasts are not quenched quickly, the growth of microlites and phenocrysts can modify the composition of the glass which will record temperatures lower than the eruption temperatures. This post-emission growth is more problematic for magmas emitted at high temperature (basaltic). Indeed, the diffusion of elements being faster, the compositional changes will be more important. It is possible to determine if the melt is properly quenched by checking the equilibrium between the rims of the olivine phenocrysts or microlites and the melt. For this purpose, one can calculate the distribution coefficient of iron and magnesium between olivine and liquid:

$$Kd\frac{Fe/Mg}{olivine/liq} = \frac{(\frac{FeO}{MgO})olivine}{(\frac{FeO}{MgO})liq} \quad [2.9]$$

The value of this coefficient is 0.30 ± 0.03 (Roeder and Emslie 1970) for most magma liquids. Other thermometers can be used for differentiated magmas that do not contain olivine, but they are generally less accurate (Putirka 2008).

2.6.4. *Crystals in equilibrium: pre-eruptive conditions in magma reservoirs*

Many other thermometers and barometers, based on equilibria between two minerals, or between a mineral and a silicate liquid, have been developed from experimental calibrations (see, e.g., the review by Putirka (2008)). The application of thermometers (and barometers) requires a knowledge of the chemistry of the minerals, usually brought by in situ analyses using an electron microprobe. Two minerals in contact with each other, or the rim of an euhedral mineral and the neighboring glass, are analyzed, and the crystallization temperature (or pressure) is derived from an equation that can be purely empirical or based on thermodynamic constraints. In addition to the determination of pre-eruptive temperatures developed in the previous paragraph, thermometers and barometers can be applied to phenocrysts to determine the storage conditions of magmas in magma reservoirs.

Quantifying the crystallization pressures allows us to obtain the storage depths of the magmas, and thus to locate the active reservoirs. It is a complementary approach to geophysical approaches (seismology, magnetotelluric, deformation, etc.) that can confirm the results with an independent methodology. The main barometers based on mineral composition rely on the incorporation of aluminum in silicates, particularly in pyroxenes and amphiboles. Barometers based on clinopyroxene composition are widely used for basaltic systems (Neave et al. 2019; Ubide et al. 2019), although their accuracy remains limited (6–8 km at best; Putirka 2008). However, they can identify the presence of different storage levels and possible magma re-injections. For dacitic or rhyolitic differentiated magmas, amphibole is also a good barometer (Johnson and Rutherford 1989; Ridolfi and Renzulli 2012; Mutch et al. 2016). By assuming that magmas are saturated with volatile elements in the magma reservoir, it is also possible to use the volatile element concentrations of magmatic inclusions to determine their minimum storage pressure. Magmatic inclusions are small volumes of magma trapped during the growth of phenocrysts (see Chapter 1). Completely embedded in the minerals, they preserve the volatile element contents prior to degassing (Anderson 1974; Dixon et al. 1995; Newman and Lowenstern 2002). Entrapment pressures can be calculated by measuring H_2O and CO_2 contents in the inclusions and comparing to experimentally determine solubility models (Iacono-Marziano et al. 2012). Measurements of H_2O and CO_2 contents are performed either by Raman spectrometry, infrared spectrometry (FTIR) or using an ion probe (SIMS).

Numerous thermometers are also available to determine temperatures in magmatic reservoirs and their variations (Holland and Blundy 1994; Putirka 2008). This makes it possible to identify recharge processes that do not result in mixing with the recharge component. The process of "self-mixing" (Couch et al. 2001) occurs when a cold, non-eruptible magma is heated by the intrusion of a hotter magma. The heating results in crystal melting and increased proportions of liquid, which can make the magma eruptible and trigger an eruption (Calbuco in Chile, Morgado et al. 2019). Such cryptic mixing cannot be deciphered from the geochemistry of pyroclasts, nor is it evidenced by the presence of enclaves. However, an increase in temperature from the rim to the core of the phenocrysts is typical of self-mixing and allows anticipation of a possible eruption of the intruding magma. According to some theories, this mixing mechanism is a key mechanism in the evolution of large silicic magma reservoirs.

2.6.5. *Crystals in disequilibrium: diffusion and kinetics of magmatic processes*

Chemical variations in phenocrysts can also tell us about the processes involved in magmatic reservoirs, particularly the triggering mechanisms and their durations (Cashman and Blundy 2013; Kilgour et al. 2014). In a phenocryst-poor magma, phenocrysts are generally homogeneous. If the rate of crystallization is higher, chemical elements are removed from the liquid during phenocryst formation, and the composition of the melt changes, in turn causing a change in the composition of the new layers of the phenocrysts that form. Thus, the composition gradually changes from the core to the rim of the crystals, recording the progressive crystallization and cooling of the magma. This is growth zoning (see Figure 2.32(a)). In general, ferromagnesian minerals (olivine, pyroxenes, amphiboles, micas) will be enriched in Fe (and to a lesser extent Mn), and depleted in Mg. The feldspars will be enriched in Na (and to a lesser extent K), and will be depleted in Ca. This is called normal zoning. During a re-injection of less differentiated magma, the change in composition of the magma will lead to a disequilibrium and sometimes a dissolution of crystals that results in anhedral/rounded forms. When growth resumes, the new crystals will crystallize from the less differentiated magma, giving rise to a so-called reverse zoning (e.g. for an olivine, an increase in Mg content and a decrease in Fe content, see Figure 2.32(c)). Reverse zoning related to a magmatic re-injection is abrupt and often present over a rounded interface resulting from dissolution (see Figures 2.32(c) and 2.32(d)). One can also imagine abrupt normal zonings, related to the injection of crystals formed in a less differentiated magma (see Figure 2.32(b)). These zonings can generally be identified from electron microscopy images and used to identify magma re-injections and mixing, critical parameters to understand the evolution of an eruption. Oscillatory zoning can also be observed and is characteristic of cyclic growth in magma reservoirs where convection is important (see Figures 2.32(e) and 2.32(f)).

Zoned crystals can also be used to determine residence times and thus characterize the timing of magma re-injections. For example, the sudden appearance of zoned crystals during an eruption may reflect a new magma re-injection if the residence time is short, or the eruption of a different portion of the reservoir if the residence time is long. A knowledge of the usual re-injection times for a given volcanic system can also help anticipate the imminence of an eruption from the triggering of precursor geophysical signals corresponding to reservoir recharge. The abrupt interfaces between zones of different compositions linked to re-injections are gradually erased thanks to the diffusion of elements across the interface. Diffusion being a kinetic phenomenon, the more the crystal is stored at high temperature after the formation of the zoning, the more the interface will fade. Knowing the temperature of the magma and the diffusion coefficients, it is possible to model compositional profiles across the interface from the diffusion equations (Fick's law) as a function of time, and to extract a crystal residence time (Costa et al. 2008) (see Figure 2.33). Diffusion profile analysis is very well developed for the simplest minerals, particularly olivine (Girona and Costa 2013; Albert et al. 2019; Sundermeyer et al. 2020) (see Figure 2.34), orthopyroxene (Solaro et al. 2020) and plagioclase (Rout and Wörner 2020).

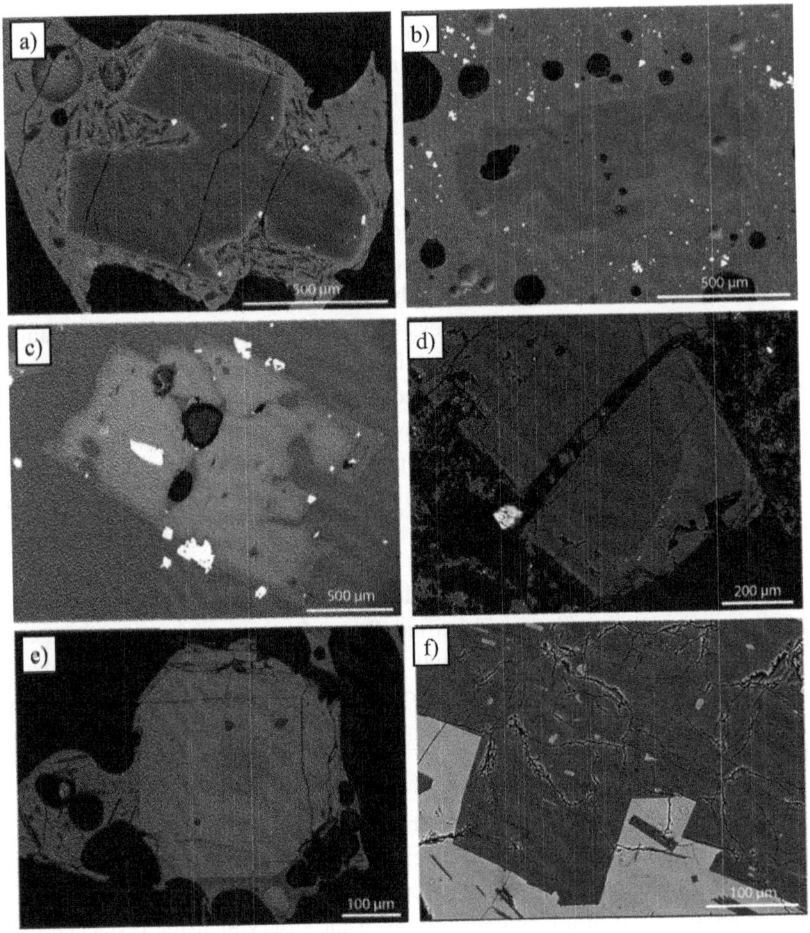

Figure 2.32. *Backscattered electron images of zoned phenocrysts*

COMMENT ON FIGURE 2.32.– *a) Normal progressive zoning of an olivine phenocryst: light rim enriched in Fe, dark core enriched in Mg. b) Abrupt normal zoning related to magma mixing, the core of the olivine crystal was resorbed and then surrounded by a growth phase. c) Reverse zoning of an olivine crystal characterized by the rounded shape of the central zone that was resorbed, and the higher Mg contents of the reverse (darker) zone. d) Multiple reverse zoning in a plagioclase crystal, showing a resorption of the internal dark zone (rich in Na), surrounded by a clear external zone (rich in Ca) corresponding to re-injection or reheating of the magmatic reservoir. e) Oscillatory zoning in a clinopyroxene. f) Oscillatory zoning in a plagioclase.*

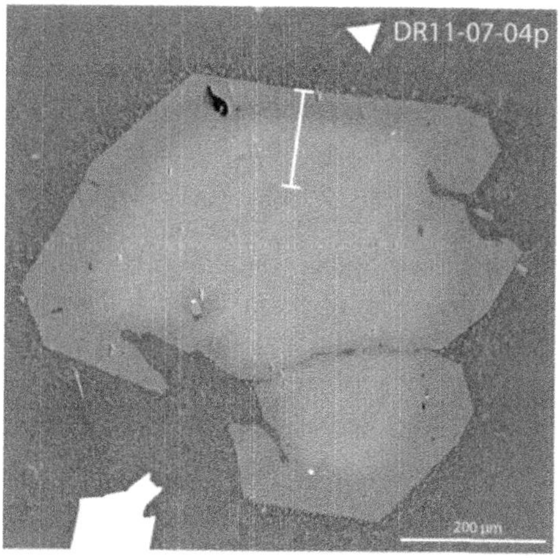

Figure 2.33. *Backscattered electron image of a reversely zoned olivine crystal from the 2018–2021 submarine eruption offshore Mayotte Island*

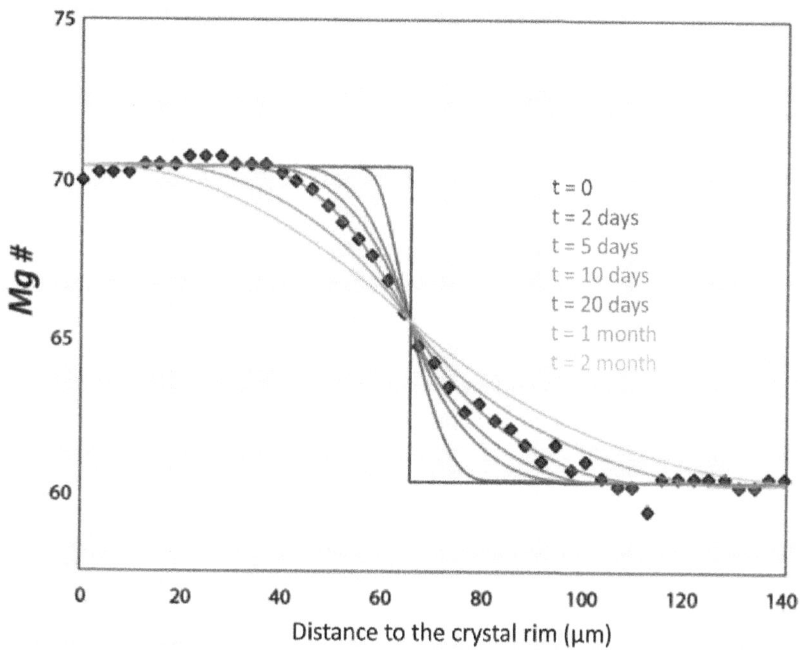

Figure 2.34. *Profile through the olivine crystal of Figure 2.33 (white line) showing the evolution of Mg# (100 Mg/(Mg+Fe) in moles) as a function of the distance to the crystal rim. For a color version of this figure, see www.iste.co.uk/lenat/hazards3.zip*

COMMENT ON FIGURE 2.34.– *The profile can be modeled by imagining a dissolution of the core of the crystal (Mg# 60), then an overgrowth of a more Mg-rich olivine (Mg# 70), followed by diffusion of the elements across the interface. The observed profile is reproduced if the crystal remains 20 days at 1,100°C after the dissolution/overgrowth step, thus a residence time in the reservoir and/or conduit of 20 days.*

Different studies have shown highly variable residence times between magma re-injection into the reservoir and an eruption: about 4 years for the 2008 eruption of Llaima in Chile (Ruth et al. 2018) and 1–10 days for an eruption of Shiveluch in Russia (Gordeychik et al. 2018). Multiple time scales can be identified in the same magmatic system, reflecting different successive recharges (e.g. 2007 and 2015 eruptions of Piton de la Fournaise) (Albert et al. 2019; Sundermeyer et al. 2020).

2.6.6. *Radiogenic isotopes*

More advanced geochemical analyses require longer lead times (weeks, months), due to the more cumbersome procedures of grinding, dissolution and analysis by mass spectrometry. Difficult to use during an eruptive crisis, these measurements allow on the other hand a fine understanding of the magmatic system and are extremely useful for monitoring long-lasting eruptive cycles (e.g. at Piton de la Fournaise) (Vlastélic et al. 2018).

2.6.6.1. *The different long-lived decay systems used as tracers*

The isotopic ratios of heavy elements are not significantly modified by the petrogenetic processes of fusion or crystallization. Their measurement in lavas gives access to the properties of the partially molten mantle from which the magmas originate. Among the different families of isotopes, the radiogenic isotopes produced by the slow decay of a parent isotope (half-life of the order of 1 billion years, see Table 2.3) are of particular importance because they record the chemical fractionation of elements over time. The large isotopic variations generated over geologic time have provided insight into the Earth's global differentiation, heterogeneity and large-scale mantle dynamics.

Measuring the isotopic composition of elements in lavas involves chemical dissolution of samples, extraction and purification of elements in clean rooms and measurement of isotopic abundances using mass spectrometers. This time-consuming, complex and infrastructure-intensive procedure has long limited the application of radiogenic isotopes and only recently have they been used to study active volcanoes. The development of this tool in the context of volcanological monitoring is motivated by:

– the identification of isotopic variations correlated to the eruptive activity, testifying to the existence and mixing of magmas of different origins during an eruption or an eruptive sequence;

– a better understanding of these variations through the development of models of melting and extraction of magmas from a heterogeneous mantle source;

– the development of plasma source mass spectrometers allowing faster and more accurate acquisition of certain isotopic ratios.

Parent isotope	Daughter isotope	Half-life (years)	Tracer (radiogenic isotope/stable isotope)
^{147}Sm	^{143}Nd	106×10^9	$^{143}Nd/^{144}Nd$
^{87}Rb	^{87}Sr	48.8×10^9	$^{87}Sr/^{86}Sr$
^{232}Th	^{208}Pb	14.01×10^9	$^{208}Pb/^{204}Pb$
^{238}U	^{206}Pb	4.468×10^9	$^{206}Pb/^{204}Pb$
^{235}U	^{207}Pb	0.738×10^9	$^{207}Pb/^{204}Pb$

Table 2.3. *Main long-lived decay systems used as tracers*

2.6.6.2. *The main tracers: lead (Pb), strontium (Sr) and neodymium (Nd)*

With three radiogenic isotopes (^{208}Pb, ^{207}Pb and ^{206}Pb) (see Table 2.3), lead is certainly the most efficient tracer. This particularity makes it possible to identify mixing processes, which are expressed by straight lines in the isotopic spaces normalized to the same isotope. The corollary is that the only stable isotope (^{204}Pb) does not allow for precisely correcting the isotopic fractionations occurring during the purification of lead on ion exchange column and during the measurement on mass spectrometer, so that the precision of the measurements is limited to approximately 100 ppm. However, the resolving power (ratio of natural variability to analytical precision) of Pb isotope ratios remains one of the highest (>1,500 for the Earth's mantle). Lead presents particularities that must be taken into account for the sampling and analysis of lavas. This chalcophile element tends to concentrate in sulfides, inducing a non-homogeneous chemical and isotopic distribution. On the other hand, there is a risk of Pb contamination by the tools used for sampling. In particular, the zinc coating (galvanization) covering the metal handles used to sample and dig the lavas is a very important potential source of Pb contamination (Harris et al. 2017). Strontium and neodymium each have a single radiogenic isotope produced by the decay of a long-lived parent isotope (see Table 2.3). On the other hand, they have several stable isotopes that allow very precise correction of the purification and measurement fractionations, and access to an accuracy of 10–20 ppm on the $^{87}Sr/^{86}Sr$

and $^{143}Nd/^{144}Nd$ ratios. The resolving powers of $^{87}Sr/^{86}Sr$ and $^{143}Nd/^{144}Nd$ (ca. 300 and 100 for the Earth's mantle, respectively) are, however, lower than for Pb because the Sr and Nd isotope variations in the mantle are smaller. Concerning the field aspects, neodymium is less sensitive than strontium to low temperature alteration processes and to contamination of lavas by sea spray rich in radiogenic Sr.

In the context of monitoring the activity of a volcano involving the measurement of a large number of samples in as short a time as possible, the ease and speed of analysis are important criteria. In this respect, it should be noted that neodymium requires a longer and more complex chemical separation procedure than strontium and lead, which can simply be separated by resin extraction chemistry (Deniel and Pin 2001). At the measurement level, Pb and Nd isotopes can be quickly and accurately measured on plasma source mass spectrometer (MC-ICPMS), while Sr requires longer measurements on solid source mass spectrometer (TIMS), implying more preparation in relation to the deposition of Sr on filament. It should be noted that Pb isotopes have several advantages (tracing capacity, resolution power, speed of analysis) but carry a more complex message than those of strontium and neodymium.

2.6.6.3. *Magma mixing, refilling and residence time*

An immediate application of isotopic tracers is the identification and quantification of magma mixing processes. For an isotopic ratio R, the simplified equation for the mixing of two magmas A and B of concentrations C_A and C_B and isotopic compositions R_A and R_B is written:

$$x_A . C_A . (R_{Ech} - R_A) + x_B . C_B . (R_{Ech} - R_B) = 0 \text{ with } x_A + x_B = 1 \qquad [2.10]$$

Knowing the compositions (R and C) of magmas A and B, it is possible to calculate their proportions x_A and x_B in any lava sample whose isotopic ratio R_{Ech} has been measured. The mixing equation can be written for n isotopic systems, offering the possibility to solve the mixing of $n + 1$ different magmas.

Magma mixing occurs most often when a deep magma is injected into a shallow reservoir. During this recharge event, the isotopic composition of the mixture (R) evolves over time toward the composition of the injected magma. Albarède (1993) showed that this temporal evolution (dR/dt) is inversely proportional to the retention time τ. In the case of a highly incompatible element for which crystallization can be neglected, we have:

$$\tau \approx \frac{C_{in}}{C_0} \times \frac{R_{in} - R_0}{dR/dt} \qquad [2.11]$$

where the index "0" is relative to the magma initially present in the magma chamber and the index "*in*" is relative to the injected magma. The volume of the magma chamber, assumed constant, can then be estimated knowing the recharge or discharge flow ϕ:

$$V = \tau \times \phi \qquad [2.12]$$

These equations have been applied to the historical volcanism of Kilauea (Hawaii) and Piton de la Fournaise (Reunion Island) (Albarède 1993; Pietruszka and Garcia 1999). The validity of the estimates of τ and V depends on the degree of knowledge/characterization:

– the composition of the injected magma;

– temporal isotopic gradients reflecting recharge phenomena;

– the flux of magma.

These parameters are known with a good degree of confidence for some recent eruptions. At Piton de la Fournaise, the March–September 1998 eruption, which followed a 5.5-year period of inactivity, is a typical example of recharging. During the 6.5 months of the eruption, the $^{87}Sr/^{86}Sr$ ratio of the lavas increased progressively from 0.70410 (composition of the 1992 magma stagnating in the magma chamber) to 0.70415. The injected magma reached the surface through a separate vent (Hudson) allowing researchers to know precisely its composition (0.70425). Knowing that the two magmas have very close Sr concentrations, we have:

$$\tau \approx \frac{(0.70425-0.70410)}{\frac{(0.70415-0.70410)}{6.5}} = 19.5 \text{ months} \qquad [2.13]$$

This estimated residence time during the recharge phase is logically much lower than the average residence time of 10–30 years estimated for the historical period (Albarède 1993; Sigmarsson et al. 2005). Knowing the average magma productivity during the 1998 eruption (60×10^6 m^3 over 6.5 months, or 3.5 m^3/s), the volume of the magma reservoir is estimated to be:

$$V = 19.5 \times 30.5 \times 24 \times 3{,}600 \times 3.5 = 180 \times 10^6 \text{ m}^3 = 0.18 \text{ km}^3 \qquad [2.14]$$

This volume is in good agreement with estimates obtained over longer periods.

At Kilauea, Pietruszka et al. (2015) conduct a more advanced approach to reproduce the evolution of isotope ratios as a function of time after a recharge event. To do so, the authors derive the expression for $R(t)$:

$$R(t) = \frac{R_0 \frac{C_0}{C_{in}} e^{-\alpha t/\tau} + \frac{R_{in}}{\alpha}(1 - e^{-\alpha t/\tau})}{\frac{C_0}{C_{in}} e^{-\alpha t/\tau} + \frac{1}{\alpha}(1 - e^{-t/\tau})} \quad [2.15]$$

The term α here includes the effects of crystallization, α being equal to $1+[f \times (D - 1)]$, where f is the fraction of injected magma that crystallizes and D is the mineral/liquid partition coefficient of the element considered. The authors thus show that the exponential decreases in the $^{206}Pb/^{204}Pb$ ratio over the periods 1959–1969 and 1970–1982 can be explained by realimentation and mixing phenomena. The decreases are consistent with a magma residence time of 1.5–3.2 years and a reservoir volume of 0.1–0.5 km^3. These two examples show that Piton de la Fournaise and Kilauea probably have similar magma reservoir sizes, although the latter has a much higher magma production. Residence time modeling can be applied to any eruption, or sequence of eruptions showing an isotopic evolution indicative of recharge. This simple approach allows a reliable estimate of the volume of the magma reservoir.

2.6.6.4. *Mantle source fusibility and magma production*

The activity of a volcano is a first-order reflection of the rate of magma supply from partially melted regions of the mantle. This rate depends not only on physical parameters, such as the size of the melt zone and the mantle ascent rate, but also on the composition of the rocks of the mantle which governs their degree of fusibility. This last parameter can vary rapidly because we know, from the study of fragments of the mantle (xenoliths) brought up by certain eruptions, that the mantle is made up of fusible lithologies (pyroxenite veins) distributed in a fine, but not homogeneous, way in a refractory matrix (peridotite). The property of fusibility (also called fertility) being directly related to the content of incompatible elements, pyroxenites generally have enriched elemental and isotopic signatures relative to peridotites, so that it is possible to trace their fusion products using isotopes such as strontium or neodymium. For example, basaltic volcanoes produce increasingly chemically and isotopically depleted lavas during eruptive sequences of typically 5–20 years (Reiners 2002; Vlastélic et al. 2018). This evolution indicates that decadal sequences are initiated by the melting of fertile regions and terminated by the extraction of liquids from refractory regions. The distribution of pyroxenite veins in the mantle thus certainly plays a major role on the cyclicity of eruptive activity. It is also thought that the large (several tens of %) melting of pyroxenite veins leads to the formation of high porosity channels that favor the extraction of liquids from the weakly melted (few %) peridotite matrix (Lundstrom et al. 2000).

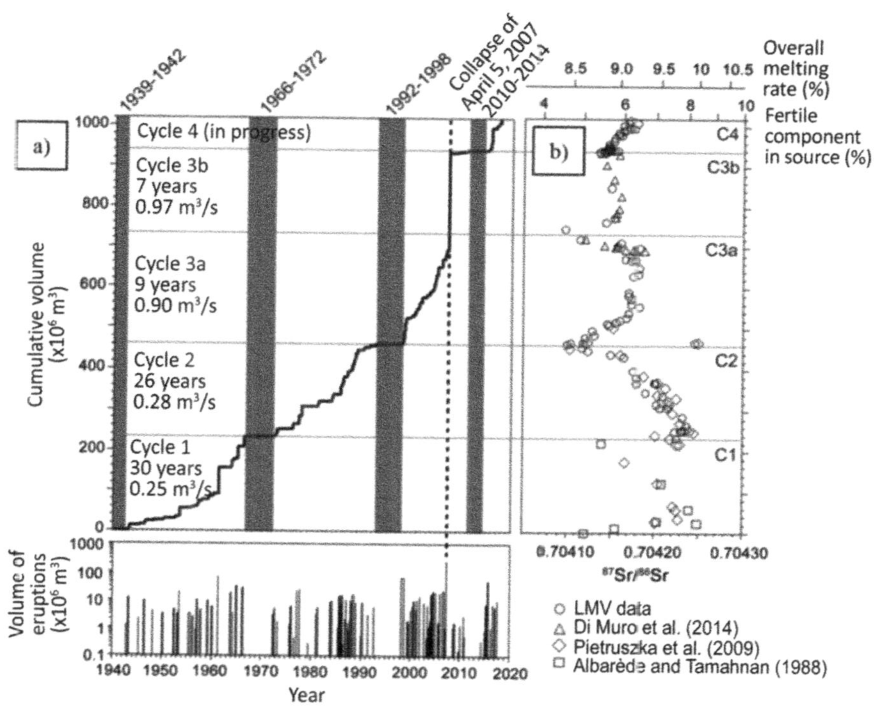

Figure 2.35. *Relationship between historical eruptive activity of Piton de la Fournaise and strontium isotopic composition of lavas. a) Volume of eruptions and cumulative volume between 1942 and 2017. b) Sr isotopic composition reported as a function of cumulative volume. The isotopic data were used to model the proportion of fertile lithology (pyroxenite) in the source and the overall melt rate (modified from Vlastelic et al. (2018))*

The melting of a mantle with a high volume concentration of pyroxenite veins will produce more magma and the geochemical signature of the latter will, paradoxically, be more depleted. We can also anticipate that the extraction of magma, and consequently the eruptive activity on the surface, will be all the more continuous as the distance between the pyroxenite veins is small. The fact that Kilauea has a richer source of pyroxenites than Piton de la Fournaise could explain its higher magmatic production, its more continuous activity and the more impoverished geochemical signature of its lavas. At Piton de la Fournaise, the systematic evolution of the $^{87}Sr/^{86}Sr$ ratio during historical eruptive cycles (1942–1966; 1972–1992; 1998–2010; 2014–) indicates that each cycle is produced by the melting of an isolated fertile domain in the refractory mantle (see Figure 2.35) (Vlastelic et al. 2018). In this case, the $^{87}Sr/^{86}Sr$ ratio presents itself as an operational tool for monitoring the progress of eruptive cycles. For example, in the current monitoring, the marked decrease in the $^{87}Sr/^{86}Sr$ ratio in 2019 and 2020 suggests that the cycle started in 2014 is nearing its end.

2.7. Measurements at the scale of pyroclastic deposits

When studying a pyroclastic deposit in order to derive the parameters of Table 2.1, the first important step is to provide an objective description of the deposit itself using a standardized method that allows everyone to understand the observations made. For this reason, and for each outcrop, a stratigraphic section must be described according to the following measurements and observations:

– Name and GPS location of the outcrop, indicating the coordinate system used.

– Thickness of the section (precautions should be taken on sections located on inclined surfaces).

– Qualitative description of the section, made after polishing the surface of the outcrop to gain access to an intact surface. This description should include:

- a sketch of the section (guidance on making these drawings was given by Houghton and Carey (2015) including the use of symbols);

- recognition of the different structures present (massive deposition or alternating beds with apparent variation in grain size);

- a description of the juvenile and non-juvenile particles;

- a possible subdivision of the whole section into different levels according to the macroscopic characteristics of the outcrop (in order to highlight different potential eruptive phases). In this case, the thickness of each subdivided level should be measured.

– If the height of the eruptive column is to be estimated from the deposits, the five largest juvenile pumice and the five largest lithic fragments should be measured from the desired levels. In this case, it is recommended to follow the procedure detailed by Bonadonna et al. (2013), which is described in section 2.9.

– If samples are to be collected for laboratory analysis (see section 2.5), each target level should be sampled on flat paleo-topography, and then a standardized sampling area should be defined (Bonadonna et al. 2013) to collect a representative amount of samples from the deposit.

– If the targeted deposit consists of large grains (and especially for pyroclastic flow deposits), image analyses of the cross-section can be used to measure the size, components and shape of the particles comprising the deposit, as reported in the studies by Gurioli et al. (2012) and Bernard et al. (2014).

Tools	Usage	Comments
Hammer (see Figure 2.36(a))	Breaking large particles to observe internal textures on unaltered surfaces	–
Foldable shovel (see Figure 2.36(b))	Polishing the surface of the section Excavation of the deposit to obtain a uniform sample surface	–
Trowel (see Figure 2.36(c))	Fine polishing of the deposit Sampling	The trowel should have a flat side for easy sampling
GPS (see Figure 2.36(d))	Annotation of geographical coordinates	–
Magnifying glass (see Figure 2.36(e))	Observations of textures (crystals and vesicles)	–
Plastic bags (see Figure 2.36(f))	Storage and transport of samples	It is best to have different-sized bags A permanent marker should be used to write the name of the sample on each bag used and also on a paper inserted in the bag itself
Compass (see Figure 2.36(g))	Inclination and azimuth of layers, particles and sections	A solar compass is preferable in the case of ferromagnetic lava
Meter (see Figure 2.36(h))	Measurements of scales and thicknesses	–
Vernier caliper (see Figure 2.36(i))	Particle measurement	–

Table 2.4. *Descriptive table of the tools shown in Figure 2.36*

In order to carry out all the measurements and observations mentioned above, we suggest using (apart from a waterproof notebook, topographic maps of the study area and a rigid board, a helmet, mountain boots, as well as appropriate clothing) the tools illustrated in Figure 2.36 and detailed in Table 2.4.

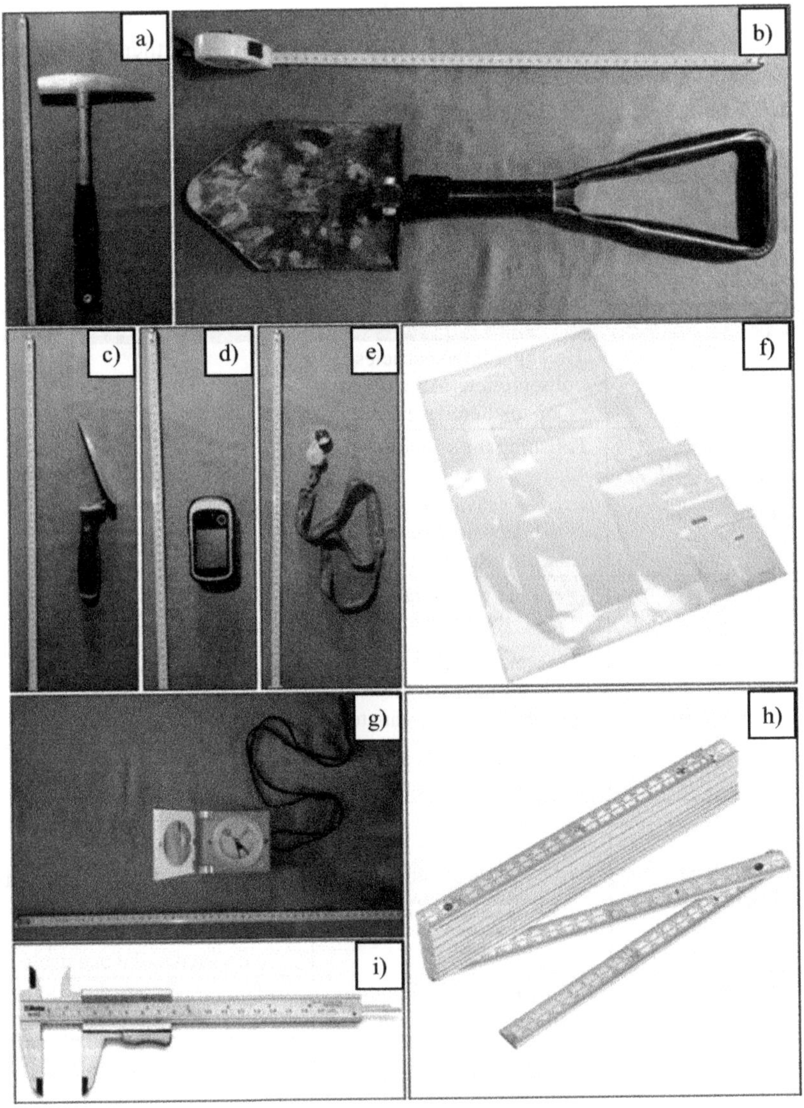

Figure 2.36. *Tools needed to perform field measurements on pyroclastic deposits (see Table 2.4)*

2.7.1. *Dispersion axis and dispersal area*

The dispersion axis of a fallout deposit is the direction of an imaginary segment starting from the volcanic vent (emission source) and indicating the maximum elongation of a fallout deposit (see Figure 2.3). This elongation is an important parameter because it provides the progression of the area potentially affected by the ash fall (syn-eruptive data) and gives insight into the main direction of the prevailing winds for former eruptions for which there are no meteorological data (post-eruptive data). The dispersion axes are angles (°) normally measured with respect to the geographic North.

The dispersal area represents the footprint of the volcanic plume on the ground, and also the area bounded by a selected isoline that characterizes the deposit (isopachs or isomasses, see next section).

2.7.1.1. *Syn-eruptive methods to determine the axis and zone of dispersion*

The dispersion axes of a volcanic plume and its relative dispersion area (which are correlated with those of the fallout deposits) are normally determined using satellite and aerial imagery. These methods are described in particular in Chapter 3 of Volume 2.

2.7.1.2. *Post-eruptive methods to determine the dispersion axis and the dispersal area*

The dispersion axis of a deposit is normally derived after drawing several isolines representing a specific feature of the deposit. The most commonly used lines are the isopachs, connecting all points in the deposit with the same thickness (see Figures 2.3(a), 2.3(b), 2.3(c) and 2.3(d)). These isolines are drawn according to the distribution of stratigraphic sections made in the field with thickness information, which are related to the entire fallout deposit or to particular fallout units. These lines have in most cases an irregular elliptical shape, reflecting the main wind direction. The volcanic vent is approximately located on one of the two foci of the resulting ellipses. From the location of the vent, the downwind and upwind parts can be distinguished (Figure 2.3(e)). Exceptional cases are those in which no or little wind is recorded at the time of the eruption, forming near-circular isopachs with the vent at their center (Figure 2.3(b)).

Except for this latter case where no dispersion axis could be derived, the dispersion axis can be easily deduced by connecting the location of the vent with the points of maximum elongation of the isopachs. The main source of uncertainty in this process is the degree of subjectivity in drawing the isopach lines. Differences in the experience and ability of each operator could lead in fact to different results

(Klawonn et al. 2014). While this could have a minor influence on the derivation of the dispersion axis, the differences could be higher if the isopach maps are used to calculate other parameters (e.g. the volume and mass of the deposit under consideration, see the following sections). To overcome this degree of subjectivity, new methods for automatic derivation of isopach lines have been implemented, such as cubic B-spline (Engwell et al. 2013) or kriging (Yang and Bursik 2016). An important measurement that is normally performed for isopachs (and also for isomasses, see next section) is to calculate their surface area, an important parameter especially for volume calculations. For cases where isopachs are drawn on topographic maps with a scale, the simplest way to obtain isopach surfaces is to approximate them to either ellipses or circles (more easily calculated). If the area of an irregular shape needs to be calculated, it is possible to use appropriate instruments, such as planimeters (see Figure 2.37) that derive the area from the perimeter of a 2D shape. In both cases, it is important to convert the obtained area using the scale of the map. If the map, on which the isopachs are plotted, has several geo-referenced points, the easiest and most accurate way to calculate the areas of the isopachs is to geo-reference the map with GIS software, digitize the individual lines and use the appropriate tool to calculate the area of a composite polygon.

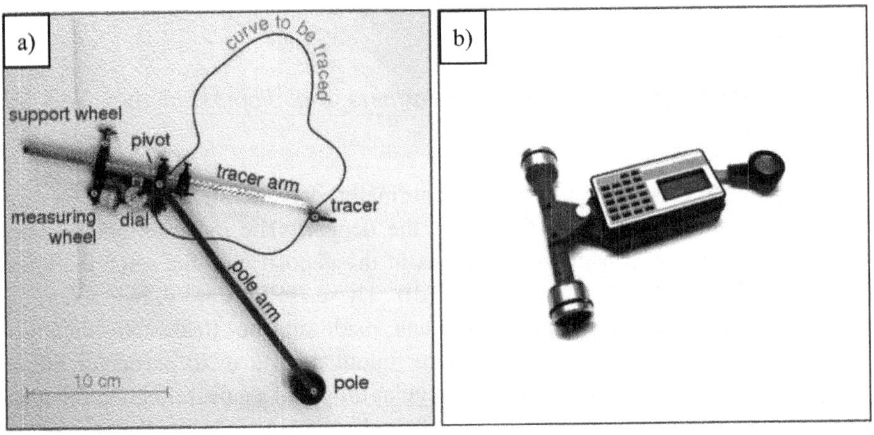

Figure 2.37. *Examples of planimeters: a) polar and b) digital*

2.7.2. *Mass load*

Mass load (sometimes called areal density) is the mass of the deposit per unit area (measured in kg/m^2, g/m^2 or g/cm^2). Point measurements of mass loading within different stratigraphic sections are typically converted to isomass maps, in which each isoline connects all points with the same mass per unit area. The process of deriving isomass maps involves the same considerations as for isopach maps, and isomass maps could also be used to obtain the dispersion axis and the dispersal area (see previous section). Mass loading is an important parameter for defining the stability of roofs under tephra load and for determining mass loading threshold values that impact different elements at risk (Biass et al. 2013), both in real time (syn-eruptive conditions) and for long-term hazard and risk studies (post-eruptive conditions).

2.7.2.1. *Syn-eruptive methods to determine the mass load*

In order to obtain mass loading measurements at different locations, the weight of a sample taken at a specific location must be divided by the sampling area (see Figures 2.4(c) and 2.4(d)). A simple example is the so-called bucket strategy, in which a large number of buckets are distributed over a discrete area of fallout for a period of time (Swanson et al. 2009). When possible, the aim is to collect a sufficient number of samples to estimate the magnitude of the event through mass loading per unit area. This measurement is of paramount importance as it provides information on the accumulation rate and will provide a first indication of the risk of infrastructure collapse due to fallout accumulation. Recently, an increasing number of devices allow syn-eruptive ash sampling and mass loading measurements (see Figures 2.4 and 2.7). Such devices should ideally be placed at several distances from the active vent. For example, ASHER (see Figure 2.7(d)) consists of a cylindrical collector that collects the ash. A scale measuring the mass of accumulated ash (between 0 and 5 kg with an accuracy of 1 g) and an ultrasonic sensor measuring the thickness of collected ash (up to 1 m thick with a resolution of 1 mm) are directly installed on the collector to measure the accumulation rate of the particles. All these parameters are measured by ASHER every 30 seconds during the whole acquisition time. A GPS receiver ensures the synchronization of the data and the precise location of the sampling. These methods are more particularly described in Chapter 4 of Volume 2.

2.7.2.2. *Post-eruptive methods to determine mass loading: isomass maps*

Mass loading can also be calculated after an eruption, based on tephra fallout deposits, especially for old eruptions without syn-eruptive sampling. In this case, the simplest way is to sample as many outcrops as possible, which should also be as far apart as possible. The number of measured sites should be decided after an initial step to identify the extent of deposition, thereby increasing the number of sections to be studied in key areas (Houghton and Carey 2015). The strategy suggested here is to space the sites at a distance no greater than approximately half the thinning distance of the deposit, that is, the average distance over which the thickness of a tephra fallout deposit reduces by half (a parameter introduced by Pyle (1989), see next section).

Once the optimum number of sections has been chosen, the mass loading for each site can be calculated by first deriving the density of the deposit. Such a measurement could be made either directly in the field or in the laboratory for each uniform portion of an outcrop. If this is done in the field, a rectangular or cylindrical volume with a predefined width, height (the thickness of the studied level) and depth should be excavated and weighed, thus obtaining the density by dividing the mass by the volume. If this is done in the laboratory, a smaller sample should be weighed and its volume measured with appropriate graduated cylinders. Once the density of the deposit has been obtained, the mass loading for a site (or for a level within the site) can be obtained simply by multiplying the density of the deposit by the thickness of the outcrop or level considered. Finally, the isomass map can be derived using the methods described in section 2.7.1.

2.7.3. *Volume*

The total volume of tephra fallout deposits from an eruption is a parameter that can only be calculated after the eruption (post-eruptive conditions). The major difficulty for this calculation is that the deposit thickness follows a nonlinear decay trend from the vent to the most distal sites. In order to address these problems, several studies have proposed methods with different assumptions and that consider different trends of deposit thickness decay. It is important to note that the volume of the deposit does not coincide with the volume of the magma that actually erupted, which is normally called Dense Rock Equivalent (DRE). To obtain the latter, the deposit volume must be corrected by subtracting the porosity of the overall deposit and the vesicularity of the juvenile material.

Once calculated, the volume of tephra fallout deposit is an extremely important parameter from a volcanic hazard assessment perspective. First, the volume of tephra fallout deposits helps determining the total volume of erupted material, which is one of the main parameters for calculating the volcanic explosivity index (VEI) described by (Newhall and Self 1982). On the other hand, the total volume of fallout deposits can be indirectly used to calculate the total mass of erupted material and thus another important parameter, namely, the eruption mass discharge rate (MDR, see section 2.10.2).

2.7.3.1. *Post-eruptive methods for determining volume*

2.7.3.1.1. Trapezoidal rule

This method proposed by Fierstein and Nathenson (1992) is based on the definition of an isopach map (see section 2.7.1.2). For each line, the area bounded by the isopach must first be calculated and this value must be plotted against the thickness of the corresponding isopach. The trapezoidal rule is the approximation of the curve described by all the points on the graph (see Figure 2.38) using a straight line joining the individual points. The formula for the volume (Δv_{n+1}) between an isopach T_n of area A_n, and an isopach T_{n+1} of area A_{n+1} (see Figure 2.38) can be written as follows:

$$\Delta v_{n+1} = \frac{1}{2}(T_{n+1} + T_n)(A_{n+1} - A_n) \quad [2.16]$$

where the first term on the right side of the equation is the area of the rectangle and the second term is the area of the triangle shown in Figure 2.38.

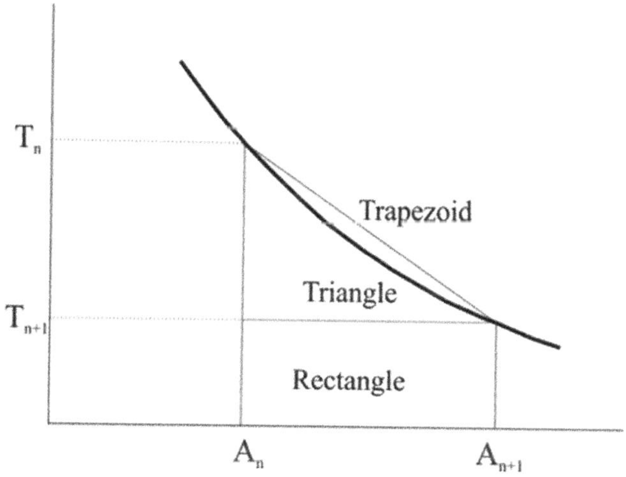

Figure 2.38. *Graphical representation of the volume calculation using the trapezoidal rule (adapted from Fierstein and Nathenson (1992))*

The total volume is therefore the sum of the entire trapezium between two consecutive isopachs. The advantage of this method is that it is quite simple in its application, but the overestimation visible in Figure 2.38 could be significant when the intervals between the isopachs are large.

2.7.3.1.2. Rose method

This method is named after the study conducted by Rose et al. (1973) proposing to plot the isopach thickness and area data on a logarithmic scale, with thickness as an independent variable on the x-axis. This allowed the nonlinearity of the data to be approximated as two segments (Figure 2.39).

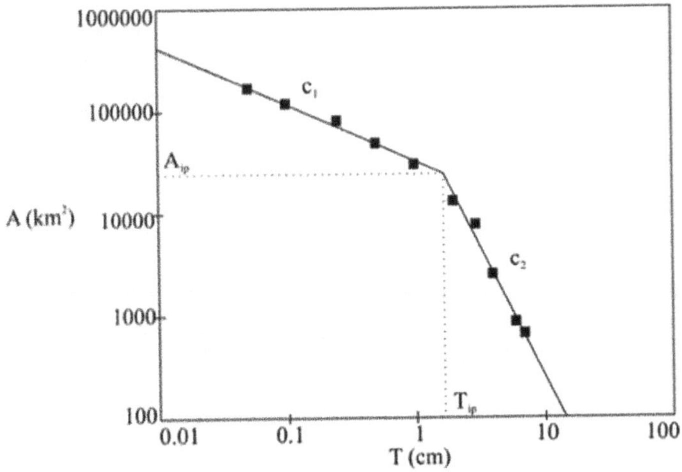

Figure 2.39. *Example of a graph associated with the May 18, 1980 Mount St. Helens eruption depositional isopachs. A_{ip} and T_{ip} indicate the coordinates of the points where lines c_1 and c_2 meet, respectively (adapted from Fierstein and Nathenson (1992))*

According to this method, a power law describes the following function:

$$A = \begin{cases} A'_0 T^{c_1} & 0 \le T \le T_{ip} \\ A''_0 T^{c2} & T_{ip} \le T \le \infty \end{cases} \quad [2.17]$$

where $A'_0 T_{ip}^{c_1} = A''_0 T_{ip}^{c_2} \equiv A_{ip}$.

A'_0 and A''_0 denote the values of the intercept area of the two lines, T_{ip} and A_{ip} the coordinates of the intercept point between the two lines and c_1 and c_2 are the angular coefficients of the two lines. When the thickness goes to infinity (in the above equation), the area goes to infinity, while when the thickness goes to zero, the area goes to infinity. To solve this problem, two constraints must be introduced, namely, $c_1 < 0$ and $c_2 < 0$. By integrating the function, we therefore obtain:

$$V = \int_0^\infty A dT = \frac{A'_0}{c_1+1}\left[T^{c_1+1}\right]_0^{T_{ip}} + \frac{A''_0}{c_2+1}\left[T^{c_2+1}\right]_{T_{ip}}^{\infty} \quad [2.18]$$

Finally, to prevent this new equation from becoming zero or infinite, two more conditions must be imposed, namely, $c_1 + 1 > 0$ and $c_2 + 1 < 0$. These two new conditions, combined with the previous one, require that:

$$0 > c_1 > -1 \quad [2.19]$$

$$c_2 < 1 \quad [2.20]$$

which implies that the volume can be given by:

$$V = A_{ip}T_{ip}\left[\frac{1}{c_1+1} - \frac{1}{c_2+1}\right] \quad [2.21]$$

Even though it is quite easy to obtain the total volume by this method, the study by Fierstein and Nathenson (1992) points out that the resulting volume is very sensitive to c_1, which in turn is sensitive to the number of isopachs available. Thus, if few isopachs with low thickness values are available, there is a high probability of volume overestimation.

2.7.3.1.3. Pyle's method

The Pyle's method was originally developed by Pyle (1989) and generalized by Fierstein and Nathenson (1992). This method is based on the observation that the thickness of tephra fallout deposits decreases exponentially with distance from the vent. Therefore, a logarithmic thickness relative to the mean distance (i.e. the radius of a circle whose area is equal to the area of the isopach) has the same exponential trend. In this way, a plot of logarithmic deposit thickness versus the square root of the area of the corresponding isopach provides a trend that is described by one or two lines (see Figure 2.40).

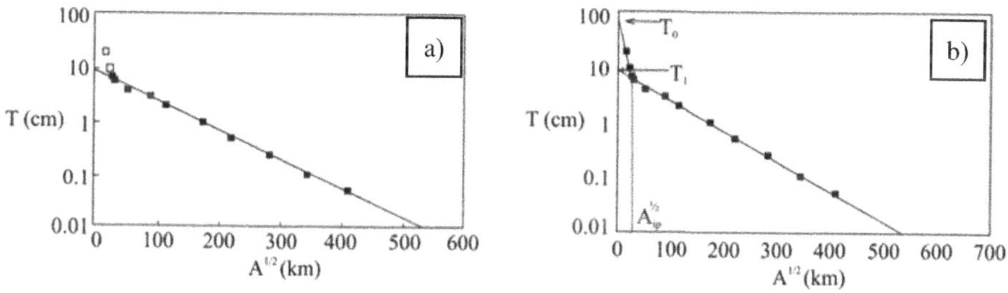

Figure 2.40. *Example of a graph associated with the May 18, 1980 Mount St. Helens eruption depositional isopachs in a log T (thickness) versus square root of the area of the associated Isopachs*

COMMENT ON FIGURE 2.40.– *a) One line (the two white squares are two proximal isopachs not included in the calculations) and b) two lines (T_0 and T_1 are the y-axis intersections of the slope breaks), adapted from Fierstein and Nathenson (1992).*

In the one-line approximation (Figure 2.40(a)), the slope (k) of the interpolation line can be calculated by knowing the coordinates of the points a and b along the line itself:

$$-k = \frac{lnT_b - lnT_a}{A_b^{1/2} - A_a^{1/2}} \quad [2.22]$$

and the function describing the exponential trend is therefore:

$$T = T_0 e^{-kA^{1/2}} \quad [2.23]$$

where T_0 is the extrapolated thickness when $A = 0$. Considering that A is an independent variable and integrating the above equations, the volume is obtained by:

$$V = \int_0^{\infty} T dA = \frac{2T_0}{k^2} \quad [2.24]$$

In the two-line approximation (Figure 2.40(b)), the exponential function can be described by:

$$T = \begin{cases} T_0 e^{-kA^{1/s}} & 0 \le A \le A_{ip} \\ T_1 e^{-k_1 A^{1/s}} & A_{ip} \le A \le \infty \end{cases} \quad [2.25]$$

where T_0 and T_1 are the intersections of the two lines with the y-axis, k and k_1 are the slopes of the two lines and A_{ip} is the meeting point of the two lines. Assuming A as an independent variable and integrating the above equations, we obtain the volume by:

$$V = \int_0^{\infty} T dA = \frac{2T_0}{k^2} + 2T_0 \left[\frac{k_1 A_{ip}^{1/2} + 1}{k_1^2} - \frac{k A_{ip}^{1/2} + 1}{k^2} \right] e^{-kA^{1/2}} \quad [2.26]$$

As an example, the volume of the Mount St. Helens eruption calculated with the one- and two-line approximations yielded values of 1.07 km^3 and 1.12 km^3, respectively. However, as pointed out by Fierstein and Nathenson (1992), the difference between the two estimates may not always be so small, depending on the location of the break in slope of Figure 2.40. In order to account for the problems associated with the lack of more distal isopachs, Sulpizio (2005) proposed that, in order to derive some of the parameters needed for the Pyle method, three empirical methods could be used, namely:

– The elliptical approach, which aims to derive distal isopachs when scattered distal data are available. It is based on the assumption that all isopachs could be approximated by ellipses, that distal isopachs have the same ellipticity (ε) as proximal isopachs and that the dispersion axis of proximal and distal isopachs is the same. In this view, when an average ellipticity is calculated for the known proximal isopachs and one of the two half-axes (a and b) of the distal isopachs is known (because the few available distal data allow for a partial plot of an isopach), then the other half-axis could be derived knowing that $\varepsilon = \sqrt{(a \wedge 2 - b \wedge 2)}$. This allows the areas of more distal isopachs to be calculated and thus reduces the uncertainty in the second line of the Pyle method.

– The method of $A_{ip}^{1/2}$ with a distal isopach, which aims to obtain the position of $A_{ip}^{1/2}$ when only one isopach is available after the break in slope (Figure 2.40). The author proposed an empirical relationship (based on data from 32 eruptions) between $A_{\mathrm{ip}}^{1/2}$ and k as $A_{ip}^{1/2} = 3.34k^{-0.96}$.

– The method of k_1 versus $A_{ip}^{1/2}$, which could be used if only proximal data are available (i.e. before the break in slope of Figure 2.40). In this case, the author did not provide a single empirical relationship, but proposed that $A_{ip}^{1/2}$ could be calculated with the previous relationship $A_{ip}^{1/2} = 3.34k^{-0.96}$ and that the relationship $k_1 = 1.61A_{ip}^{-1.02/2}$ could be used as a parameter to derive a minimum volume for the unit of interest.

2.7.3.1.4. Walker's method (or crystal concentration method)

This method was proposed by Walker (1980) to estimate the volume of the ultraplinian eruption of Taupo volcano (New Zealand). Since for this estimate the smallest available isopach line was 12.5 cm, the author inferred that most of the erupted material was not included in volume calculations based on thickness and isopach areas (see previous methods). To account for this problem, Walker (1980) developed a new method based on the following assumptions:

– The largest pumices are representative of the original magma and can therefore be used to calculate the original ratio between crystals and the glass in magma.

– During an eruption, part of the magma is broken up into particles of equal or smaller size than the largest phenocrysts.

– During eolian fractionation in the volcanic plume, crystals tend to be deposited closer to the volcanic vent than juvenile particles and, in particular, it is assumed that the crystalline part of the deposit is intact and could be accurately calculated.

Initially, for this method, it is necessary to derive different isomass maps (see section 2.7.2.2) for lithic fragments, crystals and for the total deposit. The isomass of the different components can be interpolated by multiplying the total mass per unit area by the relative proportions, respectively, of lithics and crystals, measured in as many sample sites as possible. By selecting a number of large juvenile pumice stones, the ratio of crystals to glass (i.e. the porphiricity index) could be calculated. Then, the total mass of preserved glass (M_g) is calculated by subtracting the mass of lithics (M_l) and crystals (M_c) from the total mass of the preserved deposit (M_{tot}) using the available isomass values (M/A_i) and the corresponding isomass area (A_i) for each component:

$$M_g = M_{tot} - (M_l + M_c) = \sum_{i=1}^{n} \left[\left(\frac{M}{A_i}\right) A_i\right]_{tot} - \left(\sum_{i=1}^{n}[(M/A_i)A_i]_l + \sum_{i=1}^{n}[(M/A_i)A_i]_c\right) \quad [2.27]$$

Next, the lost mass of the glassy component is obtained by subtracting M_g from the theoretical total mass of the glassy component, obtained by multiplying M_c by the porphiricity index. Finally, the mass is converted to volume by dividing it again by the density value of the deposit.

Alternatively, it is possible to use the modification of this method introduced by Fierstein and Nathenson (1992), who suggested replacing, in the Pyle's method plots, T_0 with M/A_0, which is the extrapolated intercept of a log(M/A). The main limitation of this method is that, as pointed out by Sulpizio (2005), Walker's method is only applicable to deposits that include porphyritic juvenile fragments.

2.7.3.1.5. The power-law fit

This type of integration for volume calculation was proposed by Bonadonna and Houghton (2005) and is based on the observation that the Pyle's method can significantly underestimate the total volume, especially for eruptions that produce a significant amount of fine ash. To fit the observed data, it is therefore proposed to use a power-law distribution over a thickness as a function of the square root of the area diagram, with the following formula:

$$T = T_{pl}\sqrt{A}^{(-m)} \quad [2.28]$$

where T_{pl} is a constant and m is the power-law coefficient (whose parameters are recovered when a power-law distribution is fitted to the data, using one of the available methods). Using this distribution, the volume could be obtained with:

$$V = \int_0^{\infty} T dA = \left[2T_{pl}\frac{\sqrt{A}^{(2-m)}}{2-m}\right]_0^{\infty} \quad [2.29]$$

Two integration limits (B and C, lower and upper, respectively) must be defined, however, to prevent the above equation from becoming infinite when $\sqrt{A} = 0$ and $\sqrt{A} = \infty$. The equation thus becomes:

$$V = 2T_{pl}\frac{\sqrt{A}^{(2-m)}}{2-m}\left(C^{(2-m)} - B^{(2-m)}\right) \quad [2.30]$$

This method has the advantage, as Bonadonna and Houghton (2005) point out, that the adjustment of the data does not have to be done by choosing an arbitrary number of segments as was done, for example, with the Pyle's method, although the choice of the integration limits is not simple. As an example, to estimate the volume of the Ruapehu eruption (New Zealand) in 1996, Bonadonna and Houghton (2005) chose the lower boundary B as the distance of the maximum calculated thickness, while the upper boundary C was the downwind limit of significant volcanic cloud propagation, as shown in satellite images.

2.7.3.1.6. Weibull fit

Bonadonna and Costa (2012) proposed to obtain the volume by fitting a Weibull distribution over a thickness as a function of the square root of the area diagram. A Weibull function has the general form:

$$T = \theta\left(\frac{\sqrt{A}}{\lambda}\right)^{n-2} e^{-\left(\sqrt{A}/\lambda\right)^{n}} \quad [2.31]$$

where λ represents the decay length scale characteristic of deposit thinning, θ represents a thickness and n is a shape parameter. These three parameters could be obtained when fitting a Weibull distribution to the data: Bonadonna and Costa (2012) propose to choose a range for the three parameters, derive a series of Weibull distributions, fit each distribution to the data and finally choose the distribution (and thus the triplet of parameters) that minimizes the following function:

$$\sigma^2 = \sum_{i=1}^{N} w_i[T_i(obs) - T_i(calc)]^2 \quad [2.32]$$

where w_i are weighting factors (could be = 1 if a uniform error is assigned to all data, or could be different if some data points have different uncertainties), N is the number of data points, and $T_i(obs)$ and $T_i(calc)$ represent the observed and calculated thickness values, respectively. Once λ, θ and n are determined, the total volume can be obtained with:

$$V = \int_0^{\infty} T dA = \frac{2\theta\lambda^2}{n} \quad [2.33]$$

Note that in this case, the Weibull function does not depend on the integration limits because it can be integrated between zero and infinity. In addition, Bonadonna and Costa (2012) point out that the Weibull function is less sensitive when proximal, medial or distal data are missing.

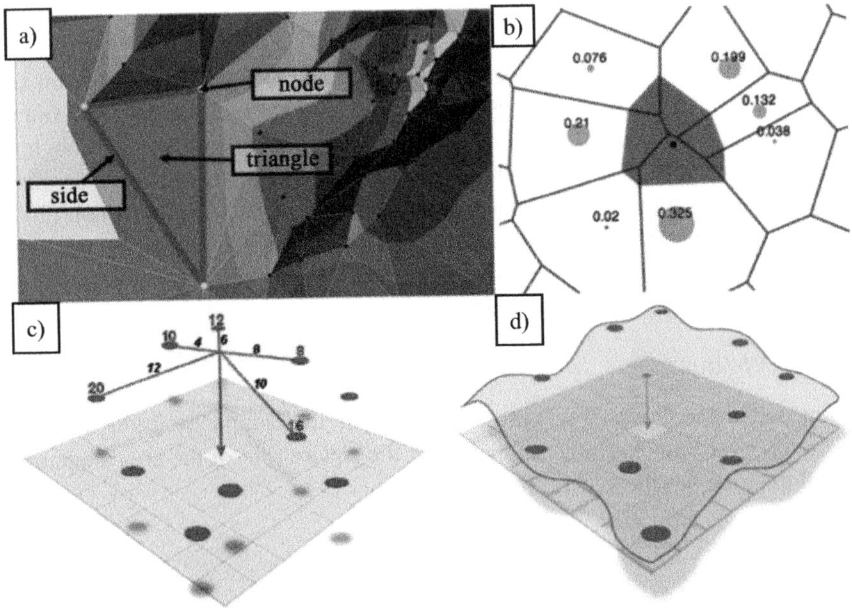

Figure 2.41. *Examples of different interpolators. a) TIN; b) NNI; c) IDW and d) SPLINE. For a color version of this figure, see www.iste.co.uk/lenat/hazards3.zip*

2.7.3.1.7. Spatial interpolation methods

The increasing availability of the GIS tool for spatial data processing allowed Bisson and Del Carlo (2013) to test different spatial interpolators available from ArcGIS software to calculate the volume of a tephra fallout deposit using geo-referenced thickness data from several stratigraphic sections to create a digital model of the deposit itself. The authors tested four different interpolators:

– The TIN (see Figure 2.41(a)), which is a vector surface model consisting of a network of Delaunay triangles. Thickness values are assigned to points on the edges of the triangles according to linear interpolations between the nodes of each triangle.

– The NNI (see Figure 2.41(b)), an interpolator that creates around any unmeasured point a Voronoi polygon, calculates the proportion of that polygon's area in each of the surrounding Voronoi polygons created in the measured points and assigns a thickness value to the unmeasured points that depends on the weighted average with respect to the overlap area.

– The IDW (see Figure 2.41(c)), which determines the values of the unmeasured points using a linearly weighted combination of a set of sample points, where the weight is a function of the inverse distance.

– The SPLINE (see Figure 2.41(d)), which instead interpolates the values with a mathematical function that minimizes the global curvature of the surface.

Bisson and Del Carlo (2013) compared volume estimates made with these interpolators to a volume estimate made using the Pyle method and concluded that the TIN volume estimate is closest to the Pyle estimate.

2.7.3.1.8. Statistical methods

The purely statistical method proposed by Burden et al. (2013) does not rely on generating isopach or isomass maps, and also includes a means of quantifying the uniform distribution of error in measured thicknesses using Bayesian analysis. The basic assumption of this method is that the thickness y_i of each point in the deposit can be modeled by considering a linear regression of a natural logarithmic transformation ($z_i = \ln(y_i)$) in the form:

$$E(Z_i|d_i) = \alpha + \beta d_i \quad [2.34]$$

where d_i is the distance from point i to the vent, α is the modeled thickness of the deposit at the vent location and β is the change in thickness as the distance increases by 1 km from the vent. Since each measured thickness point is subject to an error that depends on many factors (Engwell et al. 2013), each point is also assumed to have an error ϵ_i that is normally distributed. Moreover, assuming that $\beta < 0$ (as the thickness decreases away from the vent), the volume V can be calculated:

$$V(\alpha,\beta) = \int_0^{exp(\alpha)} \pi \left(\frac{logy-\alpha}{\beta}\right)^2 dy \quad [2.35]$$

α and β are calculated by the authors with Bayesian linear regression analysis, and interested readers can refer to the aforementioned article for more details.

2.8. Total grain size distribution

Total grain size distribution (TGSD) represents the theoretical eruptive mixture injected into the atmosphere during volcanic explosive eruptions. In the literature,

TGSD can either be determined for the products of the entire eruption (Murrow et al. 1980; Durant and Rose 2009) or for individual eruptive units (Bonadonna et al. 2015a; Janebo et al. 2018). While the majority of methods for determining TGSD are applied under post-eruptive conditions as a result of different techniques for integrating single grain size measurements (see section 2.5.1), several recent attempts have been made to characterize the PSD in real time. It should be noted, however, that the TGSD obtained from single measurements needs to be integrated over time to obtain a complete TGSD representative of the entire eruption. Furthermore, the accuracy of the current instruments is not yet optimal for obtaining complete information on the full-size range of particles dispersed in the atmosphere and deposited on the ground (Pioli et al. 2019), implying that the vast majority of TGSDs available from literature are derived from post-eruptive conditions. In addition, TGSD calculations are affected by several problems, as pointed out by Bonadonna and Houghton (2005), which are the sometimes sparse and/or uneven distribution of data points (especially for older and smaller eruptions) and difficulties in detecting and measuring particle sizes smaller than 63 μm (4Φ).

As pointed out in the study by Pioli et al. (2019), TGSD is important for gaining insight into the fragmentation mechanism of an eruption and for obtaining an important input parameter for numerical models of tephra transport and deposition and pyroclastic density currents. In the same study, the authors also provide some useful observations regarding the effect that the number of samples and the extent of the sampling area have on the TGSD calculation. In particular, the reconstructed and actual TGSD show small differences in median (MdΦ) and sorting (σΦ) (see Table 2.2) if the sample sites cover most of the depositional area and are mainly concentrated in the part of the deposition with the largest portion of the total mass. In addition to these observations, the study by Tsunematsu and Bonadonna (2015) suggests that tephra deposits associated with large explosive eruptions (plume height greater than 30 km) should be sampled at least 200 km from the vent to ensure that TGSD is not depleted of fine particles.

2.8.1. *Syn-eruptive methods for determining TGSD*

Chapters 3 and 4 of Volume 2 report on satellite and ground-based remote sensing methods for estimating this parameter.

2.8.2. *Post-eruptive methods for determining TGSD*

2.8.2.1. *Averaging techniques*

The most commonly used method for calculating TGSD from multiple grain size analyses is to average these samples by different criteria.

The simplest approach is to derive the TGSD as an unweighted average of all available samples, as done by Walker (1981). Despite its simplicity, this approach has the disadvantage that individual samples representative of a small (or large) portion of the deposit may be over- or under-represented in the calculated TGSD.

In another study, Walker (1980) proposed calculating TGSD using isomass maps (see section 2.7.2.2) of each grain size class (see Figure 2.42(a)), and integrating TGSD over the surface of all maps. In detail, as summarized by Cioni et al. (2020), for each stratigraphic section sampled, the mass per unit area (M/A) is obtained by multiplying the average depositional density measured on that section by the thickness of the section. For each site, the M/A value for each Φ grain size class (M/A Φ) is then extracted by multiplying the total M/A value by the mass percentage of each grain size at that site. The isomass map of each grain size class is finally plotted and integrated over the area to extract the total mass (within the deposit) of each grain size class. One problem with this technique was raised by Bonadonna and Houghton (2005), who point out that some bias in the final TGSD calculation could be introduced if the samples are not uniformly distributed.

Similar to Walker's isomass method, Murrow et al. (1980) used the isopachs (see section 2.7.1.2) as the averaging factor. In detail, all samples in each isopach contour are averaged, and then the resulting grain size distributions for each isopach interval are weighted against the volumetric contribution of each interval. Parfitt (1998) refined this method by using the mass of each size in each isopach grain size distributions interval calculated with volume estimates and clast density data as a weighting factor.

Another way in which TGSD has historically been derived is to divide the depositional area into different segments/polygons (see Figure 2.42(b)) and associate each segment/polygon with a different parameter to be used as a weighting factor. This can be either volume (Sparks et al. 1981), mass (Carey and Sigurdsson 1982) or mass per unit area (Bonadonna and Houghton 2005). The limitation of this method, however, is the subjectivity through which the segments/polygons are defined.

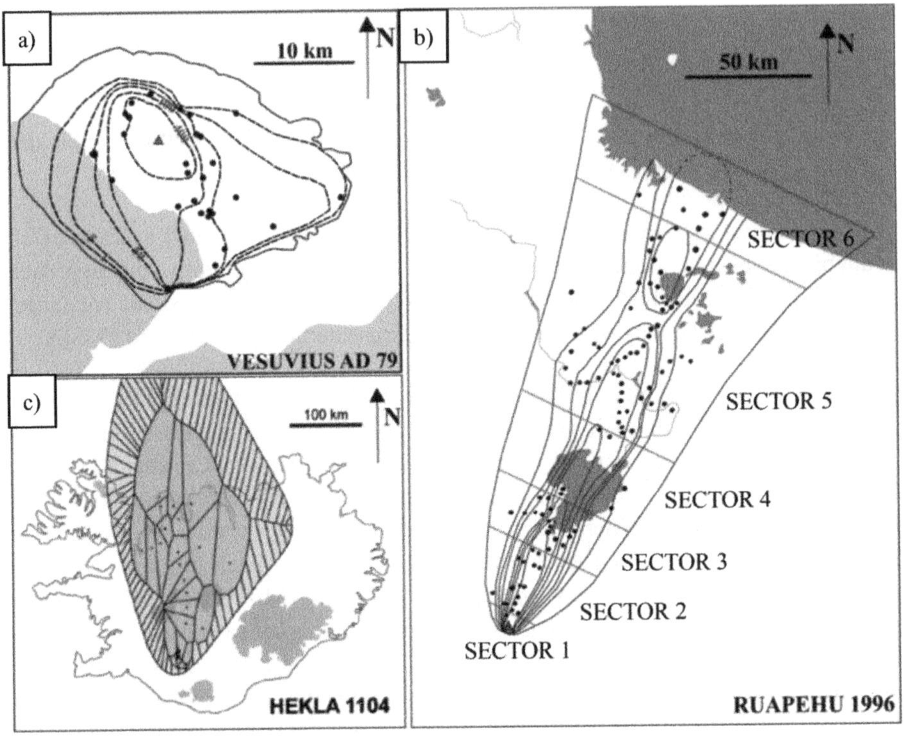

Figure 2.42. *Methods of calculating the TGSD. For a color version of this figure, see www.iste.co.uk/lenat/hazards3.zip*

COMMENT ON FIGURE 2.42.– *a) Isomasses of single Φ units (modified from Cioni et al. (2020)). Dashed lines: isomasses for $\Phi = -1$; blue solid line: deposit boundary; solid black dots: sample points; red triangle: volcanic vent. b) Subdivision of deposits (modified from Bonadonna and Houghton (2005)). The outermost black solid line is the depositional boundary; solid black dots are sample points; red solid lines are sector boundaries; red triangle is the volcanic vent. c) Voronoi tessellation (modified from Janebo et al. (2018)). The outermost black solid line is the depositional boundary, the black solid dots are the sample points and the yellow triangle is the volcano vent.*

The most recently used method for deriving TGSD is the "Voronoi tessellation" method introduced by Bonadonna and Houghton (2005). The basic idea of this method is to divide the deposit into as many polygons as sample points (see Figure 2.42(c)), with an outer integration boundary defined by the smallest available isopach line (ideally the 0-thickness isopach). Each polygon is constructed such that its edges represent points equidistant from two sample points. The mass per unit area and the grain size distribution of each sample point are assigned to each polygon. The TGSD is finally obtained as the area-weighted average of all Voronoi cells over the entire deposit.

2.8.2.2. *Data adjustment techniques*

Recent attempts have been made to derive TGSD distributions from independent eruptive parameters and known density functions. This has been possible due to the increased availability of TGSD estimates and in order to provide the TGSDs needed for numerical simulations aimed at real-time forecasting and long-term hazard assessment. In particular, Costa et al. (2016) analyzed published TGSDs from 20 eruptions of different magnitudes, and proposed to derive TGSDs by using a statistical distribution given by the sum of two log-normal (Bi-Gaussian) distributions, whose main parameters are well correlated with the height of the eruptive plume (itself related to the MDR) and the viscosity of the magma. The proposed distribution is:

$$f(\Phi) = p\frac{1}{\sigma_1\sqrt{2\pi}}exp\left[-\frac{(\Phi-\mu_1)^2}{2\sigma_1^2}\right] + (1-p)\frac{1}{\sigma_2\sqrt{2\pi}}exp\left[-\frac{(\Phi-\mu_2)^2}{2\sigma_2^2}\right] \qquad [2.36]$$

where μ_1, μ_2 and σ_1, σ_2 denote the mean and standard deviation of the two Gaussian distributions in Φ units, respectively, while p is a parameter determining the weight of each distribution (and thus of each subpopulation). These five parameters (μ_1, μ_2, σ_1, σ_2 and p) can be determined empirically from the plume height (H, in km) and magma viscosity (η, in Pa.s) with the following relationships:

$$\sigma_1 \approx a_1 + b_1H \text{ with } a_1 = 0.67 \pm 0.32 \text{ and } b_1 = 0.07 \pm 0.02 \qquad [2.37]$$

$$\mu_1 + 3\sigma_1 \approx a_2 + b_2H \text{ with } a_2 = 0.96 \pm 0.97 \text{ and } b_2 = 0.20 \pm 0.06 \qquad [2.38]$$

$$\mu_2 - \mu_1 \approx a_3(log_{10}\eta)^{b_3} \text{ with } a_3 = 1.62 \pm 0.60 \text{ and } b_3 = 0.66 \pm 0.10 \qquad [2.39]$$

$$\sigma_2 \approx 1.46 \pm 0.41\Phi \qquad [2.40]$$

$$p - a_4e^{(b_4log_{10}\eta)} \text{ with } a_4 = 1.61 \pm 0.47 \text{ and } b_4 = 0.31 \pm 0.07 \qquad [2.41]$$

2.9. Plume heights

In volcanic hazard assessment, plume (or column) height is a key parameter because it can be related to dispersion and used to derive other higher level parameters. These latter allow for better classification of the eruption and a better understanding of the associated eruption dynamics, such as the VEI and the MDR (mass flux per unit time in kg/s).

Here, we refer to the plume height measured relative to sea level (meters above sea level), where this height is the height of the center of mass of the volcanic cloud (maximum plume height, at spreading level, see Figure 2.43) (Bonadonna et al. 2015a, 2015b). Depending on the ratio of horizontal wind speed to plume rise speed, volcanic plumes can develop as strong (subvertical) or weak (curved) plumes and eventually reach the neutral buoyancy level (see Figure 2.43). In this level, their density is equal to that of the surrounding atmosphere, and they begin to spread laterally, as turbulent gravity currents, around this level and developing the umbrella-shaped cloud. As this cloud spreads away from the source, detaches from the source and is diluted and then passively transported by wind advection and atmospheric turbulence, it is called a volcanic cloud.

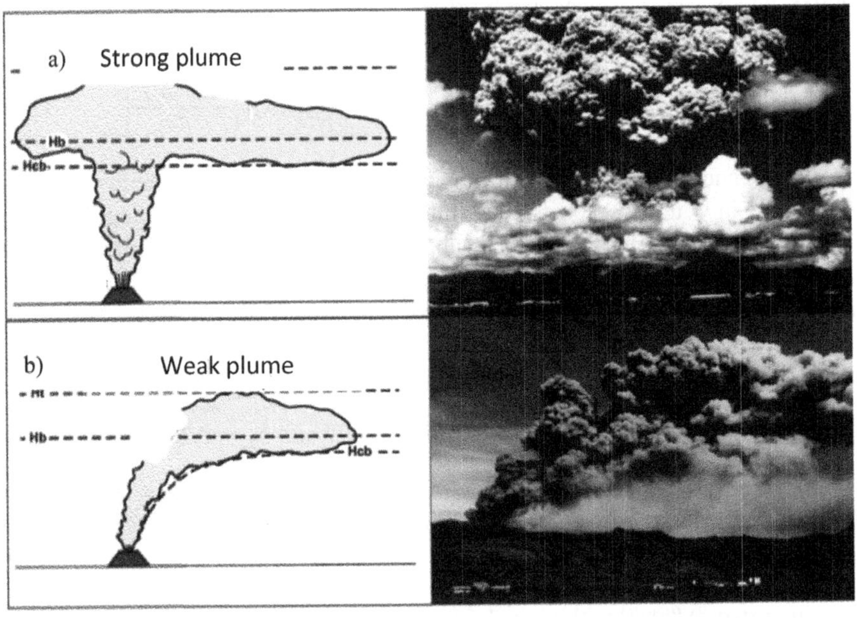

Figure 2.43. *Sketch showing the main characteristics of a) a strong volcanic plume (18-km-high volcanic plume from one of the series of explosive eruptions of Mount Pinatubo beginning on June 12, 1991; photograph by David H. Harlow, USGS) and b) a weak volcanic plume (Eyjafjallajökull plume extending toward southeast Iceland on May 4, 2010)*

COMMENT ON FIGURE 2.43.– *Figure adapted from Degruyter and Bonadonna (2012). Ht, Hb and Hcb indicate the maximum plume height, the neutral buoyancy level (NBL) height and the height of the base of the turbulent current, respectively (modified from Bonadonna et al. (2015b)).*

2.9.1. *Syn-eruptive methods for determining plume heights*

In Chapters 3 and 4 of Volume 2, satellite or ground-based remote sensing methods for estimating plume heights are reported.

2.9.2. *Post-eruptive methods for determining plume heights*

2.9.2.1. *Carey and Sparks' method (1986)*

In general, an eruptive column is formed by the thrust zone, the convective ascent zone (which is absent in weak columns) and the expansion zone described above (see Figure 2.43). The vertical velocity of particles within the plume decreases linearly with height, and has a cross-sectional profile with the shape of a Gaussian function having its maximum along the column centerline. For each clast, of a certain size and density, it is possible to define support envelopes. Therefore, considering the two boundaries *Hb* and *Ht* (see Figure 2.43), clast entrainment and transport are controlled by the average vertical velocity, which can be represented as a series of particle support envelopes where the vertical velocity of the column is equal to the terminal sedimentation velocity of a specific clast size and density. The maximum dispersion of a specific clast size is determined by the location of the edges of its mushroom-shaped support envelope under windless conditions. Additional radial transport in the region of lateral plume extension is imposed by the lateral intrusion of plume material (see Figure 2.44).

Based on this model, Carey and Sparks (1986) proposed that clast dispersal is determined by two parameters: column height and wind velocity (see Figure 2.44), which can be inferred from isopleth maps of juvenile pumice and lithic fragments. The model first calculates the velocity field (vertical and radial velocities) within the plume, and the TFV of clasts of a given diameter (see Figure 2.45). These values are then used to construct a particle support envelope, the area within which these clasts are entrained by the convective column. When a clast reaches the boundary of its envelope, it falls, following a trajectory governed by the radial velocity of the umbrella-shaped cloud and the downwind velocity. In the wind, the expansion of the cloud stops when both are equal (stagnation point).

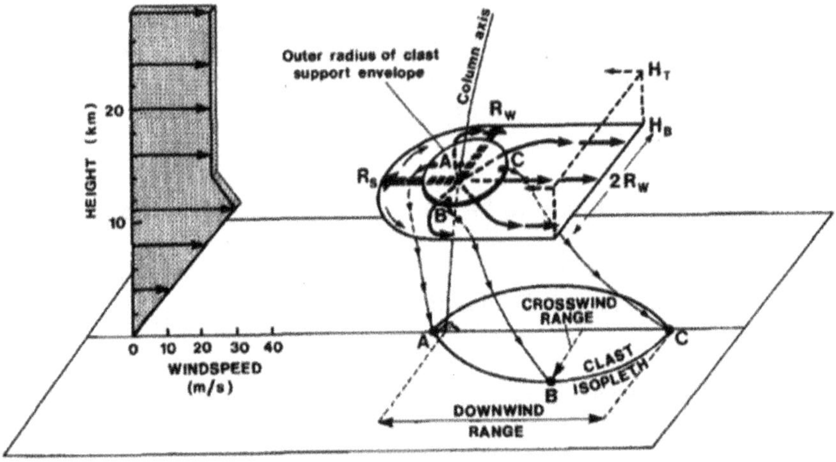

Figure 2.44. *Schematic representation of the influence of a crosswind on the deposition of particles within an eruptive plume*

COMMENT ON FIGURE 2.44.– *A windward stagnation point develops at Rs where the radial velocity in the umbrella region is equal to the crosswind velocity. The radial flow is deflected by the crosswind and passes through the leeward plane defined by Ht, Hb and two times Rw (radial expansion width). The trajectories of three clasts are described by A (upwind), B (sideways) and C (upwind) and define the geometry of the isopleth measured on the ground. The vertical wind profile adopted for the calculations is shown on the left. Clasts A, B and C are freed from the outer radius of a support envelope that lies between the two horizontal planes Ht and Hb (modified from Carey and Sparks (1986)).*

This model was used to construct diagrams where the maximum tailwind range and crosswind range of a given clast size can be plotted to determine the column height and wind speed (see Figure 2.45). Because the height is derived from the distribution of the largest clasts, the method provides the maximum column height. It is still widely used today, but it has two weaknesses:

– The user has to draw the data "by hand" on the diagram, which may lead to some inaccuracy.

– A single diagram cannot encompass all clast sizes at once.

Therefore, Biass et al. (2015) improve the method with a MatLab implementation that interpolates the Carey and Sparks (1986) diagrams into a continuous range of clast sizes, and provides well-defined estimates of plume height and wind speed. They also evaluate the influence of particle shape on the drag coefficient (hence terminal velocity) based on experiments by Bagheri and Bonadonna (2016).

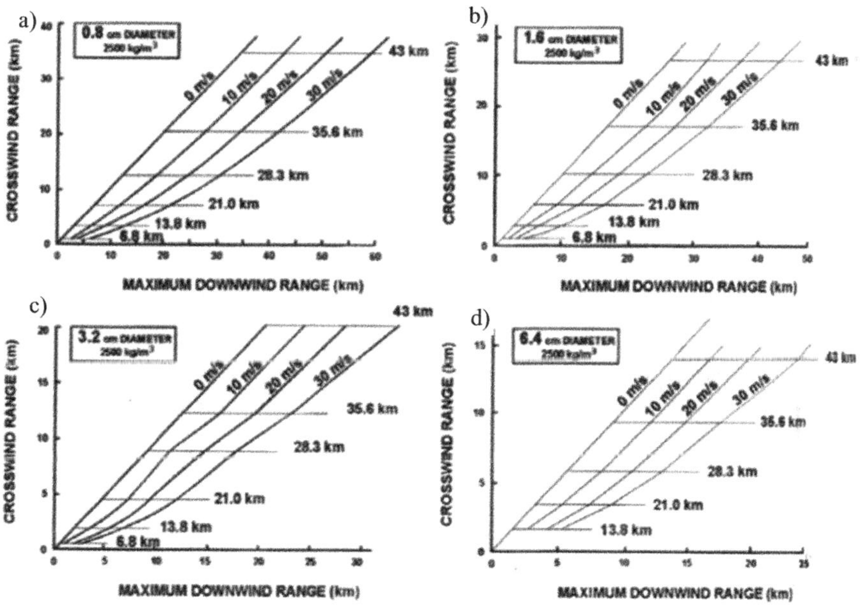

Figure 2.45. *a) Crosswind range versus maximum tailwind range for different clast sizes with a density of 2,500 kg/m³*

COMMENT ON FIGURE 2.45.– *A single isopleth geometry defines a unique combination of plume height (between 7 and 43 km) and wind speed (10, 20 and 30 m/s), for a) 0.8-cm-diameter clasts; b) 1.6-cm-diameter clasts; c) 3.2-cm-diameter clasts; and d) 6.4-cm-diameter clasts (modified from Carey and Sparks (1986)).*

2.9.2.2. *Bonadonna and Costa's method (2013)*

This method also relies on the relationship between maximum lithic clast size (ML) and vent distance. The resulting empirical relationship is:

$$H\ (\text{km}) = 5.01\lambda_{ML}^{0.55} \qquad [2.42]$$

with a typical error of 10–15% and a maximum error of 25%.

2.9.2.3. *Rossi et al. method (2019)*

Finally, the model of Rossi et al. (2019) accounts for the interaction of the plume and the wind. This new strategy refines both the definition of the support envelope of a clast, accounting for the effect of wind advection on an upwelling plume, and the determination of particle trajectories, incorporating the drag force into the sedimentation equation. By introducing a drag coefficient, this model takes into account the effect of particle shape on the sedimentation rate.

From a depositional perspective, the issue is the measurement of a maximum number of lithic clasts. According to Bonadonna et al. (2013), dense, non-juvenile fragments of similar density (e.g. pre-existing lava fragments) should be selected from sections in a specified area (see Figure 2.46) based on a grain size and lithic content given by an outcrop. In the case of sections that cannot be excavated, the resulting evaluation of the largest clasts should be considered a minimum estimate. In the case of lithic-poor deposits (e.g. basaltic tephra), only the densest juveniles should be used and the associated density should be measured. Clasts should be characterized on the basis of the geometric mean of its three orthogonal axes with the minimum ellipsoid approximation (see Figure 2.46). Only clasts with a form factor $F > 0.3$ (where $F = (b + c) / 2a$) should be considered to avoid large discrepancies with the spherical shape assumption. The 50th percentile method of the 20 largest clasts is considered the best way to evaluate the largest clasts.

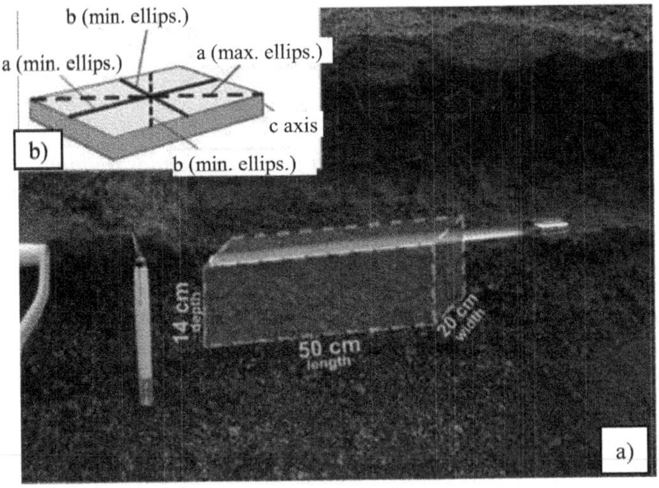

Figure 2.46. *a) Example of sampling a section of a specific area, where the depth is equal to the thickness of the deposit. b) Two different basic approaches to measurements of the three axes of a clast based on the maximum (max. ellips.) and minimum (min. ellips.) ellipsoid. Adapted from Bonadonna et al. (2013)*

2.10. Mass discharge rate

The mass discharge rate (MDR) – aka mass eruption rate (MER), expressed in kg/s – represents the maximum intensity of an eruption and can be used as a scale for explosive eruptions based on a logarithmic index of magma discharge rate, similar to the Richter scale used to describe earthquake magnitude, as reported by Carey and Sigurdsson (1989).

2.10.1. *Syn-eruptive methods to determine the MDR*

MDRs can be estimated by satellite methods (Volume 2, Chapter 3) or by ground-based remote sensing (Volume 2, Chapter 4).

2.10.2. *Post-eruptive methods to determine the MDR*

The MDR can be derived from the height of the eruptive column using the analytical model of Sparks (1986) which is based on the buoyant plume theory (Morton et al. 1956; Settle 1978; Wilson et al. 1978) and improved by allowing for an adiabatic variation within the column and a variation in atmospheric temperature (see Figure 2.47).

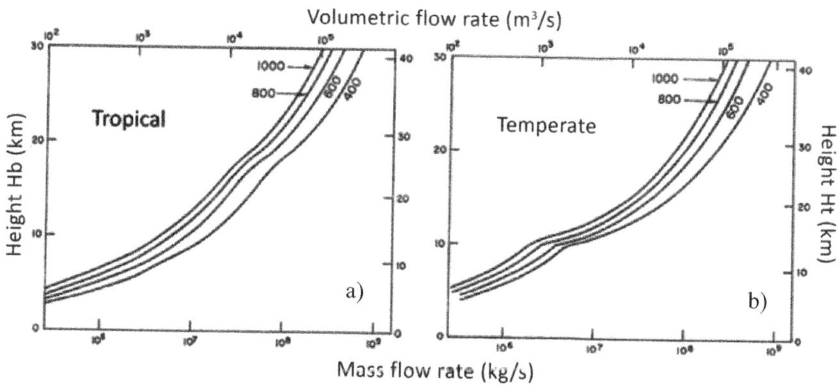

Figure 2.47. *Variation of eruptive spout heights (Hb and Ht, see Figure 2.42) with magma mass discharge rate*

COMMENT ON FIGURE 2.47.– *Each curve shows calculations for a different magma eruption temperature. The equivalent volume flow rate for a magma density of*

2,500 kg/m³ is also shown. Calculations are shown for a) tropical and b) temperate atmosphere (modified from Sparks (1986)).

The study by Carey and Sigurdsson (1989) interpreted the correlation between plume height and volume to link conduit enlargement and associated increases in MDR with time. Large-volume magma bodies can sustain relatively high eruption rates (forming relatively high plumes) before magmatic pressure is exhausted. A similar mechanism may contribute to the correlation in Figure 2.48.

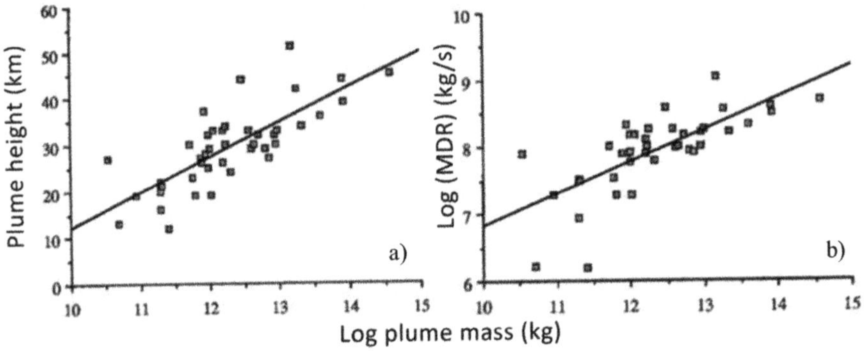

Figure 2.48. *a) Plume height versus log of deposit mass associated with Plinian fallout, for Pleistocene and recent Plinian eruptions. b) Log of magma discharge rate (intensity) versus log of deposit mass associated with Plinian fallout of the same eruptions shown in (a)*

COMMENT ON FIGURE 2.48.– *Plume heights were calculated using the equations of Sparks (1986). Magma discharge rates were calculated from the model of Carey and Sparks (1986). Plume heights and magma discharge rates show a positive correlation with the mass of material erupted during the Plinian fallout phase (adapted from Carey and Sigurdsson (1989)).*

Wilson and Walker (1987) relate MDR to the maximum plume height H with the empirical relationship:

$$\mathrm{MDR} = \left(\frac{H}{k}\right)^4 \qquad [2.43]$$

where k is a constant depending on the chemistry of the magma. For example, for the siliceous magma of Haynaputina (Peru), a factor of 0.236 was used (Prival et al. 2020). This simple equation is only valid for plumes with circular vents below 35 km in height.

Mastin et al. (2009) implemented a method based on the best fit between the volumetric flow rate $\dot{V}$ ($m^3{}_{DRE}/s$) with the maximum plume height H, calculated on a database of about 30 eruptions (see Figure 2.49):

$$\dot{V} = \left(\frac{H}{k}\right)^{4.15} \qquad [2.44]$$

where k is a constant set to 2. Knowing the DRE density of the magma, the volumetric flow rate can then be converted to MDR.

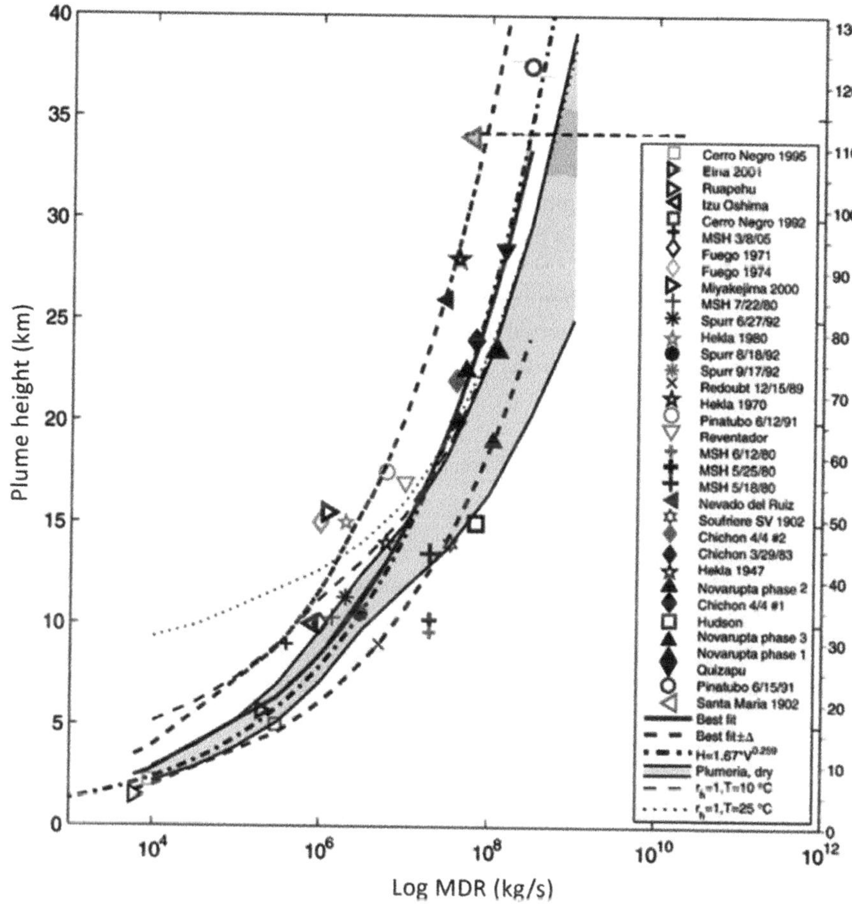

Figure 2.49. *Plume height above blowhole versus mass discharge rate for eruptions listed in legend*

COMMENT ON FIGURE 2.49.– *The symbols in the legend are ordered from lowest to highest eruption rate. The plotted height H includes both Ht and Hb depending on the estimation method. Isopleth-based estimates, denoted by solid symbols, give the cloud height Hs while all other methods are expected to give Ht (adapted from Mastin et al. (2009)).*

Finally, Degruyter and Bonadonna (2012) developed a new formulation taking into account wind:

$$\text{MDR} = \pi \frac{\rho_{a0}}{g'} \left(\frac{\alpha^2 \bar{N}}{10.9} H^4 + \frac{\beta^2 \bar{N}^2 \bar{v}}{6} H^3 \right) \qquad [2.45]$$

where ρ_{a0} is the reference density of the surrounding atmosphere (kg/m^3), g' is measured in m/s^2 and defined as $g \cdot \left((c_0\theta_0 - c_{a0}\theta_{a0})/c_{a0}\theta_{a0}\right)$ with g being the gravitational acceleration, α is the radial drag coefficient, $\bar{N}$ is the average buoyancy frequency (s^{-1}), H is the height of the plume above the vent (m), β is the wind entrainment coefficient and $\bar{v}$ is the average wind speed across the plume (m/s).

The MDR estimates are then used to calculate the minimum eruption duration, with all volume estimates converted to mass. However, if the MDR is obtained from the maximum column height, the relative duration of the eruption could be underestimated.

2.11. Temperature

As previously shown in several studies (Thomas and Sparks 1992; Tait et al. 1998; Hort and Gardner 2000), juvenile pumice larger than 6 cm in diameter undergoes little heat loss during its fall and can be deposited at temperatures between 10% and 20% of its magmatic temperature. Thus, plinian deposits, depending on their grain size, thickness and distance from the vent, can remain hot enough to present hazards to life and infrastructure (Thomas and Sparks 1992).

2.11.1. *Syn-eruptive methods to determine the temperature*

Several syn-eruptive methods can be used for the determination of the temperature of the clasts or even the eruptive plume (see Table 2.1). We refer the reader to Chapters 3 and 4 of Volume 2.

2.11.2. *Post-eruptive methods for determining temperature*

The temperature of a deposit at the time of deposition can be recovered from the melting point of natural or anthropogenic objects that are found covered by the deposits themselves and that have been partially or totally burned. In Table 2.5, we have reported the temperature estimate from objects found within pyroclastic flow deposits.

Eruption	Location	Year	Temperature	Reference
Karkar	Papua New Guinea	1979	300°C, burnt trees	(McKee et al. 1981)
Lamington	Papua New Guinea	1951	200°C, melting of plastics for 2 minutes	(Taylor 1958)
Ngauruhoe	New Zealand	1975	90°C, measured 90 minutes after the eruption	(Nairn et al. 1976)
Mayon	Philippines	1968	98°C, at 60 cm deep and 5 days after the eruption; 240°C in the blocks	(Moore and Melson 1969)
Taal	Philippines	1965	< 100°C	–
Asama	Japan	1783	900°C, strongly welded deposit	(Aramaki 1956)
Augustine	Alaska (United States)	1975	600°C, measured at 5 m depth; 400°C a few days after	(BVE 1978)
Mt. St. Helens	Washington State (USA)	1980	350°C, charred trees	(Moore and Sisson 1981)
Mt. Pelée	Martinique	1902	1,075°C, deformed gold coins 700°C, deformed glass bottles > 300°C, burning clothes	(Trechmann 1938) (Will 1903)
Soufrière	Saint-Vincent	1902	< 300°C, unburned clothing on the east coast	(Will 1903)
Soufrière	Saint-Vincent	1902	300–305°C infrared spectrometry of carbonized wood	(Bardintzeff and Boudon 1979)

Table 2.5. *Temperature measured in pyroclastic flow deposits and from natural or anthropogenic objects trapped within the deposits (adapted from Blong (1984), for references listed in the table)*

Figure 2.50. *a) Roman amphorae from the archeological site of Pompeii, covered by white and gray fallout from the 79 CE eruption. b) Collapsed tiles found within the fallout deposit (a and b) from Zanella et al. (2007). c) General view of the volcanic deposits from the Megalochori quarry. The white line and arrow indicate the contact of the deposits with the paleosol from which the studied samples were taken. Inset: photo of collected pottery fragments. From Tema et al. (2013). For a color version of this figure, see www.iste.co.uk/lenat/hazards3.zip*

The thermal effect of pyroclastic fallout deposits can also be estimated using standardized paleomagnetic techniques on lithic clasts or anthropogenic fragments embedded in volcanic deposits.

For the Pompeii fallout deposits, Zanella et al. (2007) found that the white pyroclastic (phonolitic) fallout deposit had a temperature high enough to heat the tile to 120–140°C (see Figure 2.50(a)), while the gray (tephri-phonolitic) fallout deposit contained lithic fragments that reached a minimum temperature of 180–220°C and 220–260°C (see Figure 2.50(b)). Similar temperatures have been estimated for the Minoan Plinian fallout deposits (Santorini, Greece, see Figure 2.50(c)), ranging from 180°C to 240°C from lithic clasts and 140°C to 180°C from ceramic fragments (Tema et al. 2013). In their study, Tema et al. (2013) showed that ceramic fragments covered and warmed by Minoan deposits can reliably represent the equilibrium temperature between the cold paleosol and warmer pumice fallout. Compared to lithic clasts, traditionally used in most temperature estimation studies of pyroclastic deposits, ceramic fragments have the advantage of being more thermally stable and having a less complex thermal history (Zanella et al. 2007).

2.12. Ballistic parameters

Ballistic projectiles ejected during explosive eruptions follow a parabolic trajectory in the air that is little affected by wind and eventually land on the ground (Wilson 1972). These blocks/bombs can cause significant damage, such as roof penetration (Blong 1984), demolition of mountain huts, injury to humans (Blong 1984; Baxter and Gresham 1997) and initiation of fires (Pistolesi et al. 2011). Therefore, inferring the ejection conditions of ballistic projectiles and estimating their landing velocity and energy are useful for reducing their risk. In addition, to avoid high casualties related to ballistic projectiles, it is important to make risk maps (Crandell et al. 1984; Alatorre-Ibargüengoitia et al. 2012; Fitzgerald et al. 2014), build shelters around active craters and reinforce the roofs of mountain shelters (Pomonis et al. 1999). Past ballistic landing, point maps are therefore a useful tool for hazard characterization, but also for determining the precise position of the active crater when this information is not available.

2.12.1. *Syn-eruptive methods for determining ballistic trajectories*

The syn-eruptive methods (see Figure 2.1(d)) are described in Chapter 4 of Volume 2.

2.12.2. *Post-eruptive methods for determining the ballistic landing position map and ballistic flight parameters*

Detailed maps of ballistic landing points have been produced for several explosive events (Minakami 1942; Self et al. 1980; Adams et al. 2006; Kilgour et al. 2010; Swanson et al. 2010; Gurioli et al. 2013), as scattered bomb marks (see Figure 2.13) and/or as isopleth maps (see Figure 2.51). The general strategy is to map each bomb in the field or from aerial photos, recording the GPS position and collecting all useful measurements (size, mass, direction of bomb/block impacts). The direction of impact (see Figure 2.52) can be used to retrieve information about the ejection site of each ejected bomb (see Figure 2.53).

Ballistic parameters have been measured for the mapped bombs (Lorenz 1970; Fudali and Melson 1971; Steinberg 1976; Self et al. 1980, Fagents and Wilson 1993; Mastin 2001; Kilgour et al. 2010; Pistolesi et al. 2011; Gurioli et al. 2013) as shown in Figure 2.53.

To eject a bomb at a horizontal distance d and elevation h, the initial velocity (V_0) and angle (θ) (under no-drag conditions) are:

$$V_0 = \sqrt{\frac{gd^2}{d\sin(2\theta)-2\Delta h\cos^2\theta}} \qquad [2.46]$$

where the minimum speed required to reach any point on a distribution map is now obtained with a θ of 45° + $\alpha/2$, α being the angle between the launch and landing points. All bombs will also be subject to the drag force (F_D), and will therefore experience a deceleration (a_D) given by:

$$a_D = \frac{F_D}{m} = -\frac{p_{air}AC_v v^2}{2m} \qquad [2.47]$$

where m is the mass, A is the cross-sectional area and C_v is the shape-dependent drag coefficient (= 0.47 for a sphere); ρ_{air} is the density of the air and v is the instantaneous velocity of the bomb. This means that a higher V_0 is needed to launch a bomb the same distance as can be achieved under no-drag conditions, and the mass, area and shape of the bomb are needed to reconstruct the angles and initial velocities of the bombs. Holding everything else constant, we see that the deceleration due to drag increases as the velocity increases or the mass of the bomb decreases.

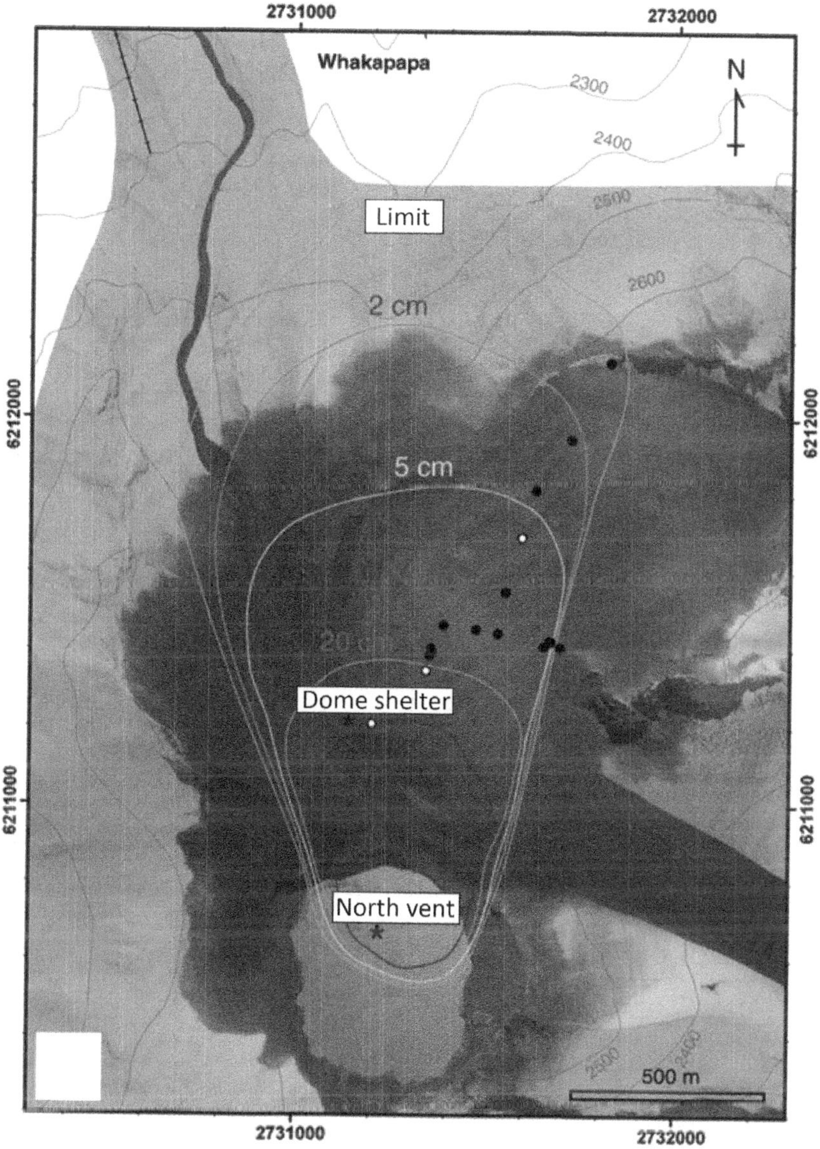

Figure 2.51. *Isopleth map of ballistic fallout from the proximal area of Ruapehu's September 25, 2007 eruption. Black dots: measured and sampled ballistic blocks; white dots: counted, measured and sampled ballistic fragments (modified from Kilgour et al. (2010)). For a color version of this figure, see www.iste.co.uk/lenat/hazards3.zip*

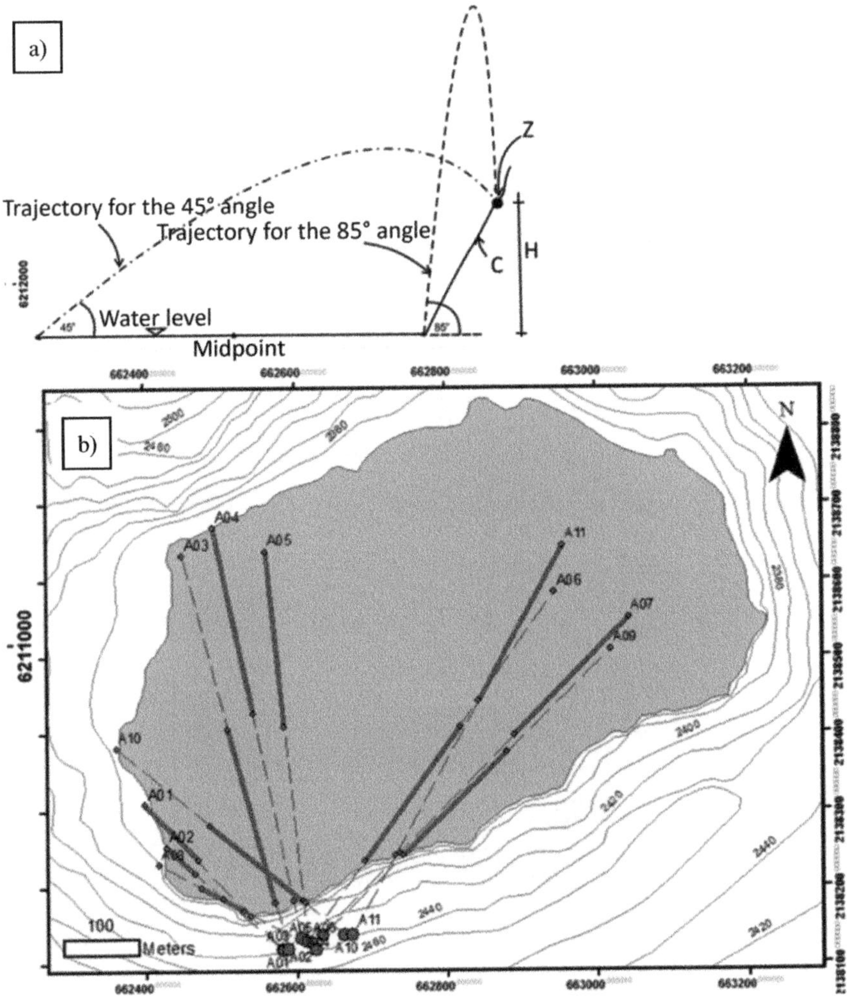

Figure 2.52. *a) Schematic diagram showing two probable areas of ballistic projectile ejection (ejection angles of 45° and 85°). b) Map of ballistic block distribution (red dots) within the Atexcac Axalapazco maar deposits. For a color version of this figure, see www.iste.co.uk/lenat/hazards3.zip*

COMMENT ON FIGURE 2.52.– *Thin red dashed lines: ballistic directions, based on the model shown in (a) and solid red lines: probable explosion zone for each ballistic clast in this unit. Figures (a) and (b) are adapted from López-Rojas and Carrasco-Núñez (2015).*

Figure 2.53. *a) and b) Strombolian ballistic bombs of mafic compositions originating from the January 2010 major explosion at Stromboli (Gurioli et al. 2013). The corresponding impact map is presented within Figure 2.13. c) Limestone boulder. d) Basalt boulder. e) Juvenile boulder with bread crust texture. f) Microdioritic blocks*

COMMENT ON FIGURE 2.53.– *c)–f) ballistic fragments are from the Atexcac Axalapazco maar deposits in the eastern Mexican volcanic belt (adapted from López-Rojas and Carrasco-Núñez (2015)).*

2.13. References

Adams, N.K., Houghton, B.F., Fagents, S.A., Hildreth, W. (2006). The transition from explosive to effusive eruptive regime: The example of the 1912 Novarupta eruption, Alaska. *Bull. Geol. Soc. Am.*, 118, 620–634.

Alatorre-Ibargüengoitia, M.A., Delgado-Granados, H., Dingwell, D.B. (2012). Hazard map for volcanic ballistic impacts at Popocatépetl volcano (Mexico). *Bull. Volcanol.*, 74, 2155–2169.

Albarède, F. (1993). Residence time analysis of geochemical fluctuations in volcanic series. *Geochim. Cosmochim. Acta*, 57, 615–621.

Albert, H., Costa, F., Di Muro, A., Herrin, J., Métrich, N., Deloule, E. (2019). Magma interactions, crystal mush formation, timescales, and unrest during caldera collapse and lateral eruption at ocean island basaltic volcanoes (Piton de la Fournaise, La Réunion). *Earth Planet. Sci. Lett.*, 515, 187–199.

Alfano, F., Bonadonna, C., Delmelle, P., Costantini, L. (2011). Insights on tephra settling velocity from morphological observations. *J. Volcanol. Geotherm. Res.*, 208, 86–98.

Anderson, A.T. (1974). Evidence for a picritic, volatile-rich magma beneath Mt. Shasta, California. *J. Petrol.*, 15, 243–267.

Andrews, B.J. and Befus, K.S. (2020). Supersaturation nucleation and growth of plagioclase: A numerical model of decompression-induced crystallization. *Contrib. to Mineral. Petrol.*, 175, 23.

Aschenbrenner, B.C. (1956). A new method of expressing particle sphericity. *J. Sediment. Petrol.*, 26, 15–31.

Avery, M.R., Panter, K.S., Gorsevski, P.V. (2017). Distinguishing styles of explosive eruptions at Erebus, Redoubt and Taupo volcanoes using multivariate analysis of ash morphometrics. *J. Volcanol. Geotherm. Res.*, 332, 1–13.

Bagheri, G. and Bonadonna, C. (2016). Aerodynamics of volcanic particles: Characterization of size, shape, and settling velocity. In *Volcanic Ash: Hazard Observation*, volume 1, Mackie, S., Cashman, K., Ricketts, H., Rust, A., Watson, M. (eds). Elsevier, Amsterdam.

Bagheri, G.H., Bonadonna, C., Manzella, I., Vonlanthen, P. (2015). On the characterization of size and shape of irregular particles. *Powder Technol.*, 270, 141–153.

Bai, L., Baker, D.R., Hill, R.J. (2010). Permeability of vesicular Stromboli basaltic glass: Lattice Boltzmann simulations and laboratory measurements. *J. Geophys. Res.*, 115, B07201–B07201.

Barberi, F., Cioni, R., Rosi, M., Santacroce, R., Sbrana, A., Vecci, R. (1989). Magmatic and phreatomagmatic phases in explosive eruptions of Vesuvius as deduced by grain-size and component analysis of the pyroclastic deposits. *J. Volcanol. Geotherm. Res.*, 38, 287–307.

Barbero, B.R. and Ureta, E.S. (2011). Comparative study of different digitization techniques and their accuracy. *CAD Comput. Aided Des.*, 43, 188–206.

Baxter, P.J. and Gresham, A. (1997). Deaths and injuries in the eruption of Galeras Volcano, Colombia, 14 January 1993. *J. Volcanol. Geotherm. Res.*, 77, 325–338.

Bernard, B. (2013). Homemade ashmeter: A low-cost, high-efficiency solution to improve tephra field-data collection for contemporary explosive eruptions. *J. Appl. Volcanol.*, 2, 1.

Bernard, J., Kelfoun, K., Le Pennec, J.L., Vallejo Vargas, S. (2014). Pyroclastic flow erosion and bulking processes: Comparing field-based vs. modeling results at Tungu-rahua volcano, Ecuador. *Bull. Volcanol.*, 76, 1–16.

Biass, S., Frischknecht, C., Bonadonna, C. (2013). A fast GIS-based risk assessment for tephra fallout: The example of Cotopaxi volcano, Ecuador-Part II: Vulnerability and risk assessment. *Nat. Hazards*, 64, 615–639.

Biass, S., Bagheri, G., Bonadonna, C. (2015). A Matlab implementation of the Carey and Sparks (1986) model. https://vhub.org/resources/3922.

Bisson, M. and Del Carlo, P. (2013). A GIS-based application for volume estimation and spatial distribution analysis of tephra fallout: A case study of the 122 BC Etna eruption. *Ann. Geophys.*, 56, 105.

Blong, R.J. (1984). *Volcanic Hazards: A Sourcebook on the Effects of Eruptions*. Academic Press, Sydney.

Bloomfield, L.J. and Kerr, R.C. (2000). A theoretical model of a turbulent fountain. *J. Fluid Mech.*, 424, 197–216.

Blott, S.J. and Pye, K. (2008). Particle shape: A review and new methods of characterization and classification. *Sedimentology*, 55, 31–63.

Bombrun, M., Harris, A., Gurioli, L., Battaglia, J., Barra, V. (2015). Anatomy of a Strombolian eruption: Inferences from particle data recorded with thermal video. *J. Geophys. Res. Solid Earth*, 120, 2367–1687.

Bonadonna, C. (2014). Future developments and new directions in modeling and monitoring volcanic ash clouds. *Eos (Washington, DC).*, 95, 64.

Bonadonna, C. and Costa, A. (2013). Plume height, volume, and classification of explosive volcanic eruptions based on the Weibull function. *Bull. Volcanol.*, 75, 742.

Bonadonna, C. and Houghton, B.F. (2005). Total grain-size distribution and volume of tephra-fall deposits. *Bull. Volcanol.*, 67, 441–456.

Bonadonna, C. and Phillips, J.C. (2003). Sedimentation from strong volcanic plumes. *J. Geophys. Res. Solid Earth*, 108, 2340.

Bonadonna, C., Cioni, R., Pistolesi, M., Connor, C., Scollo, S., Pioli, L., Rosi, M. (2013). Determination of the largest clast sizes of tephra deposits for the characterization of explosive eruptions: A study of the IAVCEI commission on tephra hazard modelling. *Bull. Volcanol.*, 75, 1–15.

Bonadonna, C., Cioni, R., Pistolesi, M., Elissondo, M., Baumann, V. (2015a). Sedimentation of long-lasting wind-affected volcanic plumes: The example of the 2011 rhyolitic Cordón Caulle eruption, Chile. *Bull. Volcanol.*, 77, 13.

Bonadonna, C., Costa, A., Folch, A., Koyaguchi, T. (2015b). Tephra dispersal and sedimentation. In *The Encyclopedia of Volcanoes*, 2nd edition, Sigurdsson, H. (ed.). Academic Press, Amsterdam.

Bottinga, Y. and Weill, D.F. (1970). Densities of liquid silicate systems calculated from partial molar volumes of oxide components. *Am. J. Sci.*, 269, 169–182.

Brenot, H., Theys, N., Clarisse, L., Van Geffen, J., Van Gent, J., Van Roozendael, M., Van Der, A.R., Hurtmans, D., Coheur, P.F., Clerbaux, C. et al. (2014). Support to aviation control service (SACS): An online service for near-real-time satellite monitoring of volcanic plumes. *Nat. Hazards Earth Syst. Sci.*, 14, 1099–1123.

Brown, R.J., Bonadonna, C., Durant, A.J. (2012). A review of volcanic ash aggregation. *Phys. Chem. Earth*, 45–46, 65–78.

Brunauer, S., Emmett, P.H., Teller, E. (1938). Adsorption of gases in multimolecular layers. *J. Am. Chem. Soc.*, 60, 309–319.

Buckland, H.M., Eychenne, J., Rust, A.C., Cashman, K.V. (2018). Relating the physical properties of volcanic rocks to the characteristics of ash generated by experimental abrasion. *J. Volcanol. Geotherm. Res.*, 349, 335–350.

Burden, R.E., Phillips, J.C.C., Hincks, T.K. (2011). Estimating volcanic plume heights from depositional clast size. *J. Geophys. Res.*, 116. http://dx.doi.org/10.1029/2011JB008548.

Burden, R.E., Chen, L., Phillips, J.C. (2013). A statistical method for determining the volume of volcanic fall deposits. *Bull. Volcanol.*, 75, 1–10.

Bursik, M.I., Sparks, R.S.J., Gilbert, J.S., Carey, S.N. (1992). Sedimentation of tephra by volcanic plumes I: Theory and its comparison with a study of the Fogo A plinian deposit, Sao Miguel (Azores). *Bull. Volcanol.*, 54, 329–344.

Büttner, R., Dellino, P., Zimanowski, B. (1999). Identifying magma-water interaction from the surface features of ash particles. *Nature*, 401, 688–690.

Carey, S. and Bursik, M. (2015). Volcanic plumes. In *The Encyclopedia of Volcanoes*, 2nd edition, Sigurdsson, H. (ed.). Academic Press, Amsterdam.

Carey, S. and Sigurdsson, H. (1982). Influence of particle aggregation on deposition of distal tephra from the May 18, 1980, eruption of Mount St. Helens volcano. *J. Geophys. Res.*, 87, 7061–7072.

Carey, S. and Sigurdsson, H. (1989). The intensity of plinian eruptions. *Bull. Volcanol.*, 51, 28–40.

Carey, S. and Sparks, R.S.J. (1986). Quantitative models of the fallout and dispersal of tephra from volcanic eruption columns. *Bull. Volcanol.*, 48, 109–125.

Cas, R.A.F. and Wright, J.V. (1987). *Volcanic Successions, Modern and Ancient*. Chapman & Hall, London.

Casalbore, D., Romagnoli, C., Chiocci, F., Frezza, V. (2010). Morpho-sedimentary characteristics of the volcaniclastic apron around Stromboli volcano (Italy). *Mar. Geol.*, 269, 132–148.

Cashman, K. and Blundy, J. (2013). Petrological cannibalism: The chemical and textural consequences of incremental magma body growth. *Contrib. to Mineral. Petrol.*, 166, 703–729.

Cerminara, M., Esposti Ongaro, T., Valade, S., Harris, A.J.L. (2014). Volcanic plume vent conditions retrieved from infrared images: A forward and inverse modeling approach. *J. Volcanol. Geotherm. Res.*, 300, 129–147.

Chouet, B., Hamisevicz, N., McGetchin, T.R. (1974). Photoballistics of volcanic jet activity at Stromboli, Italy. *J. Geophys. Res.*, 79, 4961–4976.

Cioni, R., Sbrana, A., Vecci, R. (1992). Morphologic features of juvenile pyroclasts from magmatic and phreatomagmatic deposits of Vesuvius. *J. Volcanol. Geotherm. Res.*, 51, 61–78.

Cioni, R., D'Oriano, C., Bertagnini, A. (2008). Fingerprinting ash deposits of small scale eruptions by their physical and textural features. *J. Volcanol. Geotherm. Res.*, 177, 277–287.

Cioni, R., Tadini, A., Gurioli, L., Bertagnini, A., Mulas, M., Bevilacqua, A., Neri, A. (2020). Estimating eruptive parameters and related uncertainties for pyroclastic density currents deposits: Worked examples from Somma-Vesuvius (Italy). *Bull. Volcanol.*, 82(65), 1–20.

Colò, L., Ripepe, M., Gurioli, L., Harris, A.J.L. (2020). Fragmentation processes during Strombolian explosions revealed using particle size distribution mapping. *Frontiers Earth Science*, 8(356), 1–10.

Colombier, M., Gurioli, L., Druitt, T.H., Shea, T., Boivin, P., Miallier, D., Cluzel, N. (2017a). Textural evolution of magma during the 9.4-ka trachytic explosive eruption at Kilian Volcano, Chaîne des Puys, France. *Bull. Volcanol.*, 79(17), 1–24.

Colombier, M., Wadsworth, F.B., Gurioli, L., Scheu, B., Kueppers, U., Di Muro, A., Dingwell, D.B. (2017b). The evolution of pore connectivity in volcanic rocks. *Earth Planet. Sci. Lett.*, 462, 99–109.

Costa, A., Macedonio, G., Folch, A. (2006). A three-dimensional Eulerian model for transport and deposition of volcanic ashes. *Earth Planet. Sci. Lett.*, 241, 634–647.

Costa, F., Dohmen, R., Chakraborty, S. (2008). Time scales of magmatic processes from modeling the zoning patterns of crystals. *Reviews in Mineralogy & Geochemistry*, 69, 545–594.

Costa, A., Pioli, L., Bonadonna, C. (2016). Assessing tephra total grain-size distribution: Insights from field data analysis. *Earth Planet. Sci. Lett.*, 443, 90–107

Couch, S., Sparks, R.S.J., Carroll, M.R. (2001). Mineral disequilibrium in lavas explained by convective self-mixing in open magma chambers. *Nature*, 411, 1037–1039.

Crandell, D.R., Booth, B., Kusumadinata, K., Shimozuru, D., Walker, G.P.L., Westercamp, D. (1984). *Source Book for Volcanic Hazards Zonation*. Unesco Press, Paris.

Dartevelle, S., Ernst, G.G.J., Bernard, A. (2002). Origin of the Mount Pinatubo climactic eruption cloud: Implications for volcanic hazards and atmospheric impacts. *Geology*, 30, 663–666.

Degruyter, W. and Bonadonna, C. (2012). Improving on mass flow rate estimates of volcanic eruptions. *Geophys. Res. Lett.*, 39(L16308), 1–6.

Dellino, P. and La Volpe, L. (1996). Image processing analysis in reconstructing fragmentation and transportation mechanisms of pyroclastic deposits. The case of Monte Pilato-Rocche Rosse eruptions, Lipari (Aeolian Islands, Italy). *J. Volcanol. Geotherm. Res.*, 71, 13–29.

Dellino, P., Mele, D., Bonasia, R., Braia, G., La Volpe, L., Sulpizio, R. (2005). The analysis of the influence of pumice shape on its terminal velocity. *Geophys. Res. Lett.*, 32, 1–4.

Deniel, C. and Pin, C. (2001). Single-stage method for the simultaneous isolation of lead and strontium from silicate samples for isotopic measurements. *Anal. Chim. Acta*, 426, 95–103.

Dioguardi, F., Mele, D., Dellino, P., Dürig, T. (2017). The terminal velocity of volcanic particles with shape obtained from 3D X-ray microtomography. *J. Volcanol. Geotherm. Res.*, 329, 41–53.

Dixon, J.E., Stolper, E.M., Holloway, J.R. (1995). An experimental study of water and carbon dioxide solubilities in mid-ocean ridge basaltic liquids. Part I: Calibration and solubility models. *J. Petrol.*, 36, 1607–1631.

Durant, A.J. and Rose, W.I. (2009). Sedimentological constraints on hydrometeor-enhanced particle deposition: 1992 eruptions of Crater Peak, Alaska. *J. Volcanol. Geotherm. Res.*, 186, 40–59.

Dürig, T., Sonder, I., Zimanowski, B., Beyrichen, H., Büttner, R. (2012). Generation of volcanic ash by basaltic volcanism. *J. Geophys. Res. Solid Earth*, 117, 1204.

Edwards, M.J., Pioli, L., Harris, A.J.L., Gurioli, L., Thivet, S. (2020). Magma fragmentation and particle size distributions in low intensity mafic explosions: The July/August 2015 Piton de la Fournaise eruption. *Sci. Rep.*, 10. https://doi.org/10.1038/s41598-020-69976-y.

Engwell, S.L., Sparks, R.S.J., Aspinall, W.P. (2013). Quantifying uncertainties in the measurement of tephra fall thickness. *J. Appl. Volcanol.*, 2(5), 1–12.

Ersoy, O., Şen, E., Aydar, E., Tatar, I., Çelik, H.H. (2010). Surface area and volume measurements of volcanic ash particles using micro-computed tomography (micro-CT): A comparison with scanning electron microscope (SEM) stereoscopic imaging and geometric considerations. *J. Volcanol. Geotherm. Res.*, 196, 281–286.

Eshel, G., Levy, G.J., Mingelgrin, U., Singer, M.J. (2004). Critical evaluation of the use of laser diffraction for particle-size distribution analysis. *Soil Sci. Soc. Am. J.*, 68, 736–743.

Eychenne, J. and Le Pennec, J.L. (2012). Sigmoidal particle density distribution in a subplinian scoria fall deposit. *Bull. Volcanol.*, 74, 2243–2249.

Eychenne, J., Houghton, B.F., Swanson, D.A., Carey, R.J., Swavely, L. (2015). Dynamics of an open basaltic magma system: The 2008 activity of the Halema'uma'u Overlook vent, Kīlauea Caldera. *Earth Planet. Sci. Lett.*, 409, 49–60.

Fagents, S.A. and Wilson, L. (1993). Explosive volcanic eruptions VII. The ranges of pyroclasts ejected in transient volcanic explosions. *Geophys. J. Int.*, 113, 359–370.

Fagents, S.A., Gregg, T.K.P., Lopes, R.M.C. (2013). *Modeling Volcanic Processes: The Physics and Mathematics of Volcanism*. Cambridge University Press, Cambridge.

Faure, F., Trolliard, G., Nicollet, C., Montel, J.M. (2003). A developmental model of olivine morphology as a function of the cooling rate and the degree of undercooling. *Contrib. to Mineral. Petrol.*, 145, 251–263.

Fierstein, J. and Nathenson, M. (1992). Another look at the calculation of fallout tephra volumes. *Bull. Volcanol.*, 54. http://dx.doi.org/10.1007/BF00278005.

Fisher, R.V. (1966). Rocks composed of volcanic fragments and their classification. *Earth Sci. Rev.*, 1, 287–298.

Fisher, R.V. and Schmincke, H.U. (1984). *Pyroclastic Rocks*. Springer, Berlin.

Fitzgerald, R.H., Tsunematsu, K., Kennedy, B.M., Breard, E.C.P., Lube, G., Wilson, T.M., Jolly, A.D., Pawson, J., Rosenberg, M.D., Cronin, S.J. (2014). The application of a calibrated 3D ballistic trajectory model to ballistic hazard assessments at Upper Te Maari, Tongariro. *J. Volcanol. Geotherm. Res.*, 286, 248–262.

Folk, R.L. and Ward, W.C. (1957). Brazos River bar [Texas]: A study in the significance of grain size parameters. *J. Sediment. Res.*, 27, 3–26.

Formenti, Y., Druitt, T.H., Kelfoun, K. (2003). Characterisation of the 1997 Vulcanian explosions of Soufrière Hills Volcano, Montserrat, by video analysis. *Bull. Volcanol.*, 65, 587–605.

Freret-Lorgeril, V., Donnadieu, F., Eychenne, J., Soriaux, C., Latchimy, T. (2019). In situ terminal settling velocity measurements at Stromboli volcano: Input from physical characterization of ash. *J. Volcanol. Geotherm. Res.*, 374, 62–79.

Freret-Lorgeril, V., Gilchrist, J., Donnadieu, F., Jellinck, A.M., Delanoë, J., Latchimy, T., Vinson, J.P., Caudoux, C., Peyrin, F., Hervier, C. et al. (2020). Ash sedimentation by fingering and sediment thermals from wind-affected volcanic plumes. *Earth Planet. Sci. Lett.*, 534, 116072.

Freundt, A. and Rosi, M. (1998). *Explosive Volcanism. A Physical Description*. Springer Verlag, Berlin.

Fudali, R.F. and Melson, W.G. (1971). Ejecta velocities, magma chamber pressure and kinetic energy associated with the 1968 eruption of arenal volcano. *Bull. Volcanol.*, 35, 383–401.

Gansecki, C., Lopaka, L.R., Shea, T., Lundblad, S.P., Hon, K., Parcheta, C. (2019). The tangled tale of Kīlauea's 2018 eruption as told by geochemical monitoring. *Science*, 366, 147.

Ganser, G.H. (1993). A rational approach to drag prediction of spherical and nonspherical particles. *Powder Technol.*, 77, 143–152.

Gardner, J.E., Thomas, R.M.E., Jaupart, C., Tait, S. (1996). Fragmentation of magma during Plinian volcanic eruptions. *Bull. Volcanol.*, 58, 144–162.

Girona, T. and Costa, F. (2013). DIPRA: A user-friendly program to model multi-element diffusion in olivine with applications to timescales of magmatic processes. *Geochemistry, Geophys. Geosystems*, 14, 422–431.

Gordeychik, B., Churikova, T., Kronz, A., Sundermeyer, C., Simakin, A., Wörner, G. (2018). Growth of, and diffusion in, olivine in ultra-fast ascending basalt magmas from Shiveluch volcano. *Sci. Rep.*, 8, 1–15.

Gouhier, M. and Donnadieu, F. (2008). Mass estimations of ejecta from Strombolian explosions by inversion of Doppler radar measurements. *J. Geophys. Res. Solid Earth*, 113, 5383.

Gurioli, L., Zanella, E., Pareschi, M.T., Lanza, R. (2007). Influences of urban fabric on pyroclastic density currents at Pompeii (Italy) 1: Flow direction and deposition. *J. Geophys. Res.*, 112, B05213.

Gurioli, L., Harris, A.J.L., Houghton, B.F., Polacci, M., Ripepe, M. (2008). Textural and geophysical characterization of explosive basaltic activity at Villarrica volcano. *J. Geophys. Res. Solid Earth*, 113(B08206), 1–16.

Gurioli, L., Zanella, E., Gioncada, A., Sbrana, A. (2012). The historic magmatic-hydrothermal eruption of the Breccia di Commenda, Vulcano, Italy. *Bull. Volcanol.*, 74, 1235–1254.

Gurioli, L., Harris, A.J.L., Colò, L., Bernard, J., Favalli, M., Ripepe, M., Andronico, D. (2013). Classification, landing distribution, and associated flight parameters for a bomb field emplaced during a single major explosion at Stromboli, Italy. *Geology*, 41, 559–562.

Gurioli, L., Colò, L., Bollasina, A.J., Harris, A.J.L., Whittington, A., Ripepe, M. (2014). Dynamics of Strombolian explosions: Inferences from field and laboratory studies of erupted bombs from Stromboli volcano. *J. Geophys. Res. Solid Earth*, 119, 319–345.

Gurioli, L., Di Muro, L., Vlastélic, I., Bachèlery, P., Harris, A., Thivet, S., Calabro', L., Rivet, S., Delage, E. (2016). Les apports d'une vision intégrée des données volcanologiques (DYNVOLC). *Rev. Auvergne "Obs. Phys. du globe"*, 1(620–621), 299–335.

Gurioli, L., Di Muro, A., Vlastélic, I., Moune, S., Thivet, S., Valer, M., Villeneuve, N., Boudoire, G., Peltier, A., Bachèlery, P. et al. (2018). Integrating field, textural, and geochemical monitoring to track eruption triggers and dynamics: A case study from Piton de la Fournaise. *Solid Earth*, 9, 431–455.

Hammer, J.E. and Rutherford, M.J. (2002). An experimental study of the kinetics of decompression-induced crystallization in silicic melt. *J. Geophys. Res. Solid Earth*, 107(ECV 8-1-ECV), 8–24.

Hammer, J.E., Cashman, K.V., Hoblitt, R.P., Newman, S. (1999). Degassing and microlite crystallization during pre-climactic events of the 1991 eruption of Mt. Pinatubo, Philippines. *Bull. Volcanol.*, 60, 355–380.

Harris, A.J.L., Battaglia, J., Donnadieu, F., Gurioli, L., Kelfoun, K., Labazuy, P., Sawyer, G., Valade, S., Bombun, M., Barra, V. et al. (2013). Full bandwidth remote sensing for total parameterization of volcanic plumes. *Eos (Washington, DC)*, 94, 321–322.

Harris, A.J.L., Villeneuve, N., Di Muro, A., Ferrazzini, V., Peltier, A., Coppola, D., Favalli, M., Bachèlery, P., Froger, J.L., Gurioli, L. et al. (2017). Effusive crises at Piton de la Fournaise 2014–2015: A review of a multi-national response model. *J. Appl. Volcanol.*, 6, 11.

Heiken, G. (1972). Morphology and petrography of volcanic ashes. *Bull. Geol. Soc. Am.*, 83, 1961–1988.

Heiken, G. and Wohletz, K.H. (2016). *Volcanic Ash.* University of California Press, Berkeley.

Helz, R.T. and Thornber, C.R. (1987). Geothermometry of Kilauea Iki lava lake, Hawaii. *Bull. Volcanol.*, 49, 651–668.

Higgins, M.D. (2000). Measurement of crystal size distributions. *Am. Mineral.*, 85, 1105–1116.

Higgins, M.D. (2002). Reply to comment on "closure in crystal size distributions (CSD), verification of CSD calculations and the significance of CSD fans". *Am. Mineral.*, 87, 1244.

Holland, T. and Blundy, J. (1994). Non-ideal interactions in calcic amphiboles and their bearing on amphibole-plagioclase thermometry. *Contrib. to Mineral. Petrol.*, 116, 433–447.

Hort, M. and Gardner, J. (2000). Constraints on cooling and degassing of pumice during Plinian volcanic eruptions based on model calculations. *J. Geophys. Res. Solid Earth*, 105, 25981–6001.

Horwell, C.J. and Baxter, P.J. (2006). The respiratory health hazards of volcanic ash: A review for volcanic risk mitigation. *Bull. Volcanol.*, 69, 1–24.

Houghton, B.F. and Carey, R.J. (2015). Pyroclastic fall deposits. *The Encyclopedia of Volcanoes*, 2nd edition, Sigurdsson, H. (ed.). Academic Press, Amsterdam.

Houghton, B.F. and Wilson, C.J.N. (1989). A vesicularity index for pyroclastic deposits. *Bull. Volcanol.*, 51, 451–462.

Hurst, A.W. and Turner, R. (1999). Performance of the program ASHFALL for forecasting ashfall during the 1995 and 1996 eruptions of Ruapehu volcano. *New Zeal. J. Geol. Geophys.*, 42, 615–622.

Iacono-Marziano, G., Morizet, Y., Le Trong, E., Gaillard, F. (2012). New experimental data and semi-empirical parameterization of H_2O-CO_2 solubility in mafic melts. *Geochim. Cosmochim. Acta*, 97, 1–23.

Inman, D.L. (1952). Measures for describing the size distribution of sediments. *J. Sed. Pet.*, 22, 125–145.

Irvine, T.N. and Baragar, W.R.A. (1971). A guide to the chemical classification of the common volcanic rocks. *Can. J. Earth Sci.*, 8, 523–548.

Janebo, M.H., Houghton, B.F., Thordarson, T., Bonadonna, C., Carey, R.J. (2018). Total grain-size distribution of four subplinian–Plinian tephra from Hekla volcano, Iceland: Implications for sedimentation dynamics and eruption source parameters. *J. Volcanol. Geotherm. Res.*, 357, 25–38.

Johnson, M.C. and Rutherford, M.J. (1989). Experimental calibration of the aluminum-in-hornblende geobarometer with application of Long Valley caldera (California) volcanic rocks. *Geology*, 17, 837–841.

Jones, T.J., McNamara, K., Eychenne, J., Rust, A.C., Cashman, K.V., Scheu, B., Edwards, R. (2016). Primary and secondary fragmentation of crystal-bearing intermediate magma. *J. Volcanol. Geotherm. Res.*, 327, 70–83.

Jordan, S.C., Dürig, T., Cas, R.A.F., Zimanowski, B. (2014). Processes controlling the shape of ash particles: Results of statistical IPA. *J. Volcanol. Geotherm. Res.*, 288, 19–27.

Jordan, S.C., Le Pennec, J.L., Gurioli, L., Roche, O., Boivin, P. (2016). Highly explosive eruption of the monogenetic 8.6 ka BP La Vache et Lassolas scoria cone complex (Chaîne des Puys, France). *J. Volcanol. Geotherm. Res.*, 313, 15–28.

Kaminski, E. and Jaupart, C. (1998). The size distribution of pyroclasts and the fragmentation sequence in explosive volcanic eruptions. *J. Geophys. Res. Solid Earth*, 103, 29759–29779.

Kilgour, G., Manville, V., Della Pasqua, F., Graettinger, A., Hodgson, K.A., Jolly, G.E. (2010). The 25 September 2007 eruption of Mount Ruapehu, New Zealand: Directed ballistics, surtseyan jets, and ice-slurry lahars. *J. Volcanol. Geotherm. Res.*, 191, 1–14.

Kilgour, G.N., Saunders, K.E., Blundy, J.D., Cashman, K.V., Scott, B.J., Miller, C.A. (2014). Timescales of magmatic processes at Ruapehu volcano from diffusion chronometry and their comparison to monitoring data. *J. Volcanol. Geotherm. Res.*, 288, 62–75.

Klawonn, M., Houghton, B.F., Swanson, D.A., Fagents, S.A., Wessel, P., Wolfe, C.J. (2014). From field data to volumes: Constraining uncertainties in pyroclastic eruption parameters. *Bull. Volcanol.*, 76, 1–16.

Krumbein, W.C. (1934). Size frequency distributions of Sediments. *SEPM J. Sediment. Res.*, 4, 65–77.

Krumbein, W.C. (1938). Size frequency distributions of sediments and the normal phi curve. *SEPM J. Sediment. Res.*, 8, 84–90.

Kunii, D. and Levenspiel, O. (1991). *Fluidisation Engineering*. Wiley, New York.

Kylling, A., Kahnert, M., Lindqvist, H., Nousiainen, T. (2014). Volcanic ash infrared signature: Porous non-spherical ash particle shapes compared to homogeneous spherical ash particles. *Atmos. Meas. Tech.*, 7, 919–929.

Launeau, P. and Robin, P.Y.F. (1996). Fabric analysis using the intercept method. *Tectonophysics*, 267, 91–119.

Lautze, N., Taddeucci, J., Andronico, D., Houghton, B.F., Niemeijer, A., Scarlato, P. (2013). Insights into explosion dynamics and the production of ash at Stromboli from samples collected in real-time, October 2009. *Spec. Pap. Geol. Soc. Am.*, 498, 125–139.

Le Maitre, R.W., Streckeisen, A., Zanettin, B., Le Bas, M., Bonin, B., Bateman, P., Bellieni, G., Dudek, A., Efremova, S., Keller J. et al. (2002). *Igneous Rocks – A Classification and Glossary of Terms*. Cambridge University Press, Cambridge.

Leduc, L., Gurioli, L., Harris, A., Colò, L., Rose-Koga, E.F. (2015). Types and mechanisms of strombolian explosions: Characterization of a gas-dominated explosion at Stromboli. *Bull. Volcanol.*, 77(8), 1–16.

Leibrandt, S. and Le Pennec, J.L. (2015). Towards fast and routine analyses of volcanic ash morphometry for eruption surveillance applications. *J. Volcanol. Geotherm. Res.*, 297, 11–27.

Liu, Y., Anderson, A.T., Wilson, C.J.N. (2007). Melt pockets in phenocrysts and decompression rates of silicic magmas before fragmentation. *J. Geophys. Res. Solid Earth*, 112(B06204), 1–12.

Liu, E.J., Cashman, K.V., Rust, A.C. (2015a). Optimising shape analysis to quantify volcanic ash morphology. *GeoResJ*, 8, 14–30.

Liu, E.J., Cashman, K.V., Rust, A.C., Gislason, S.R. (2015b). The role of bubbles in generating fine ash during hydromagmatic eruptions. *Geology*, 43, 239–242.

Lloyd, A.S., Ruprecht, P., Hauri, E.H., Rose, W., Gonnermann, H.M., Plank, T. (2014). NanoSIMS results from olivine-hosted melt embayments: Magma ascent rate during explosive basaltic eruptions. *J. Volcanol. Geotherm. Res.*, 283, 1–18.

Lockwood, J.P. and Hazlett, R.W. (2010). *Volcanoes: Global Perspectives*. Wiley, Hoboken.

López-Rojas, M. and Carrasco-Núñez, G. (2015). Depositional facies and migration of the eruptive loci for Atexcac axalapazco (Central Mexico): Implications for the morphology of the crater. *Rev. Mex. Ciencias Geológicas*, 32, 377–394.

Lorenz, V. (1970). Some aspects of the eruption mechanism of the Big Hole Maar, central Oregon. *Geological Society of America*, 81(6), 1823–1830.

Lundstrom, C.C., Gill, J., Williams, Q. (2000). A geochemically consistent hypothesis for MORB generation. *Chem. Geol.*, 162, 105–126.

Manga, M., Patel, A., Dufek, J. (2011). Rounding of pumice clasts during transport: Field measurements and laboratory studies. *Bull. Volcanol.*, 73, 321–333.

Mastin, L.G. (2001). A simple calculator of ballistic trajectories for blocks ejected during volcanic eruptions. US Department of the Interior, US Geological Survey, 18.

Mastin, L.G., Guffanti, M., Servranckx, R., Webley, P., Barsotti, S., Dean, K., Durant, A., Ewert, J.W., Neri, A., Rose, W.I. et al. (2009). A multidisciplinary effort to assign realistic source parameters to models of volcanic ash-cloud transport and dispersion during eruptions. *J. Volcanol. Geoth. Res.*, 186, 10–21.

Mattsson, H.B. (2010). Textural variation in juvenile pyroclasts from an emergent, Surtseyan-type, volcanic eruption: The Capelas tuff cone, São Miguel (Azores). *J. Volcanol. Geotherm. Res.*, 189, 81–91.

Mele, D. and Dioguardi, F. (2018). The grain size dependency of vesicular particle shapes strongly affects the drag of particles. First results from microtomography investigations of Campi Flegrei fallout deposits. *J. Volcanol. Geotherm. Res.*, 353, 18–24.

Mele, D., Dioguardi, F., Dellino, P. (2018). A study on the influence of internal structures on the shape of pyroclastic particles by X-ray microtomography investigations. *Ann. Geophys.*, 61(6), 1–11.

Michon, L., Di Muro, A., Villeneuve, N., Saint-Marc, C., Fadda, P., Manta, F. (2013). Explosive activity of the summit cone of Piton de la Fournaise volcano (La Réunion Island): A historical and geological review. *J. Volcanol. Geotherm. Res.*, 264, 117–133.

Minakami, T. (1942). 5. On the distribution of volcanic ejecta. Part 1: The distributions of volcanic bombs ejected by the recent explosion of Asama. *Bull. Earthq. Res. Inst.*, 20, 65–92.

Morgado, E., Morgan, D.J., Harvey, J., Parada, M.Á., Castruccio, A., Brahm, R., Gutiérrez, F., Georgiev, B., Hammond, S.J. (2019). Localised heating and intensive magmatic conditions prior to the 22–23 April 2015 Calbuco volcano eruption (Southern Chile). *Bull. Volcanol.*, 81, 24.

Morgan, D.J. and Jerram, D.A. (2006). On estimating crystal shape for crystal size distribution analysis. *J. Volcanol. Geotherm. Res.*, 154, 1–7.

Morton, B.R., Taylor, G.I., Turner, J.S. (1956). Turbulent gravitational convection from maintained and instantaneous sources. *Proc. R. Soc. London Ser. A: Math. Phys. Sci.*, 234, 1–23.

Moussallam, Y., Rose-Koga, E.F., Koga, K.T., Médard, E., Bani, P., Devidal, J.L., Tari, D. (2019). Fast ascent rate during the 2017–2018 Plinian eruption of Ambae (Aoba) volcano: A petrological investigation. *Contrib. to Mineral. Petrol.*, 174, 90.

Murrow, P.J., Rose, W.I., Self, S. (1980). Determination of the total grain size distribution in a Vulcanian eruption column, and its implications to stratospheric aerosol perturbation. *Geophys. Res. Lett.*, 7, 893–896.

Mutch, E.J.F., Blundy, J.D., Tattitch, B.C., Cooper, F.J., Brooker, R.A. (2016). An experimental study of amphibole stability in low-pressure granitic magmas and a revised Al-in-hornblende geobarometer. *Contrib. to Mineral. Petrol.*, 171, 85.

Neave, D.A., Bali, E., Guðfinnsson, G.H., Halldórsson, S.A., Kahl, M., Schmidt, A.S., Holtz, F. (2019). Clinopyroxene-liquid equilibria and geothermobarometry in natural and experimental tholeiites: The 2014–2015 Holuhraun eruption, Iceland. *J. Petrol.*, 60, 1653–1680.

Newhall, C.G. and Self, S. (1982). The volcanic explosivity index (VEI): An estimate of explosive magnitude for historical volcanism. *J. Geophys. Res.*, 87, 1231–1238.

Newman, S. and Lowenstern, J.B. (2002). VOLATILECALC: A silicate melt-H_2O-CO_2 solution model written in Visual Basic for excel. *Comput. Geosci.*, 28, 597–604.

Nurtiani, D. and Bouvet de Maisonneuve, C. (2018). Furthering the investigation of eruption styles through quantitative shape analyses of volcanic ash particles. *J. Volcanol. Geotherm. Res.*, 354, 102–114.

Parfitt, E.A. (1998). A study of clast size distribution, ash deposition and fragmentation in a Hawaiian-style volcanic eruption. *J. Volcanol. Geotherm. Res.*, 84, 197–208.

Pfeiffer, T., Costa, A., Macedonio, G. (2005). A model for the numerical simulation of tephra fall deposits. *J. Volcanol. Geotherm. Res.*, 140, 273–294.

Pietruszka, A.J. and Garcia, M.O. (1999). The size and shape of Kilauea Volcano's summit magma storage reservoir: A geochemical probe. *Earth Planet. Sci. Lett.*, 167, 311–320.

Pietruszka, A.J., Heaton, D.E., Marske, J.P., Garcia, M.O. (2015). Two magma bodies beneath the summit of Kilauea Volcano unveiled by isotopically distinct melt deliveries from the mantle. *Earth Planet. Sci. Lett.*, 413, 90–100.

Pioli, L. and Harris, A.J.L. (2019). Real-time geophysical monitoring of particle size distribution during volcanic explosions at Stromboli volcano (Italy). *Front. Earth Sci.*, 7, 1–13.

Pioli, L., Bonadonna, C., Pistolesi, M. (2019). Reliability of total grain-size distribution of tephra deposits. *Sci. Rep.*, 9, 1–15.

Pistolesi, M., Donne, D.D., Pioli, L., Rosi, M., Ripepe, M. (2011). The 15 March 2007 explosive crisis at Stromboli volcano, Italy: Assessing physical parameters through a multidisciplinary approach. *J. Geophys. Res. Solid Earth*, 116(B12206), 1–18.

van der Plas, L. and Tobi, A.C. (1965). A chart for judging the reliability of point counting results. *Am. J. Sci.*, 263, 87–90.

Pomonis, A., Spence, R., Baxter, P. (1999). Risk assessment of residential buildings for an eruption of Furnas Volcano, Sao Miguel, the Azores. *J. Volcanol. Geotherm. Res.*, 92, 107–131.

Porritt, L.A., Russell, J.K., Quane, S.L. (2012). Pele's tears and spheres: Examples from Kilauea Iki. *Earth Planet. Sci. Lett.*, 333/334, 171–180.

Potuzak, M., Nichols, A.R.L., Dingwell, D.B., Clague, D.A. (2008). Hyperquenched volcanic glass from Loihi Seamount, Hawaii. *Earth Planet. Sci. Lett.*, 270, 54–62.

Prata, A.J. and Bernardo, C. (2009). Retrieval of volcanic ash particle size, mass and optical depth from a ground-based thermal infrared camera. *J. Volcanol. Geotherm. Res.*, 186, 91–107.

Prival, J.M., Thouret, J.C., Japura, S., Gurioli, L., Bonadonna, C., Mariño, J., Cueva, K. (2020). New insights into eruption source parameters of the 1600 CE Huaynaputina Plinian eruption, Peru. *Bull. Volcanol.*, 82(1), 1–19.

Putirka, K.D. (2008). Thermometers and barometers for volcanic systems. *Rev. Mineral. Geochemistry*, 69, 61–120.

Pyle, D.M. (1989). The thickness, volume and grainsize of tephra fall deposits. *Bull. Volcanol.*, 51, 1–15.

Rausch, J., Grobéty, B., Vonlanthen, P. (2015). Eifel maars: Quantitative shape characterization of juvenile ash particles (Eifel Volcanic Field, Germany). *J. Volcanol. Geotherm. Res.*, 291, 86–100.

Reiners, P.W. (2002). Temporal-compositional trends in intraplate basalt eruptions: Implications for mantle heterogeneity and melting processes. *Geochemistry, Geophys. Geosystems*, 3, 1–30.

Rhéty, M., Harris, A., Villeneuve, N., Gurioli, L., Médard, E., Chevrel, O., Bachélery, P. (2017). A comparison of cooling-limited and volume-limited flow systems: Examples from channels in the Piton de la Fournaise April 2007 lava-flow field. *Geochemistry, Geophys. Geosystems*, 18, 3270–3291.

Ridolfi, F. and Renzulli, A. (2012). Calcic amphiboles in calc-alkaline and alkaline magmas: Thermobarometric and chemometric empirical equations valid up to 1,130°C and 2.2 GPa. *Contrib. to Mineral. Petrol.*, 163, 877–895.

Riley, C.M., Rose, W.I., Bluth, G.J.S. (2003). Quantitative shape measurements of distal volcanic ash. *J. Geophys. Res. Solid Earth*, 108, B10, 2504, 1–15.

Ripepe, M., Rossi, M., Saccorotti, G. (1993). Image processing of explosive activity at Stromboli. *J. Volcanol. Geotherm. Res.*, 54, 335–351.

Robert, B., Harris, A., Gurioli, L., Médard, E., Sehlke, A., Whittington, A. (2014). Textural and rheological evolution of basalt flowing down a lava channel. *Bull. Volcanol.*, 76, 1–21.

Roeder, P.L. and Emslie, R.F. (1970). Olivine-liquid equilibrium. *Contrib. to Mineral. Petrol.*, 29, 275–289.

Rose, W.I., Bonis, S., Stoiber, R.E., Keller, M., Bickford, T. (1973). Studies of volcanic ash from two recent Central American eruptions. *Bull. Volcanol.*, 37, 338–364.

Rossi, E., Bonadonna, C., Degruyter, W. (2019). A new strategy for the estimation of plume height from clast dispersal in various atmospheric and eruptive conditions. *Earth Planet. Sci. Lett.*, 505, 1–12.

Rout, S.S. and Wörner, G. (2020). Constraints on the pre-eruptive magmatic history of the Quaternary Laacher See volcano (Germany). *Contrib. to Mineral. Petrol.*, 175, 73.

Rust, A.C. and Cashman, K.V. (2004). Permeability of vesicular silicic magma: Inertial and hysteresis effects. *Earth Planet. Sci. Lett.*, 228, 93–107.

Rust, A.C. and Cashman, K.V. (2011). Permeability controls on expansion and size distributions of pyroclasts. *J. Geophys. Res. Solid Earth*, 116, 1–17.

Ruth, D.C.S., Costa, F., Bouvet de Maisonneuve, C., Franco, L., Cortés, J.A., Calder, E.S. (2018). Crystal and melt inclusion timescales reveal the evolution of magma migration before eruption. *Nat. Commun.*, 9, 1–9.

Sahagian, D.L. and Proussevitch, A.A. (1998). 3D particle size distributions from 2D observations: Stereology for natural applications. *J. Volcanol. Geotherm. Res.*, 84, 173–196.

Schipper, C.I., White, J.D.L., Houghton, B.F. (2011). Textural, geochemical, and volatile evidence for a strombolian-like eruption sequence at lōihi seamount, Hawaii. *J. Volcanol. Geotherm. Res.*, 207, 16–32.

Schipper, C.I., Castro, J.M., Tuffen, H., James, M.R., How, P. (2013). Shallow vent architecture during hybrid explosive-effusive activity at Cordón Caulle (Chile, 2011–12): Evidence from direct observations and pyroclast textures. *J. Volcanol. Geotherm. Res.*, 262, 25–37.

Schmid, R. (1981). Descriptive nomenclature and classification of pyroclastic deposits and fragments. *Neues Jahrb. fur Mineral. Monatshefte*, 70, 190–196.

Schmith, J., Höskuldsson, Á., Holm, P.M. (2017). Grain shape of basaltic ash populations: Implications for fragmentation. *Bull. Volcanol.*, 79(2), 1–16.

Scollo, S., Prestifilippo, M., Spata, G., D'Agostino, M., Coltelli, M. (2009). Monitoring and forecasting Etna volcanic plumes. *Nat. Hazards Earth Syst. Sci.*, 9, 1573–1585.

Self, S., Kienle, J., Huot, J.P. (1980). Ukinrek Maars, Alaska, II. Deposits and formation of the 1977 craters. *J. Volcanol. Geotherm. Res.*, 7, 39–65.

Settle, M. (1978). Volcanic eruption clouds and the thermal power output of explosive eruptions. *J. Volcanol. Geotherm. Res.*, 3, 309–324.

Shea, T. (2017). Bubble nucleation in magmas: A dominantly heterogeneous process? *J. Volcanol. Geotherm. Res.*, 343, 155–170.

Shea, T. and Hammer, J.E. (2013). Kinetics of cooling- and decompression-induced crystallization in hydrous mafic-intermediate magmas. *J. Volcanol. Geotherm. Res.*, 260, 127–145.

Shea, T., Larsen, J.F., Gurioli, L., Hammer, J.E., Houghton, B.F., Cioni, R. (2009). Leucite crystals: Surviving witnesses of magmatic processes preceding the 79 AD eruption at Vesuvius, Italy. *Earth Planet. Sci. Lett.*, 281, 88–98.

Shea, T., Gurioli, L., Larsen, J.F., Houghton, B.F., Hammer, J.E., Cashman, K.V. (2010a). Linking experimental and natural vesicle textures in Vesuvius 79AD white pumice. *J. Volcanol. Geotherm. Res.*, 192, 69–84.

Shea, T., Houghton, B.F., Gurioli, L., Cashman, K.V., Hammer, J.E., Hobden, B.J. (2010b). Textural studies of vesicles in volcanic rocks: An integrated methodology. *J. Volcanol. Geotherm. Res.*, 190, 271–289.

Shea, T., Gurioli, L., Houghton, B.F., Cioni, R., Cashman, K.V. (2011). Column collapse and generation of pyroclastic density currents during the A.D. 79 eruption of Vesuvius: The role of pyroclast density. *Geology*, 39, 695–698.

Shea, T., Gurioli, L., Houghton, B.F. (2012). Transitions between fall phases and pyroclastic density currents during the AD 79 eruption at Vesuvius: Building a transient conduit model from the textural and volatile record. *Bull. Volcanol.*, 74, 2363–2381.

Shea, T., Hellebrand, E., Gurioli, L., Tuffen, H. (2014). Conduit- to localized-scale degassing during plinian eruptions: Insights from major element andvolatile (Cl and H_2O) analyses within vesuvius ad 79 pumice. *J. Petrol.*, 55, 315–344.

Sheridan, M.F. and Marshall, J.R. (1983). Interpretation of pyroclast surface features using SEM images. *J. Volcanol. Geotherm. Res.*, 16, 153–159.

Sigmarsson, O., Condomines, M., Bachèlery, P. (2005). Magma residence time beneath the Piton de la Fournaise Volcano, Reunion Island, from U-series disequilibria. *Earth Planet. Sci. Lett.*, 234, 223–234.

Solaro, C., Balcone-Boissard, H., Morgan, D.J., Boudon, G., Martel, C., Ostorero, L. (2020). A system dynamics approach to understanding the deep magma plumbing system beneath Dominica (Lesser Antilles). *Front. Earth Sci.*, 8, 1–17.

Sparks, R.S.J. (1986). The dimensions and dynamics of volcanic eruption columns. *Bull. Volcanol.*, 48, 3–15.

Sparks, R.S.J., Wilson, L., Sigurdsson, H. (1981). The pyroclastic deposits of the 1875 eruption of Askja, Iceland. *Philos. Trans. R. Soc. London Ser. A: Math. Phys. Sci.*, 299, 241–273.

Steinberg, G.S. (1976). To the determination of the energy and the depth of volcano Sheveluch explosion. *Mod. Geol.*, 5, 255–257.

Steinberg, G.S. and Lorenz, V. (1983). External ballistic of volcanic explosions. *Bull. Volcanol.*, 46, 333–348.

Sugar, P., Sugarova, J., Morovic, L. (2008). Application of 3D optical scanning for the shape accuracy analysis of machine parts produced by multi-pass metal spinning. *Annals of DAAAM and Proceedings of the International DAAAM Symposium*, 1331–1332.

Sugawara, T. (2000). Empirical relationships between temperature, pressure, and MgO content in olivine and pyroxene saturated liquid. *J. Geophys. Res. Solid Earth*, 105, 8457–8472.

Sulpizio, R. (2005). Three empirical methods for the calculation of distal volume of tephra-fall deposits. *J. Volcanol. Geotherm. Res.*, 145, 315–336.

Sundermeyer, C., Di Muro, A., Gordeychik, B., Wörner, G. (2020). Timescales of magmatic processes during the eruptive cycle 2014–2015 at Piton de la Fournaise, La Réunion, obtained from Mg–Fe diffusion modelling in olivine. *Contrib. to Mineral. Petrol.*, 175(1), 1–16.

Swanson, D., Wooten, K., Orr, T. (2009). Buckets of ash track tephra flux from Halema'uma'u crater, Hawai'i. *Eos Trans. AGU*, 90, 427.

Swanson, D., Zolkos, S., Haravitch, B. (2010). Ballistic blocks surrounding Kilauea's caldera [abs.]. *Am. Geophys. Union, Fall Meet. Abstr.*, 91:V33B–V0, abstract no., V11D-2313.

Taddeucci, J., Pompilio, M., Scarlato, P. (2002). Monitoring the explosive activity of the July–August 2001 eruption of Mt. Etna (Italy) by ash characterization. *Geophys. Res. Lett.*, 29, 1029–1032.

Tait, S., Thomas, R., Gardner, J., Jaupart, C. (1998). Constraints on cooling rates and permeabilities of pumice in an explosive eruption jet from colour and magnetic mineralogy. *J. Volcanol. Geotherm. Res.*, 86, 79–91.

Tema, E., Kondopoulou, D., Pavlides, S. (2013). Palaeotemperature estimation of the pyroclastic deposit covering the pre-Minoan palaeosol at Megalochori Quarry, Santorini (Greece): Evidence from magnetic measurements. *Stud. Geophys. Geod.*, 57, 627–646.

Textor, C., Graf, H.F., Herzog, M., Oberhuber, J.M., Rose, W.I., Ernst, G.G.J. (2006). Volcanic particle aggregation in explosive eruption columns. Part I: Parameterization of the microphysics of hydrometeors and ash. *J. Volcanol. Geotherm. Res.*, 150, 359–377.

Thivet, S., Gurioli, L., Di Muro, A., Derrien, A., Ferrazzini, V., Gouhier, M., Coppola, D., Galle, B., Arellano, S. (2020a). Evidences of plug pressurization enhancing magma fragmentation during the September 2016 basaltic eruption at Piton de la Fournaise (La Réunion Island, France). *Geochemistry, Geophys. Geosystems*, 21, 1–30.

Thivet, S., Gurioli, L., Di Muro, A., Eychenne, J., Besson, P., Nedelec, J.M. (2020b). Variability of ash deposits at Piton de la Fournaise (La Reunion Island): Insights into fragmentation processes at basaltic shield volcanoes. *Bull. Volcanol.*, 82(63), 1–20.

Thivet, S., Gurioli, L., Muro, D. (2020c). A basaltic dyke eruptions at Piton de La Fournaise: Quantification of magmatic overpressure, conduit and eruptive dynamics. *Contrib. to Mineral. Petrol.*, 175(26), 1–24.

Thivet, S., Harris, A.J., Gurioli, L., Bani, P., Barnie, T., Bombrun, M., Marchetti, E. (2021). Multi-parametric field experiment links explosive activity and persistent degassing at Stromboli. *Frontiers in Earth Science*, 9, 431.

Thomas, R.M.E. and Sparks, R.S.J. (1992). Cooling of tephra during fallout from eruption columns. *Bull. Volcanol.*, 54, 542–553.

Thomas, N., Jaupart, C., Vergniolle, S. (1994). On the vesicularity of pumice. *J. Geophys. Res.*, 99, 15633–15644.

Thornber, C.R., Heliker, C., Sherrod, D.R., Kauahikaua, J.P., Miklius, A., Okubo, P.G., Trusdell, F.A., Budahn, J.R., Ridley, W.I., Meeker, G.P. (2003). Kilauea east rift zone magmatism: An episode 54 perspective. *J. Petrol.*, 44, 1525–1559.

Tomlinson, E.L., Smith, V.C., Menzies, M.A. (2020). Chemical zoning and open system processes in the Laacher See magmatic system. *Contrib. to Mineral. Petrol.*, 175, 19.

Toramaru, A. (2006). BND (bubble number density) decompression rate meter for explosive volcanic eruptions. *J. Volcanol. Geotherm. Res.*, 154, 303–316.

Toramaru, A., Noguchi, S., Oyoshihara, S., Tsune, A. (2008). MND (microlite number density) water exsolution rate meter. *J. Volcanol. Geotherm. Res.*, 175, 156–167.

Tsunematsu, K. and Bonadonna, C. (2015). Grain-size features of two large eruptions from Cotopaxi volcano (Ecuador) and implications for the calculation of the total grain-size distribution. *Bull. Volcanol.*, 77, 64.

Ubide, T., Caulfield, J., Brandt, C., Bussweiler, Y., Mollo, S., Di Stefano, F., Nazzari, M., Scarlato, P. (2019). Deep magma storage revealed by multi-method elemental mapping of clinopyroxene megacrysts at Stromboli volcano. *Front. Earth Sci.*, 7, 239.

Vernon, R. (2004). *A Practical Guide to Rock Microstructure*. Cambridge University Press, Cambridge.

Verolino, A., White, J.D.L., Dürig, T., Cappuccio, F. (2019). Black Point – Pyroclasts of a Surtseyan eruption show no change during edifice growth to the surface from 100 m water depth. *J. Volcanol. Geotherm. Res.*, 384, 85–102.

Villeneuve, N., Neuville, D.R., Boivin, P., Bachèlery, P., Richet, P. (2008). Magma crystallization and viscosity: A study of molten basalts from the Piton de la Fournaise volcano (La Réunion island). *Chem. Geol.*, 256, 242–251.

Vlastélic, I., Di Muro, A., Bachèlery, P., Gurioli, L., Auclair, D., Gannoun, A. (2018). Control of source fertility on the eruptive activity of Piton de la Fournaise volcano, La Réunion. *Sci. Rep.*, 8, 14478.

Vonlanthen, P., Rausch, J., Ketcham, R.A., Putlitz, B., Baumgartner, L.P., Grobéty, B. (2015). High-resolution 3D analyses of the shape and internal constituents of small volcanic ash particles: The contribution of SEM micro-computed tomography (SEM micro-CT). *J. Volcanol. Geotherm. Res.*, 293, 1–12.

Walker, G.P.L. (1971). Grain-size characteristics of pyroclastic deposits. *J. Geol.*, 79, 696–714.

Walker, G.P.L. (1973). Explosive volcanic eruptions: A new classification scheme. *Geol. Rundschau*, 62, 431–446.

Walker, G.P.L. (1980). The Taupo pumice: Product of the most powerful known (ultraplinian) eruption? *J. Volcanol. Geotherm. Res.*, 8, 69–94.

Walker, G.P.L. (1981). Characteristics of two phreatoplinian ashes, and their water-flushed origin. *J. Volcanol. Geotherm. Res.*, 9, 395–407.

Welsch, B., Faure, F., Famin, V., Baronnet, A., Bachèlery, P. (2013). Dendritic crystallization: A single process for all the textures of olivine in basalts? *J. Petrol.*, 54, 539–574.

Will, J. (1903). Report on the medical relief expeditions to Martinique and St. Vincent in aid of the sufferers from the volcanic eruptions. In *Further Correspondence Relating to the Volcanic Eruptions in St. Vincent and Martinique in 1902 and 1903*, Great Britain Colonial Office (ed.). His Majesty's Stationary Office, London.

Wilson, L. (1972). Explosive volcanic eruptions II. The atmospheric trajectories of pyroclasts. *Geophys. J. R. Astron. Soc.*, 30, 381–392.

Wilson, C.J.N. and Houghton, B.F. (2000). Pyroclastic transport and deposition. In *Encyclopedia of Volcanoes*, Sigurdsson, H. (eds). Academic Press, Cambridge.

Wilson, L. and Huang, T.C. (1979). The influence of shape on the atmospheric settling velocity of volcanic ash particles. *Earth Planet. Sci. Lett.*, 44, 311–324.

Wilson, L. and Walker, G.P.L. (1987). Explosive volcanic eruptions VI. Ejecta dispersal in plinian eruptions: The control of eruption conditions and atmospheric properties. *Geophys. J. R. Astron. Soc.*, 89, 657–679.

Wilson, L., Sparks, R.S.J., Huang, T.C., Watkins, N.D. (1978). The control of volcanic column heights by eruption energetics and dynamics. *J. Geophys. Res. Solid Earth*, 83, 1829–1836.

Wohletz, K.H. (1983). Mechanisms of hydrovolcanic pyroclast formation: Grain-size, scanning electron microscopy, and experimental studies. *J. Volcanol. Geotherm. Res.*, 17, 31–63.

Wohletz, K.H., Sheridan, M.F., Brown, W.K. (1989). Particle size distributions and the sequential fragmentation/transport theory applied to volcanic ash. *J. Geophys. Res.*, 94, 15703–15721.

Wörner, G. and Schmincke, H.U. (1984). Mineralogical and chemical zonation of the Laacher See tephra sequence (East Eifel, W. Germany). *J. Petrol.*, 25, 805–835.

Xu, Z. and Zhang, Y. (2002). Quench rates in air, water, and liquid nitrogen, and inference of temperature in volcanic eruption columns. *Earth Planet. Sci. Lett.*, 200, 315–330.

Yang, Q. and Bursik, M. (2016). A new interpolation method to model thickness, isopachs, extent, and volume of tephra fall deposits. *Bull. Volcanol.*, 78, 68.

Yokoyama, T. and Takeuchi, S. (2009). Porosimetry of vesicular volcanic products by a water-expulsion method and the relationship of pore characteristics to permeability. *J. Geophys. Res. Solid Earth*, 114(B02201), 1–10.

Zanella, E., Gurioli, L., Pareschi, M.T., Lanza, R. (2007). Influences of urban fabric on pyroclastic density currents at Pompeii (Italy) 2: Temperature of the deposits and hazard implications. *J. Geophys. Res. Solid Earth*, 112(B5). https://doi.org/10.1029/2006JB004775.

3

Monitoring of Eruptive Products: Pyroclastic Density Currents and Their Deposits

Alessandro TADINI[1], Lucia GURIOLI[1], Sylvain CHARBONNIER[2], Simon THIVET[1,3] and Jean-Claude THOURET[1]

[1] *Laboratoire Magmas et Volcans, CNRS, IRD, OPGC, Université Clermont Auvergne, Clermont-Ferrand, France*

[2] *School of Geosciences, University of South Florida, Tampa, United States*

[3] *Department of Earth and Environmental Sciences, Ludwig-Maximilians-Universitat, Munich, Germany*

3.1. Introduction

Pyroclastic density currents (PDCs) are a mixture of gases and solid particles that move across the Earth's topography under the influence of gravity. PDCs are one of the most dangerous phenomena in explosive eruptions because of their mobility, high impact pressure from the current front, velocity, lethal temperature (>100°C) and ash content. The study of PDCs and their deposits is, therefore, one of the main tasks in volcanic hazard assessment. Although innovative measurements conducted recently during eruptions are promising (despite the many technical challenges), still the study of PDC deposits immediately after an eruption remains a powerful tool for assessing their hazard. Following the approach described in Chapter 2 of Volume 2, we will first describe some key features of PDCs and associated deposits (for a more detailed description, see Branney and Kokelaar

Hazards and Monitoring of Volcanic Activity 3,
coordinated by Jean-François LÉNAT.

(2002), Sulpizio et al. (2014) and Brown and Andrews (2015)). In this chapter, we will also discuss the investigative approach performed at multiple scales and detail the differences between the measurements conducted during and after the eruption. This will be followed by a detailed discussion of all the main parameters of PDCs and their deposits, considering different scales of observation and conditions of the measurements conducted during and after the eruption. Finally, we will present a brief overview of future perspectives in PDC and deposition studies.

3.2. General characteristics of pyroclastic density currents and their deposits

PDCs are mostly non-uniform phenomena (i.e. with a highly variable spatial extent and with an unsteady to quasi-steady flow regime, that is, temporally variable to quasi-stable) (Branney and Kokelaar 2002). They can be short (a few minutes) or relatively prolonged (up to several hours for the largest ignimbrites) (Bursik and Woods 1996), and with velocities ranging from 1 to 300 m/s (Cole et al. 2015). PDCs can be triggered by a wide range of explosive volcanic events, namely:

– total or partial collapse of a sustained (Plinian/sub-Plinian, see Figure 3.1(a)) or transient (Vulcanian) eruptive column (Branney and Kokelaar 2002);

– low pyroclastic fountaining ("boil-over") (Cas and Wright 1987);

– collapsing lava domes (see Figure 3.1(b)) (Rose et al. 1976; Bardintzeff 1984; Yamamoto et al. 1993; Cole et al. 2002) or loose, unwelded ignimbrites (Branney and Kokelaar 2002);

– lateral blasts (Hoblitt 1985);

– expansion of over-pressurized gas (Cas and Wright 1987);

– phreatic explosions (Sheridan 1980).

PDCs, which typically have intermediate to acidic magmatic compositions (andesites to rhyolites), contain particles with densities ranging from 200 to 2,500 kg/m^3 (Shea et al. 2012), a key element that influences current mobility and the sedimentological characteristics of associated deposits. However, some PDC deposits have a basaltic composition, as documented at Villarrica volcano (Silva Parejas et al. 2010; Costantini et al. 2011), Tonga (Caulfield et al. 2011) and Etna (Andronico et al. 2018).

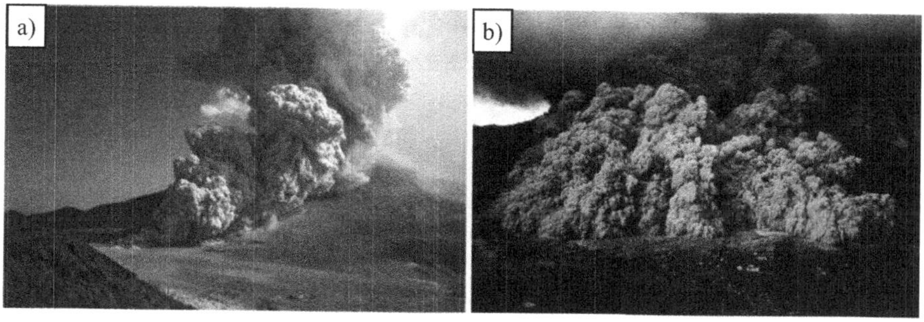

Figure 3.1. *Two examples of pyroclastic density currents (PDC): a) the August 7, 1980 Mount St. Helens (United States) eruption (photo courtesy P.W. Lipman, USGS) and b) the June 3, 1991 Mt. Unzen (Japan) dome collapse (photo courtesy Nagasaki Photo service). For a color version of this figure, see www.iste.co.uk/lenat/hazards3.zip*

From a macroscopic point of view, PDCs can be considered as a system composed of a dense mixture of pyroclastic particles (pumice, scoriae and free crystals) with gas immersed in a less dense, nearly isotropic fluid (the atmosphere). The term PDC has been proposed by several authors over the past 30 years (Valentine 1987; Druitt 1992; Wilson and Houghton 2000; Branney and Kokelaar 2002; Sulpizio et al. 2014) in an effort to describe a natural continuum between two end-members: dilute currents, in which the fluid (i.e. the gas) is prevalent, and concentrated currents, in which solid particles predominate. These two contrasting terms were named by Branney and Kokelaar (2002), respectively: "fully dilute PDCs", in which particles are carried by the turbulence of the fluid throughout the current, and "granular fluid-based PDCs", in which the particle concentration remains high enough in the lower part of the current for grain interactions and fluidization to constitute the transport mechanisms, delaying sedimentation. This terminology has the advantage of being directly related to transport processes, which can also be extracted from the lithofacies analysis of the corresponding deposits. Besides this terminology, other subdivisions have been recently proposed. Doronzo (2012) suggested distinguishing PDCs according to current velocity rather than current concentration because the former is the factor that controls transport and sedimentation. According to this distinction, PDCs are divided into convection-governed or inertia-governed currents. More recently, Palladino (2017) proposed using the more general term pyroclastic currents (PC) to avoid the term density being redundant when referring to the two particular types of currents.

Recognizing and describing the different lithofacies within PDC deposits is a key step. The lithofacies of a deposit or one of its units is characterized by stratification, grain size, grain shape, sorting, fabric and composition. Processes occurring at the "boundary zone" between the overlying current and the underlying forming deposit determine the lithofacies of associated pyroclastic deposits and ignimbrites. The latter are deposits of pumice-rich PDCs, which show characteristics interpreted as the sedimentation of products deposited under relatively high-temperature conditions. In general, PDC deposits can be described on the basis of non-genetic terms related to sedimentary structures, grain size, sorting and composition (Sulpizio et al. 2014).

To avoid ambiguity, it is important to use standard and objective terminology to describe PDC depositional lithofacies in a synthetic manner. Therefore, we suggest using different terms when referring to deposits of different PDCs. In the case of consolidated and large-volume PDC deposits (e.g. ignimbrites), we recommend the use of a lithological description of the primary components, defined by White and Houghton (2006) and shown in Figure 3.2.

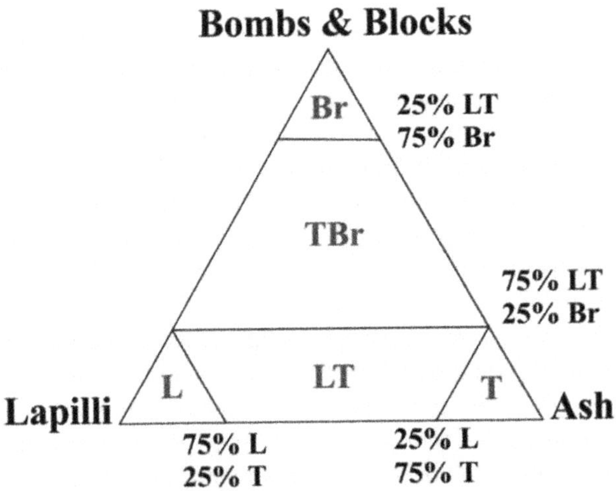

Figure 3.2. *Non-genetic classification scheme for primary tephra (modified from White and Houghton (2006)). Abbreviations: L: lapillistone; T: tuff; LT: lapilli tuff; TBr: tuff breccia; Br: breccia*

In addition to this initial characterization, several abbreviations (Branney and Kokelaar 2002) must be added in order to describe the internal architecture of the deposit: Table 3.1 shows a list of lithofacies symbols used.

Lithofacies symbol	Description
mlBr	Massive lithic breccia poor in fines
$mlBr_{poor}$	Lens of lapilli
lensL	Massive lapilli tuff
mLT	Massive lapilli tuff with inverse grading pattern
mLT_i	Massive lapilli tuff with normal grading pattern
mLT_n	Massive lapilli tuff with normal grading of lithics and inverse grading of pumices
$mLT_{(nl, ip)}$	Massive lapilli tuff with accretionary lapilli
mLT_{accr}	Eutaxitic massive lapilli tuff, with directional fabric
$emLT_f$	Thin-bedded lapilli tuff stratified
bLT	Diffuse thin-bedded lapilli tuff
dbLT	Parallel-stratified lapilli tuff
//sLT	Cross-stratified lapilli tuff
xsLT	Massive tuff with accretionary lapilli
mT_{accr}	Diffuse thin-bedded lapilli tuff with accretionary lapilli
dbT_{accr}	Stratified tuff
sT	Parallell-stratified tuff
//sT	Cross-stratified tuff
xsT	Massive lithic breccia poor in fines

Table 3.1. *List of abbreviations describing the PDC lithofacies used in this chapter*

For low-volume, unconsolidated deposits, such as block-and-ash flow deposits, the terminology in Figure 3.2 should be modified by replacing the terms "tuff" and "breccia" with "ash" (C) and "block" (B), as suggested by Sulpizio and Dellino (2008) and Sulpizio et al. (2014). The abbreviations in Table 3.1 will then be used with these new terms.

In general, PDC deposits exhibit the following characteristics, which are recognizable at different scales (as in the case of fallout deposits):

– The sub-rounded to rounded shape of the pyroclasts (see Figure 3.3(a)), especially the juvenile particles. This feature is due to abrasion and/or breakage of the larger pumiceous clasts during transport.

– Poor sorting at the scale of the outcrop, especially within massive deposits (see Figure 3.3(b)), and the presence of several structures such as grading, bedding and stratification (see Figure 3.3(b)). While poor sorting results from several processes (poor sorting at the source, abrasion and breakage during transport, en-masse deposition, etc.), the presence of an internal architecture within the deposits is a consequence of the unstable regime of PDCs, their close interaction with the underlying topography and the turbulent regime of the more dilute PDCs.

– A different interaction of the two main types of deposits with the pre-existing topography: dense, granular and fluidized PDCs tend to be valley-confined (see Figure 3.3(c)), while turbulent, dilute PDCs tend to mantle the topography, but not uniformly, being thicker in valleys but thinner on topographic highs (see Figure 3.3(c)). Unlike fallout deposits, PDC thickness does not decrease exponentially with distance from vent, but is primarily influenced by interactions with the underlying topography.

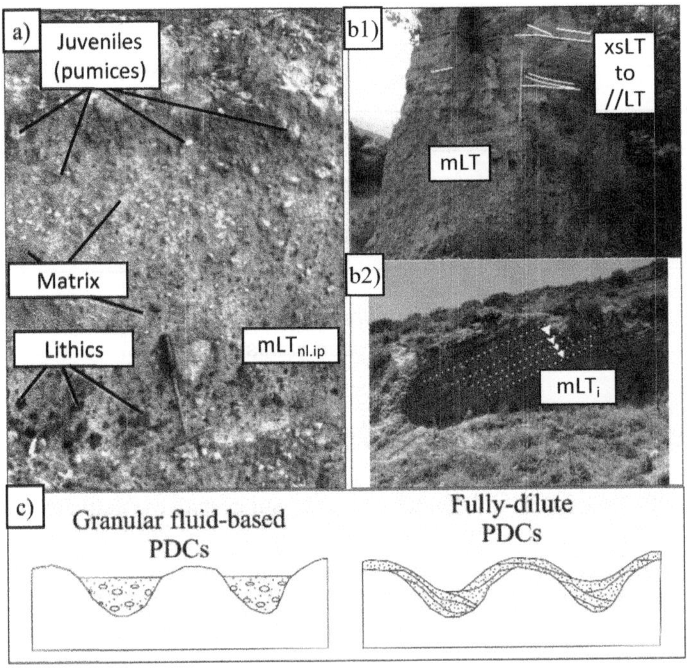

Figure 3.3. *Examples of pyroclasts, lithofacies and deposits associated with PDCs. For a color version of this figure, see www.iste.co.uk/lenat/hazards3.zip*

COMMENT ON FIGURE 3.3.– *a) Example of juvenile and lithic pyroclasts carried by an abundant ash matrix in an outcrop, which shows coarse tail grading (coarse lithic fragments toward the base and pumice toward the top. Above a certain size threshold, the matrix is not graded). b) PDC outcrops from Somma-Vesuvius volcano, Italy (b1) and on Nysiros Island, Greece (b2 – triangles highlight the inverse grading of the layers). c) Sketch of the two typical PDC deposits, on the left the "granular fluid-based" lithofacies (see b1) and, on the right, the fully diluted lithofacies of a surge (see b2) (modified after Wilson and Houghton (2000)).*

One of the main purposes of a detailed description of lithofacies association in a PDC deposit is to distinguish the number of flow units within the deposit. A "flow unit" is a genetic term meaning the deposit of a temporally discrete current. Recognizing multiple flow units within a PDC deposit requires finding evidence of flow interruption, identified through the presence of a fallout deposit surmounting that PDC (either from fine particles that have settled out of the ash cloud of the current or from the revival of a sustained column at the vent). This distinction is of great importance when dealing with numerical simulations and hazard assessment in an area covered by PDC. Using the terminology introduced above, it is then possible to propose a synthetic, non-genetic and objective description of all types of PDC deposits and then relate them to the characteristics of sedimentation induced by different processes acting at the boundary between the current and the underlying forming deposit (based on the concept of "flow-boundary zone"; Branney and Kokelaar (2002)).

3.3. The importance of scale in the joint study of deposition and individual particles

Similarly to tephra fallout deposits, PDC deposits can also provide a wide range of information if analyzed at different scales. Most of the analyses (especially at the scale of individual particle) are basically the same as those used to study fallout deposits, but can nevertheless be used to extract different information. For example, shape measurements on individual particles can shed light on the degree of rounding of the more fragile juvenile particles and, consequently, on the amount of fine particle available to feed co-ignimbritic plumes or to fluidize the granular lower portion of a dense PDC. Furthermore, the study of a PDC deposit should always take into account that an outcrop in a particular location represents only an incomplete view of a current at that location. Apart from a few cases ("plug flow"; Branney and Kokelaar 2002), where sudden mass sedimentation of a PDC results in a deposit that actually represents the current that set it up, PDC deposits must be considered (from bottom to top) as a succession of snapshots of different depositional regimes through time. From this perspective, each lithofacies is related to a different point in the temporal and spatial evolution of a PDC, which helps identify quasi-stable and unstable phases (e.g. waxing and waning) of a current. Moreover, the spatial and temporal variability, which characterizes the sedimentation parameters and regimes of PDCs, makes comparisons between lithofacies of a single PDC at different locations more difficult. Nevertheless, it is crucial to study the complete deposition of a single or series of PDCs produced during the same eruption because characteristic parameters (such as total volume or, to a lesser extent, total grain size distribution) must be identified in order to reconstruct the eruptive history and to develop input parameters for numerical models.

3.4. Syn-eruptive and post-eruptive measurements

Since PDCs are very destructive and rapidly moving phenomena, it is much more difficult, compared to the case of tephra fallout, to perform real-time measurements, which are intended to define the hazard-related parameters in Table 3.2. On the contrary, measurements conducted on PDC deposits are, in most cases, the only ways to study in detail the dynamics of these currents and to derive quantified and more relevant input parameters for numerical models. Like fallout deposits, PDCs can also provide information on the general eruptive history of a volcano (including in this case the indication of PDC types and preferred impact areas), as well as on the evolution of a singular eruption over time (fluctuations of the eruptive regime that feeds these currents). In addition, quantitative measurements of PDC deposits are also fundamental to quantify the range of variations in the parameters, listed in Table 3.2, as these in turn are essential data for numerical models.

3.4.1. *Particle scale measurements*

We have already indicated that most of the measurements performed on particles are identical to those conducted on tephra fallout deposits; therefore, we refer the reader to the detailed description of these methods in Table 2.2 and section 2.5.

Scale of observation	Desired parameter	Syn-eruptive method	Post-eruptive method
Particle	Size (Φ or mm)	Satellite, LiDAR, radar Infrared camera, ASHER (thickness measurement), disdrometer, visible camera	Screening Image analysis of the deposit Coulter Laser counter (for very fine particles)
	Components or constituents (weight %)	–	Binocular, hand sorting, image analysis of the deposit, SEM, 2D/3D microscope
	Density/porosity (kg/m^3 or g/cm^3)	–	Pycnometer (water, helium, pre-dimensioned grain flow), Archimedes method, image analysis of particles
	Geometric parameters, area (mm^2), perimeter (mm), main axes a, b, c (mm)	Infrared	2D/3D image analysis of particles, image analysis of the deposit, G3 morphology, SEM, X-ray microtomograph
	Permeability	–	Permeameter
	Settling velocity (m/s)	ASHER, Disdrometer	Equations
Deposit	Flow direction (in degrees from N)	Aerial or drone films and photos	Deposit fabric (tree trunks, nesting or sedimentary structures of clasts) Anisotropic magnetic susceptibility (AMS)
	Maximum runout (km) and invaded area (km^2)	–	Determination of the distance at which the most distant outcrop was observed, hand drawing, automatic drawing with algorithm
	Total volume (km^3 or m^3) Total mass (kg)	–	Integration from an isopach map
	Total grain distribution (TGSD) (wt%)	–	Weighted or unweighted average of each sample after sieving

	Mass flow rate (MFR- kg/s) or volumetric flow rate (VFR - m^3/s) per unit time	–	Ratio total mass (or volume)/total PDC emplacement time
	Concentration of volcanic particles ($\mu g/m^3$ or vol %)	–	Equations
	PDC velocity (m/s or km/h)	Thermal camera, infrasound sensors, Doppler radar	Characteristics of PDC-induced erosional features, equations
	Dynamic pressure (Pa)	–	Observed effects on trees, buildings, larger blocks, equations
	Temperature (°C or K)	Thermal camera Satellite	hermal effects on objects, petrological evidences, thermal remnant magnetization (TRM)

Table 3.2. *Key PDC depositional parameters relevant to hazard assessment*

PDCs are phenomena sometimes characterized by high temperatures and by particles interaction with the substratum, but also by particle-particle interaction within the current. This characteristic implies that, during the propagation of a current, the particles are subjected to a modification of their shapes, which in turn necessarily influence the measured parameters. The most important phenomena that cause these modifications are the following:

– *Bedrock erosion*: During current propagation, part of the bedrock can be eroded and incorporated into the current itself, as shown for example by Pittari et al. (2007) or Brand et al. (2014). This incorporation has a crucial impact on the definition of the constituents of a PDC deposit and on the abrasion of clasts (Bernard and Le Pennec 2016).

– *Clast abrasion*: In the case of PDCs for which interactions between clasts are one of the support/transport mechanisms (Branney and Kokelaar 2002), mechanical and thermal abrasion of the more fragile clasts can be considerable (Kueppers et al. 2012). From a numerical model perspective, it should be considered that the information inferred from a deposit does not match the composition of the initial mixture of a PDC, both in terms of particle size distribution and particle rounding (and hence shape factor).

– *Clast deformation*: Many ignimbrites (Self et al. 1988) exhibit varying degrees of deformation of pumiceous clasts due to the combined effects of lithostatic load of the deposit and high temperature, involving gas exsolution. These deformed clasts (termed "fiamme") result from the modification of the initial mixture from the collapse that generated a PDC, which can provide information on emplacement conditions (e.g. mass and volume flow rate, compaction rate and flow permeability).

3.4.2. *Deposit scale measurements*

In order to obtain hazard-related parameters from PDC deposits (see Table 3.2), it is crucial to describe the deposit as precisely as possible, following the same procedure and using the same tools as indicated in section 2.7. In addition, one must keep in mind that PDC deposits have more structural complexities; therefore, the identification of different lithofacies is crucial when inferring transport mechanisms and current types (either fully dilute or granular fluid-based in the sense of Branney and Kokelaar (2002)). Such measurements are crucial for the numerical modeling of PDCs aimed at quantifying their destructive power and forecasting their potentially invaded areas. Any study of PDC lithofacies should include both the nomenclature shown in Table 3.2 and measurements of the following parameters in addition to those described in section 2.7, namely:

– The wavelength (in cm or m) and orientation (in degrees from North) of possible dunes/antidunes and ramp-like structures in the case of dilute PDC (Douillet et al. 2013).

– The strike of imbricated clasts or the preferential orientation of individual clasts (Karátson et al. 2002).

– Oriented cores from drillings in welded deposits (e.g. some ignimbrites). In order to make these measurements in welded deposits, the necessary tools are the sun compass and a drilling equipment.

3.4.2.1. *Flow direction*

Determining the direction of a PDC (usually expressed in degrees from North) is not an easy task, especially in the case of past deposits for which there are few field evidences. This is because PDC deposits from a single current in most of the cases do not show a single main direction, but rather several radially oriented directions with respect to the crater. On the other hand, if a lobe of an individual PDC is considered (especially if it is valley-confined), a single flow direction can be identified. In this case, however, localized variations in direction can be observed due to topographic effects (Komorowski et al. 2013), including those generated by small obstacles (see Figure 3.4) (Gurioli et al. 2007).

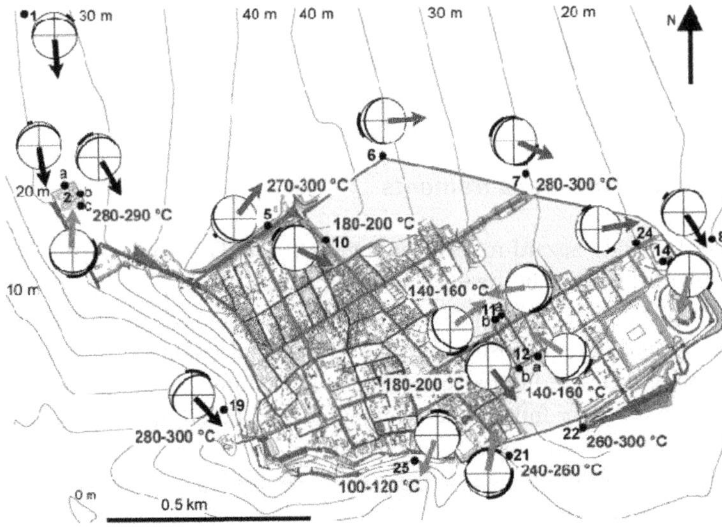

Figure 3.4. *Flow direction (anisotropic magnetic susceptibility, AMS) and temperature (thermal remnant magnetization, TRM) variations measured in the most energetic PDC deposits at Pompeii during the 79 CE eruption. For a color version of this figure, see www.iste.co.uk/lenat/hazards3.zip*

COMMENT ON FIGURE 3.4.– *The urban fabric led to the division of the lower part of the PDC into several currents that followed walls and streets. Vortices formed downstream of obstacles or in cavities. Points: sampling sites. Paleo-contours (in m) before the eruption shown in gray. AMS data: circle, projection of the magnetic foliation; large arrow, main direction of the site. TRM data: temperature, in red, defined from at least 10 samples (modified from Gurioli et al. (2005)).*

PDC flow direction is a critical measurement in the context of hazard assessment, as it can provide real-time indications about areas potentially invaded by PDC, although high PDC velocities prevent the implementation of effective mitigation strategies in syn-eruptive conditions. In addition, flow directions of past PDCs could be used to infer the location of buried, eroded or poorly located eruptive vents. This latter is the intersection point (or area) of the lines passing by the points where flow direction has been measured. Finally, deposit-derived PDC flow directions can also help better understand PDC dynamics and emplacement processes (Gurioli et al. 2007).

3.4.2.1.1. Syn-eruptive methods to determine flow direction

Tracking the direction of a PDC in real time can be of fundamental importance from a hazard perspective, especially when weather conditions prevent direct observation of the phenomenon. However, measuring the direction of a PDC in three dimensions is difficult during the eruption and, therefore, it is more correct to use the term "flow path". This latter indicates the main direction of flow propagation, which is an average of different flow directions measured in different locations. For this purpose, the most recent techniques exploit information from two different sources: thermal cameras that record the temperature of the outer envelope of a PDC (see Figure 3.5) and infrasound signals (Ripepe et al. 2009; Delle Donne et al. 2014; see Figure 3.6). The latter, in particular, are low-frequency components of sound that are below the human threshold of hearing but can be detected by appropriate instruments. These infrasound signals, common on volcanoes, are frequently used for monitoring purposes (Ripepe et al. 2007). In the case of PDCs, an array of infrasound detectors can help in recognizing the flow path by tracking the flow front with high accuracy. For example, by combining an infrasound array with thermal signals, Delle Donne et al. (2014) tracked a PDC in real time on Soufrière Hills volcano (Montserrat island, Lesser Antilles). This was done by (1) converting thermal images into binary pixels, characterized by temperatures below 0 or above 1 ambient temperature, and (2) by geo-referencing pixels with temperatures higher than the ambient temperature (related to hot volcanic material).

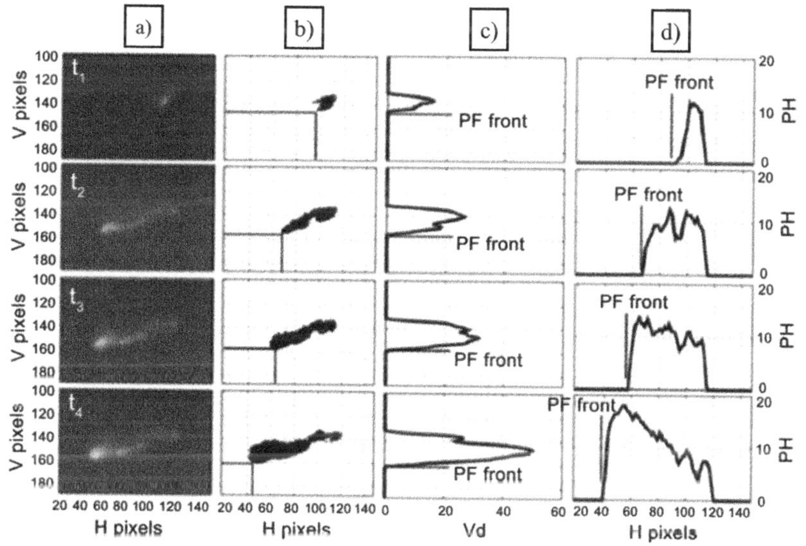

Figure 3.5. *Methodology for PDC tracking from the temperature of the outer envelope. For a color version of this figure, see www.iste.co.uk/lenat/hazards3.zip*

COMMENT ON FIGURE 3.5.– *a) Original thermal images acquired at time interval* $t_1 - t_4$ *are converted into b) binary images using a thermal threshold above the background noise, which allows detection of high-temperature areas in the camera field of view. Image deconvolution method along the c) vertical and d) horizontal axes, which plots the high temperature anomaly related to the PDC front (modified from Delle Donne et al. (2014)).*

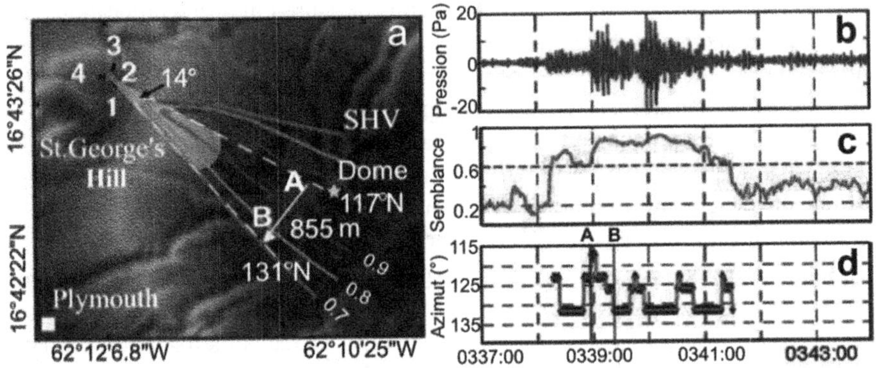

Figure 3.6. *Methodology for tracing a PDC from infrasound measurements. For a color version of this figure, see www.iste.co.uk/lenat/hazards3.zip*

COMMENT ON FIGURE 3.6.– *a) Map showing the position of an infrasound array for Soufrière Hills volcano (Montserrat island, Lesser Antilles) and the direction of the July 29, 2008 PDC as detected by the array in real time. Contour lines (0.9, 0.8 and 0.7) represent the distribution of semblance indicating the direction from which the sound originates. b) Infrasound signal (measured in Pa), associated with PDC propagation. c) Distribution of multichannel semblance as a function of the flow path (back azimuth) and time. The blue dashed line indicates the threshold used to locate the source. d) Repeated variations of the angle (azimuth) of the infrasound signals during the main PDC phase. The location of A and B on map (a) is referenced in diagram (d) (modified from Ripepe et al. (2009)).*

3.4.2.1.2. Post-eruptive methods to determine flow direction

Two main post-eruptive methods for determining flow direction are used to measure the PDC deposit fabric and anisotropic magnetic susceptibility.

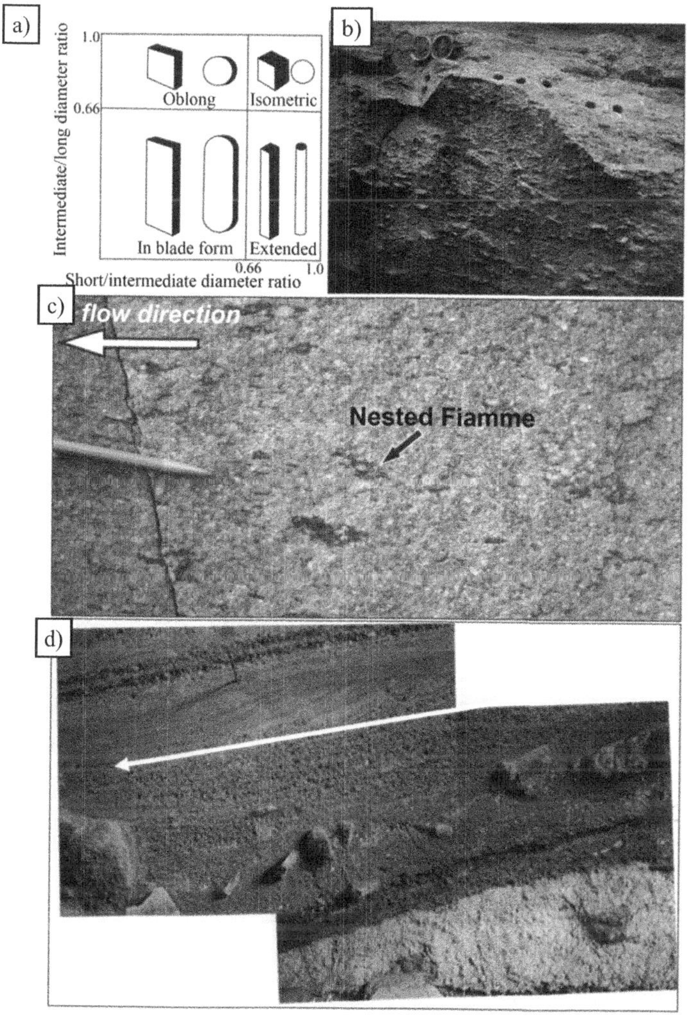

Figure 3.7. *a) Classification of clast shapes, examples of clast imbrications from b) Monte Cimino, Italy, and c) Tenerife, Spain. d) Tiles imbricated along their intermediate axes (79 CE eruption, Somma-Vesuvius): this configuration is interpreted as the result of tiles rolling within a traction carpet at the base of a PDC. For a color version of this figure, see www.iste.co.uk/lenat/hazards3.zip*

COMMENT ON FIGURE 3.7.– *a) According to Sneed and Folk (1958). b) Modified according to LaBerge et al. (2009). c) Modified according to Bryan et al. (1998). d) Modified according to Gurioli et al. (2007).*

– *Fabric*: The irregular shape of juvenile clasts allows inferring PDC direction. The principle is that (for non-turbulent flow regimes) flow direction is parallel to (1) the strike of the major axis of clasts (especially oblate and prolate ones, see Figure 3.7(a)) and (2) the direction of clast imbrication (see Figure 3.7(b)). This principle can be generally applied to all the asymmetrical elements of a PDC deposit (see Figure 3.7(c)), such as anthropic materials (Gurioli et al. 2002, 2007), elongated pumices and lithic clasts (Hughes and Druitt 1998; Karátson et al. 2002), glass shards or elongated phenocrysts (Ui et al. 1989), tree trunks (Potter and Oberthal 1987; Fisher 1990) and sedimentation structures (asymmetrical ramps of low-aspect ratio ignimbrites; Suzuki-Kamata 1988). In Figure 3.7(d), Roman tiles were deposited in a traction depositional system; therefore, differently from the above-mentioned principle, their intermediate axes and their imbrications are parallel to flow direction.

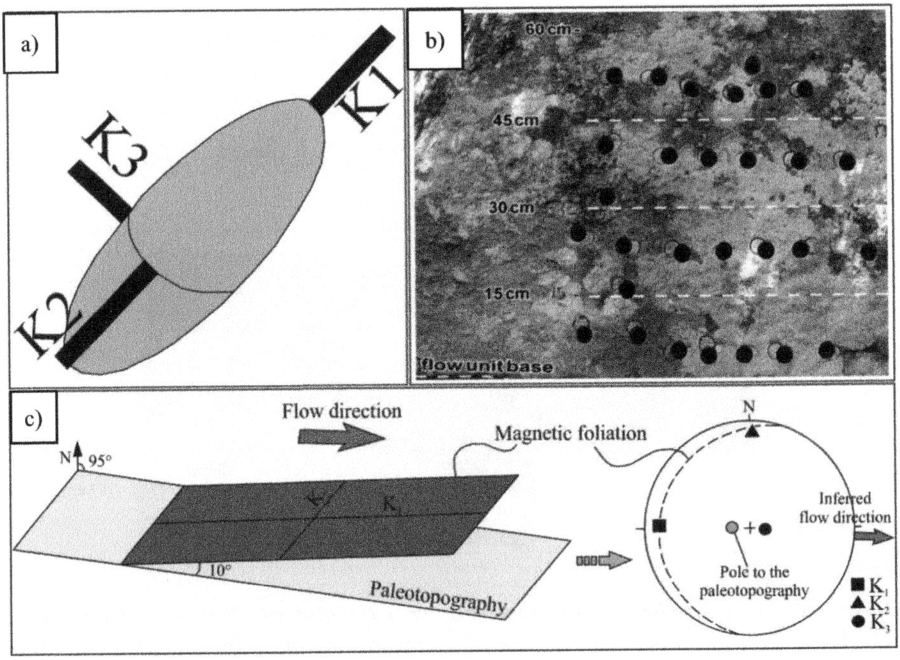

Figure 3.8. *a) Susceptibility ellipsoid (modified according to Cañón-Tapia and Mendoza-Borunda (2014)). b) Drilling to extract cores for AMS analysis (modified according to LaBerge et al. (2009)). c) Stereo-plot showing the position of K1, K2 and K3 (modified from Giordano et al. (2008))*

– *Anisotropic magnetic susceptibility (AMS)*: AMS indicates the variation in the natural magnetization of rock elements, produced by the interaction of an external magnetic field with the electrons of an element in that rock. We refer the reader to the article by Cañón-Tapia and Mendoza-Borunda (2014) for a detailed review of AMS in PDC deposits. In general, a rock placed in a magnetic field H acquires an induced magnetic momentum J_i, proportional to the field, such that:

$$J_i = K_{ij}H \quad [3.1]$$

The proportionality factor, K_{ij}, is the magnetic susceptibility represented by three components ($K_1 \geq K_2 \geq K_3$), called maximum, intermediate and minimum susceptibility, respectively. These three components define a susceptibility ellipsoid (see Figure 3.8(a)), the shape and orientation of which are controlled by the ferromagnetic mineral fabric in the rock, known as the magnetic fabric.

In order to estimate the AMS from an outcrop, several cores of identical diameter and length must be extracted (see Figure 3.8(b)). During sampling, the orientations of the cores (in degrees to the North) should be accurately measured using magnetic and solar compasses. In case the outcrop allows it, it is recommended to regularly collect samples at two or three depth levels in the same deposit to check the consistency of orientations at the same site (Cañón-Tapia and Mendoza-Borunda 2014). In the laboratory, cores are analyzed with specific instruments that provide the direction and intensities of K_1, K_2 and K_3, as well as other parameters defining magnetic susceptibility. These are total anisotropy P (K_1/K_3), lineation L (K_1/K_2), foliation F (K_2/K_3) and elongation parameter E (rounded/elongated) ($[K_2]^2/K_1K_3$).

The directions of K_1, K_2 and K_3 are represented in stereographic projections in the lower hemisphere (see Figure 3.8(c)). Analysis of the stereo-diagrams usually leads to one of the following results:

– a clustering of K_3 close to the vertical and a clustering of K_1 in the foliation plane (lineation), defining a flow direction;

– a clustering of K_3 close to the vertical and $K_1 - K_2$ determining a foliation, not defining a flow direction;

– a random arrangement of K_1, K_2 and K_3, indicating a low degree of anisotropy or a turbulent PDC;

– a clustering of $K_1 - K_2$, with K_3 inclined less than 30° from the vertical, interpreted as an imbrication (which therefore indicates the flow direction).

3.4.2.2. *Maximum runout/invaded area*

The maximum propagation distance of a PDC from its source (in m or km from the crater area) is called maximum runout. Directly dependent on this, the area invaded by the PDC (measured in m^2 or km^2) is the area constrained by the line joining all maximum runout distances. The maximum runout is usually indicated in the literature by two measurements:

– A point taken as the maximum distance reached by a PDC, chosen on the basis of the presence of primary PDC deposits at that location (De La Cruz-Reyna and Carrasco-Núñez 2002; Gurioli et al. 2010).

– An isoline enveloping the whole deposit, which should ideally be the 0-thickness isopach. This isoline can be obtained by an algorithm (e.g. Voronoi tessellation; Burt et al. 2001) or can be based on field evidences. In this case, PDC maximum runout can be either outlined immediately following an eruption (Charbonnier and Gertisser 2008), or it corresponds to the furthest possible isopach that can be traced with uncertainty bounds (Cioni et al. 2020) or without them (Wilson 1985; Gurioli et al. 2005). Recall that the Voronoi tessellation (see also section 2.8.2.1) represents a decomposition of a metric space into cells (adjacent regions), determined by the distances to a discrete set of objects in that space, typically a discrete set of points (see Figure 3.9).

The point of maximum distance reached by a PDC can sometimes be referred to as the "farthest outcrop distance" (FOD; Gurioli et al. 2010). When applying this method, one should keep in mind that the area covered by PDC deposits is always smaller than the area actually affected by such currents, since the thinnest distal part has been eroded (except in the case of deposits measured immediately following an eruption). This is also true when applying methods that express the maximum runout as an isoline without specifying the uncertainty limits. A conservative approach is therefore to consider these runout estimates as minimum values. When maximum PDC runouts are measured for ancient deposits (Ma) located in an extensional or compressional geodynamic setting, the measured distance may be under- or overestimated due to bedrock extension or compression. In this case, Roche et al. (2016) showed that it is appropriate to add a correction factor to the maximum runout to account for the paleo-morphology preceding these tectonic movements.

Estimating the maximum runout and the corresponding invaded area is obviously of great importance in the context of hazard assessment. Yet, even syn-eruptive measurements could not be used for real-time mitigation plans and, indeed, very few experiments have been attempted to date (Pallister et al. 2013). On the contrary, measurements made on past PDC deposits are of great interest as they provide key information for land management stakeholders and planners to make long-term decisions about areas previously affected by PDCs.

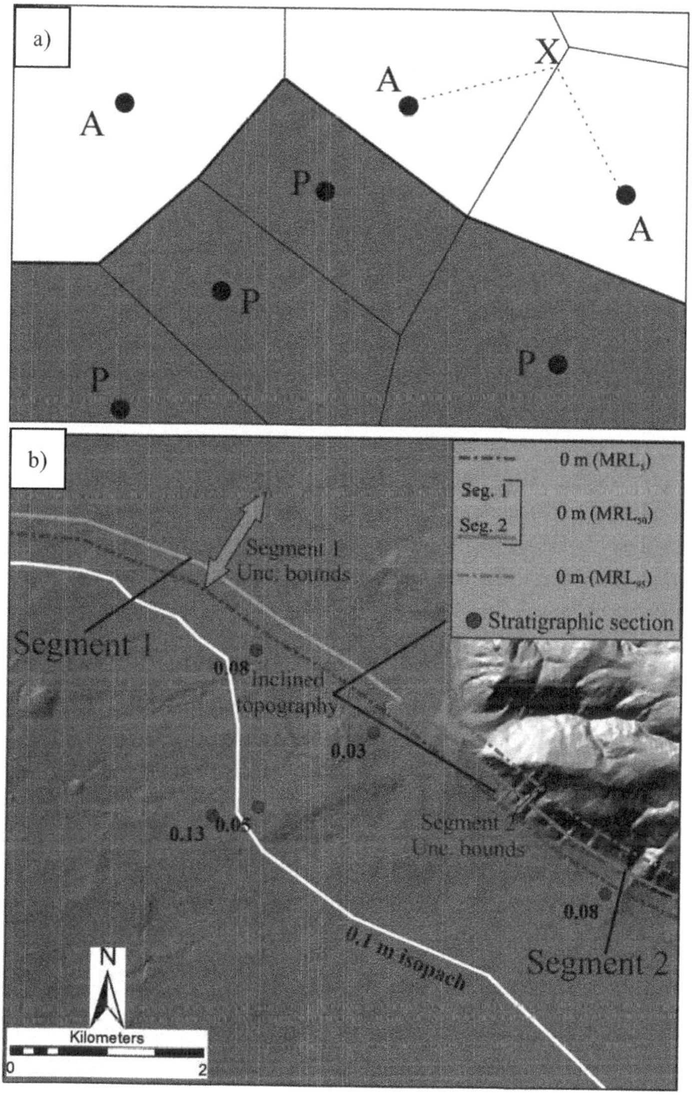

Figure 3.9. *a) Maximum runout defined by the Voronoi tessellation. b) Maximum propagation distance defined by the maximum runout lines (MRL). For a color version of this figure, see www.iste.co.uk/lenat/hazards3.zip*

COMMENT ON FIGURE 3.9.– *Point X is equidistant from the two sites where deposits are absent (A), and each of the two sites defines the connected white polygon, while the sites where deposition is present (P) define the gray polygons. The boundary between the gray and white polygons thus provides the maximum runout (modified from Burt et al. (2001)). In this example (b), two segments (1 and 2) are drawn for* MRL_{50}*, and different uncertainty limits (lower* MRL_5 *and upper* MRL_{95}*) are drawn depending on the number and quality of available data (modified from Cioni et al. (2020)).*

3.4.2.2.1. Post-eruption methods to determine the maximum runout/invaded area quotient

Farthest outcrop distance (FOD): The simplest method for inferring the maximum runout is to define the distance at which a deposit of a PDC is unambiguously identified. This measure (FOD) should actually be considered a minimum (De La Cruz-Reyna and Carrasco-Núñez 2002), or it can be integrated with qualitative indications as to the position of the maximum runout (Gurioli et al. 2010). In this regard, Gurioli et al. (2010) emphasized that the FOD and the maximum runout must coincide. However, it should be remembered that for older eruptions (whose distal deposits are poorly known) the maximum runout may be located further than the FOD, and for the case where the most distal deposits are characterized by fine ash beds (co-ignimbritic), the maximum runout is located somewhere between the FOD and these co-ignimbritic deposits.

Isoline drawn with an algorithm (Voronoi tessellation): An automatic process to obtain an isoline corresponding to the maximum runout of a PDC was proposed by Burt et al. (2001). These authors proposed to use the Voronoi tessellation method (cited above and in section 2.8.2.1), applied to a series of outcrops for which valid information on a PDC unit is available. Each of these outcrops can be associated with a Thiessen polygon, and the maximum runout of a unit can be plotted along the sides of the polygons, connected to outcrops that exhibit this PDC and those that do not (see Figure 3.9(a)).

Isoline drawn from field observations: Another way to plot the isoline corresponding to the maximum runout of a PDC is to analyze field evidences of PDC deposits. The simplest case is that of intact PDC deposits on which measurements can be made immediately after an eruption. In this favorable case, field campaigns and aerial photos or satellite images of high resolution are used to trace this boundary which thus provides the 0-thickness isoline of the maximum runout. This is not possible in the case of ancient deposits, where partial or total isolines are normally drawn based on qualitative observations of the deposits (Wilson 1985; Gurioli et al. 2005). In view of the epistemic uncertainty associated with the location of older deposits, uncertainty bounds must be associated with all maximum runout isolines. This is accomplished by placing a buffer around an isoline resulting from the "best estimate" (Neri et al. 2015), or by taking a more quantitative approach (Cioni et al. 2020). Particularly, these latter authors propose to position a maximum runout line (MRL_{50}) composed of different segments, based on field data and sedimentological characteristics of the deposits (position of the minimum observed deposit thickness, presence of topographic obstacles, etc.). The authors suggest adding an estimate of the uncertainty associated with the position of the MRL by assigning lower (MRL_5) and upper (MRL_{95}) uncertainty limits to each segment of the MRL based on the (statistical) confidence given to each segment (see Figure 3.9(b)).

3.4.2.3. *Volume*

The volume of a PDC unit (in m^3 or km^3) is a critical parameter that must be calculated in hazard assessments, with the dual purpose of reconstructing the eruptive history of a volcano and providing an input parameter for numerical models. Estimating the volume of PDC units presents greater challenges than in the case of fallout deposits, since PDC deposits generally do not show clear relationships between thickness decay and distance from the vent. This lack of relationship is primarily due to the role of interactions between the current and the underlying topography. For this reason, the volume of PDC units has been calculated only on rare occasions:

– by assigning the same thickness and slope values to homogeneous segments of the deposit (averaging values measured at individual outcrops within that segment), and calculating the volume of each segment, then summing them all (Crandell 1989; Cronin et al. 2013);

– by using techniques adapted or modified from those originally used for fallout deposits (Isaia et al. 2004; Scarpati et al. 2014).

In the first case, it should be noted that, in view of the high variability of PDC thicknesses, it is necessary to acquire a large number of measurements in order to homogenize the data over a study area, thus avoiding over- or underestimations. On the other hand, the methods used for the fallout deposits must take into account the fact that the thickness of PDC deposits does not decrease regularly with distance; therefore, methods based on the simplest postulates (Cioni et al. 2020) should be preferred. These are described in section 2.7.3.

3.4.2.4. *Total grain size distribution*

The total grain size distribution (TGSD, see Chapter 2, section 2.8.2.1) is an important parameter also for PDCs, as it can be used as an input parameter in numerical models and it allows us to deduce PDC dynamics, in particular transport and sedimentation mechanisms. Since the issues and methods are the same as those used for tephra fallout deposits, refer back to section 2.8.2.1.

In addition to the issues common to tephra fallout deposits, an additional complexity for PDC deposits is that in most cases TGSD is incomplete in fine fractions (<63 μm) due to the generation of co-ignimbritic plumes, which elutriate finer particles. Available TGSD calculations of PDCs from literature are therefore very sparse so far, and have only been performed for small currents (Lube et al. 2007) or, for larger PDC units, in cases where co-ignimbritic plume deposition could be recognized (Eychenne et al. 2012; Cioni et al. 2020).

3.4.2.5. *Mass/volumetric flow (or eruption/discharge) rate*

Mass flow rate (MFR in kg/s; sometimes referred to as mass eruption rate (MER) or mass discharge rate (MDR)) or volumetric flow rate (VFR in m^3/s) are equally important to be estimated for PDCs. MFR especially is a key input parameter for numerical models that simulate plume dynamics and PDCs, including multiphase ones (Esposti Ongaro et al. 2008). Moreover, MFRs are also critical to account for some PDCs with long runouts (Roche et al. 2009) and for the overtopping of confined PDC over valley banks, terraces and alluvial fans (Lube et al. 2011).

Note that, within a single eruption, MFR inferred from tephra fallout deposits (with reference to a sustained phase of the eruptive column, detailed in section 2.10.2) is generally less than the MFR that actually feeds PDC propagation (Bursik and Woods 1996). This is due to the different conditions in the phases of the sustained eruptive columns and PDC initiation (see section 2.2), which in most cases involves an increase in MFR from the first phase of the columns to the last, due to caldera collapse (Druitt and Sparks 1984; Cioni et al. 1992) or conduit erosion (Wilson et al. 1980). While MFRs can be estimated from data inversion techniques and numerical modeling (Dufek and Bergantz 2007; Engwell et al. 2016), the simplest approach is to estimate the average MFR by dividing the total mass of a deposit by the duration of the emplacement of a PDC. This estimate is obviously subject to several uncertainties:

– Uncertainties on the calculation of the total mass, which is usually obtained by multiplying the total volume (see sections 3.4.2.3 and 2.7.3) by an average value of deposit density.

– Uncertainties on the depositional times of PDCs, which are lower for those actually observed, but higher for those resulting from old eruptions.

However, it is crucial to estimate these MFRs in order to obtain first-order comparison values for MFRs calculated from numerical modeling. Recently, attempts have been made on PDC deposits, both observed in recent eruptions and from old eruptions:

– Roche et al. (2016) calculated MFRs and VFRs for the Peach Spring tuff (United States) PDC (~18.8 Ma). They first used the approach of Roche (2015) to calculate PDC velocity from entrained bedrock lithic clasts (see section 3.4.2.7), and then estimated the time of PDC initiation by relating velocity to the maximum distance reached by the PDC (measured from the FOD, see section 3.4.2.2). Finally, for the phase of the eruption that generated the PDC, they obtained (1) MFR by dividing the total mass of the deposit by the time of emplacement and (2) VFR by dividing the total mass by an average value of the deposit density (2,500 kg/m^3). The obtained flow rates are, respectively, $\sim 3.8 \times 10^7 - 1.5 \times 10^8$ m^3/s and $\sim 9.6 \times 10^{10} - 3.8 \times 10^{11}$ kg/s.

– Komorowski et al. (2013) calculated the MFR of the PDC phase of the 2010 eruption of Merapi volcano (Indonesia). The total volume and mass of the PDC deposit were calculated and then divided by the total duration of the sequence of explosions recorded by the seismic signals; the authors obtained a value of $\sim 10^7$ kg/s.

3.4.2.6. *Concentration of volcanic particles in PDCs*

The particle volumetric concentration in a PDC (usually named *C* and measured as volumetric percentage) is a parameter used to assess the associated risks, as even low concentrations can cause serious respiratory problems for the exposed population (Baxter et al. 1998). Despite its value, this parameter is not simple to measure, especially in syn-eruptive conditions. Part of this difficulty is also related to the stratified nature of PDCs, which implies that its properties vary with current height (Branney and Kokelaar 2002). Such stratification could imply different variations within the current from the two regime types mentioned above, namely:

– Turbulent, fully diluted currents, in which particle concentration varies abruptly between the current and the newly formed deposit, and then decreases rapidly with the height of the current itself (see Figure 3.10(a)).

– Dense, granular fluid-based currents, in which particle concentration decreases gradually from the base to the top, with no abrupt change in the transition zone from deposit to current (see Figure 3.10(b)).

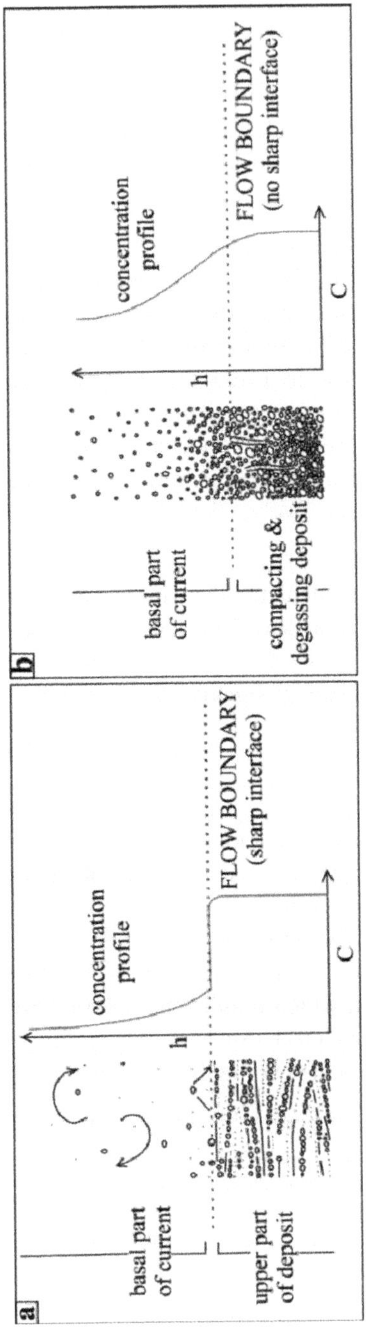

Figure 3.10. *Concentration profiles of a) a turbulent, fully diluted current and b) a dense, granular fluid-based current. Modified according to Branney and Kokelaar (2002)*

The basic conceptualization shown in Figure 3.10 was later refined with contributions from Sulpizio et al. (2014) and Dufek (2016). When the concentration of a PDC is tentatively calculated from deposit data, values are usually given as volume fractions (in %).

3.4.2.6.1. Post-eruptive methods for determining the concentration of volcanic particles in PDCs

Because of its variation with height in a PDC, particle concentration (C) is usually reported as a function of height, according to Branney and Kokelaar (2002):

$$C = C_0 \left[\frac{\eta_0(1-\eta)}{\eta(1-\eta_0)}\right]^{P_n} \quad [3.2]$$

where C_0 is the volumetric particle concentration at the non-dimensional height reference η_0, η is the non-dimensional height for which the concentration C is estimated and P_n is the particle Rouse number (Rouse 1937). Non-dimensional heights are the ratios of a certain height in the current to its total thickness, such that the top is equal to 1 and the bottom is equal to 0. η_0 is a reference height that is often set to 0.01 (Branney and Kokelaar 2002) with a reference concentration $C_0 = 75\%$ (Dellino et al. 2008). P_n (for diluted PDCs) is defined differently for each homogeneous clast population (i) of a PDC as:

$$P_{n_i} = \frac{w_i}{ku_*} \quad [3.3]$$

where w_i is the settling velocity (or terminal/fall velocity, see section 2.5.4) of the clast population i, $k \approx 0.4$ is the Von Kármán coefficient and u_* is the current shear velocity (see equation [3.10]). The term "population" here refers to a group of particles with the same sedimentation rate, regardless of their nature (juvenile or non-juvenile). In its formulation, the Rouse number indicates populations that can ($P_{n_i} < 2.5$) or cannot ($P_{n_i} > 2.5$) be maintained in suspension (Dellino et al. 2008). The total P_n number of a current is the weighted average of the P_{n_i} numbers of all clast populations, that is, according to Valentine (1987):

$$P_n = \frac{1}{\bar{S}} \sum_{i=1}^{n} \%_i \left(\frac{w_i}{ku_*}\right) \quad [3.4]$$

where $\bar{S}$ is the average volume concentration of all solid particles in a PDC and $\%_i$ is the weight/volume percentage (measured in the deposit) of clast population i at a specific stratigraphic height

Although the aforementioned equations are applicable to sustained and large-volume PDCs, other criteria must be considered for their more ephemeral and small-volume counterparts. In such cases, particle concentration must be described using the laws of granular material physics (Campbell 1990; Iverson and Vallance 2001).

3.4.2.7. *PDC velocity*

PDCs are phenomena characterized by high velocities (in m/s or km/h), sometimes reaching several hundreds of m/s (Roche et al. 2009). Estimating PDC velocity (noted as u in equation [3.4]) is therefore extremely useful for hazard assessment and can provide insight into its potential impacts. Due to the influence of topography exerted on PDCs, an averaged velocity does not apply to the entire deposit, but different velocity measurements should be made at as many locations as possible. In addition, velocity is a stratum-dependent property in most PDCs (similar to particle concentration), whereas many velocity measurements are often related only to the basal portion of the current. Two conceptual vertical velocity profiles have been used to describe a gradual variation in velocity profiles in stratified PDCs, based on Branney and Kokelaar (2002) (see Figure 3.11).

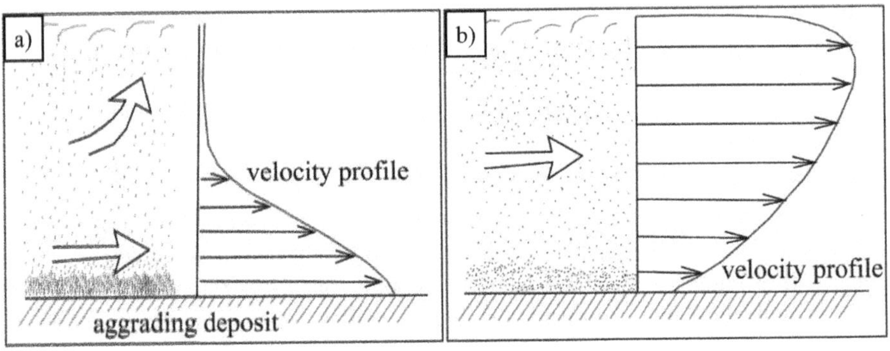

Figure 3.11. *Two conceptual vertical velocity profiles in PDCs with a lower dense current, a) faster and b) slower. Modified according to Branney and Kokelaar (2002)*

As pointed out by Branney and Kokelaar (2002), velocity profiles also vary with changes in slope, sedimentation rate, air ingestion and instability of the source supply.

3.4.2.7.1. Syn-eruptive methods to determine the PDC velocity

The velocity at the front of a PDC can usually be inferred from the combined analysis of thermal cameras and infrasound data (Delle Donne et al. 2014), the principles of which are described in section 3.4.2.1. Recently, Scharff et al. (2019) sought to measure particle velocities within a moving PDC at Colima volcano (Mexico) using Doppler radar (see Figure 3.12).

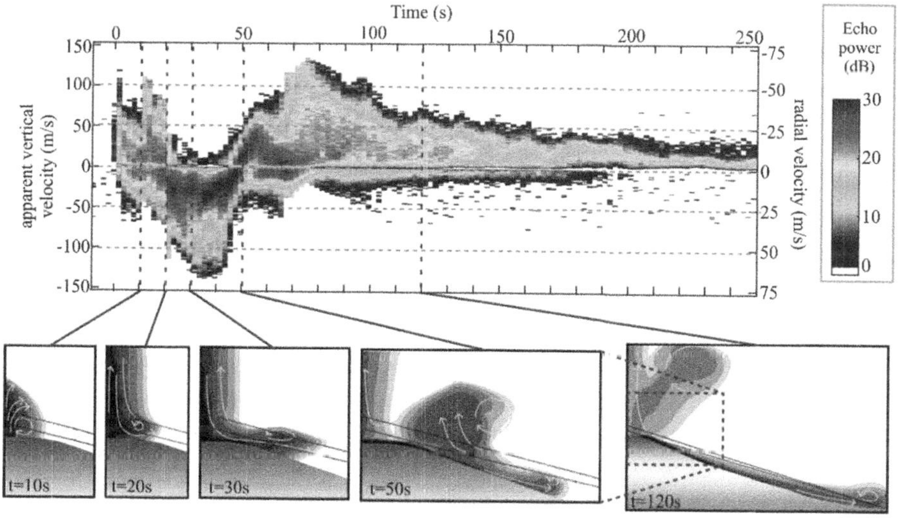

Figure 3.12. *Doppler radar data displayed as a velocigram (modified according to Scharff et al. (2019)). For a color version of this figure, see www.iste.co.uk/lenat/hazards3.zip*

COMMENT ON FIGURE 3.12.– *Each pixel shows the approximate amount of material (echo power in dB decibels, color-coded) moving at a certain speed (y-axis) and over time (x-axis). The echo power is in dB and relative to the noise level. For the apparent vertical velocity and the measured radial velocity, + and – indicate the velocities toward and away from the radar, respectively.*

3.4.2.7.2. Post-eruptive methods for determining PDC velocity

In the absence of direct syn-eruptive measurements of PDC velocity, it is necessary to extract this parameter from the deposits. Performing such calculations is not without difficulties. First, PDC stratification implies that measurements made at a particular location and height are related to the properties of the basal portion of the current as it passes through that location, and not to the PDC as a whole (Branney and Kokelaar 2002). Second, as pointed out by Sulpizio et al. (2014), there are only a few methods available for determining PDC velocity, some of which are not tested or validated by laboratory experiments or observations. Nevertheless, it is important to extract at least first-order values of PDC velocity (possibly with uncertainty margins) and compare them with experimental and simulated velocities from numerical models.

PDC velocities can be measured from field data collected after an eruption in two ways:

– Through the effects induced by PDC erosion, paleo-topography and macroscopic characteristics of the deposits.

– Through equations to be applied once the rheology of the PDC has been estimated from lithofacies analysis at several outcrops. Such equations can be used either to deduce the velocity of the basal part of the current (in the flow-boundary zone) or to obtain the total vertical velocity profile of a PDC.

– *Erosion effects, role of paleo-topography and macroscopic characteristics of deposits*: The small number of PDC velocity measurements made this way is typical of a single type of deposit and could not be applied to other deposits and/or their parent currents. A few examples are mentioned here for comparison:

- Kieffer and Sturtevant (1988) used the erosional groove geometry (see Figure 3.13(a)) of the blast, a directed lateral explosion from the 1980 eruption of Mount St. Helens (United States), to derive $u = 235$ m/s;

- Wilson (1985) used the heights of obstacles overcome by the Taupo (New Zealand) ignimbrite to estimate its velocity at various points with the equation relating potential energy to kinetic energy, and he extracted $u \geq 200$ m/s;

- Freundt and Schmincke (1986) used the cross-sectional surface slope of a current from Laacher See volcano (Germany) as it passed through a valley bend to derive $u = 10$–25 m/s (see Figure 3.13(b));

- Roche (2015) calculated PDC velocities resulting from accidental lithic particles entrainment for a variety of dilute and dense PDCs worldwide, with values up to ~100 m/s for dilute currents and ~25–30 m/s for dense currents.

Wilson's (1985) approach can be generalized by considering that if a PDC overcomes a morphological obstacle, its kinetic energy should be greater than the potential energy needed to overcome it:

$$\frac{mu^2}{2} = mgh \rightarrow u = \sqrt{2gh} \quad [3.5]$$

with m being the total mass of a PDC, g the gravitational acceleration, u the velocity of the PDC and h the height of the obstacle. This is an estimate of the minimum velocity of the current before it overcomes the morphological height, assuming that this current has a thickness less than the height of the obstacle and that the kinetic energy is completely converted into potential energy. Alternatively, the formula of Valentine (1987) can be used:

$$\left(\frac{1}{2}\rho_c u^2\right)_{y_{ds}} = g\int_{y_{ds}}^{y_m}(y_{ds} - y)\left(\frac{-d\rho_c}{dy}\right)dy \qquad [3.6]$$

where ρ_c is the current density (see equation [3.14]), y_m is the height of the morphological obstacle and y_{ds} is the height below which the particles cannot ascend the morphological obstacle.

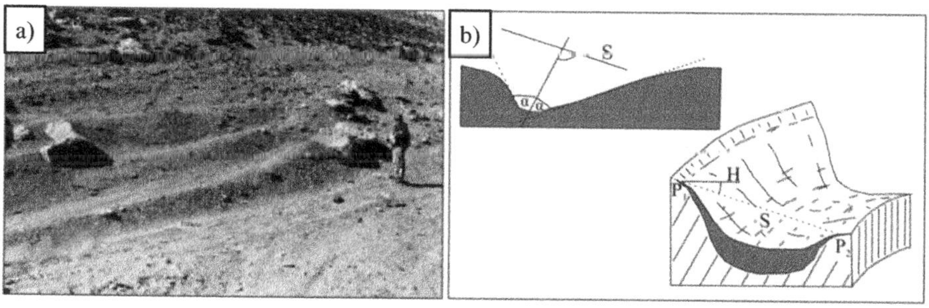

Figure 3.13. *a) Example of erosional furrows from Lascar volcano (modified according to Sparks et al. (1997)). b) Estimated asymmetric (upward) erosion channel PDC velocity and cross-section in a valley bend*

COMMENT ON FIGURE 3.13.– *S is the slope of the reconstructed current surface, which is perpendicular to the bisector between the two flanks of an asymmetric (upward) erosion channel, and connects the outer (P1) and inner (P2) perimeter of a valley curve. Modified according to Freundt and Schmincke (1986).*

Roche's (2015) approach, in contrast, starts from measurements of the maximum size (D in m) of entrained particles.

For dilute PDCs, the critical velocity profile required for the onset of motion of these particles is:

$$u_{(z)} = \frac{1}{k}\sqrt{\frac{\theta_a(\rho_p - \rho_c)gD}{\rho_c}}\, ln\left(\frac{z}{z_0}\right) \qquad [3.7]$$

where k is the Von Kármán coefficient (0.4), ρ_p is the density of the accidental lithic particles, ρ_c is the current density (see equation [3.14]), z is the height in the PDC where the velocity is estimated and z_0 is the hydrodynamic roughness considered equal to $D/30$. The parameter θ_a is the critical Shields number at the time the onset of motion occurs, which depends on the substrate slope α:

$$\theta_a = \theta_0 \left(cos\alpha + \frac{sin\alpha}{tan\Phi} \right) \quad [3.8]$$

where θ_0 is the critical Shields number for a horizontal bedrock (0.052–0.086) and Φ is the internal friction angle (30°).

For dense PDCs, it is instead necessary to measure the minor axis (C, in m) of the entrained particles and determine whether they look roughly like parallelepipeds or ellipsoids.

In both cases, the form factor is $\xi = 1$ or $\xi = 2/3$, respectively. The velocity at the dense PDC front is then calculated with:

$$u = \sqrt{\frac{\xi(\rho_p - \rho_f)gC}{\gamma\rho_c}} \quad [3.9]$$

where ρ_f is the density of the interstitial fluid (air) and $\gamma = 0.057$.

Equations to extract the basal velocity of a turbulent or fully diluted PDC: These types of currents are those that primarily deposit //sLT and xsLT lithofacies (see Table 3.1). For such cases, it can be applied the suspension criterion (Brissette and Lajoie 1990) which states that, in turbulent currents, the settling velocities of most particles (w_s, see section 2.5.4) are less than the shear rate of the current (u^* being the shear stress normalized by the density acting at the base of the current).

The first approach to calculate u^* is to extract the grain size distribution of a deposit (see section 2.5.1) and choose a grain size representative of the suspension criterion. The sedimentation rate of this grain size is then assumed to be equal to the shear rate. This representative grain size can be set equivalent to: (1) the coarsest grain size (Sparks 1983); (2) the 5th percentile of the cumulative grain size distribution (Lajoie et al. 1989); (3) the grain size at the break in slope, in the cumulative distribution, between the coarsest and midsize subpopulations (Middleton 1976); or (4) the average grain size (Brissette and Lajoie 1990).

The second approach to calculate shear rate is based on equations that relate u^* to shear stress (Wilcock 1996):

$$u^* = \sqrt{\frac{\tau_0}{\rho_g}} = \sqrt{\frac{\Delta\rho g h \sin\theta}{\rho_g}} \qquad [3.10]$$

where τ_0 is the shear stress exerted by the current, $\Delta\rho$ is the difference in density between the current (see equation [3.14]) and the atmosphere, ρ_g is the density of the gas phase and θ is the slope.

The shear rate can be considered equal to the current velocity. In the case where the actual current velocity must be determined, experimental relationships must be used to obtain u from u^* (Middleton and Southard 1984).

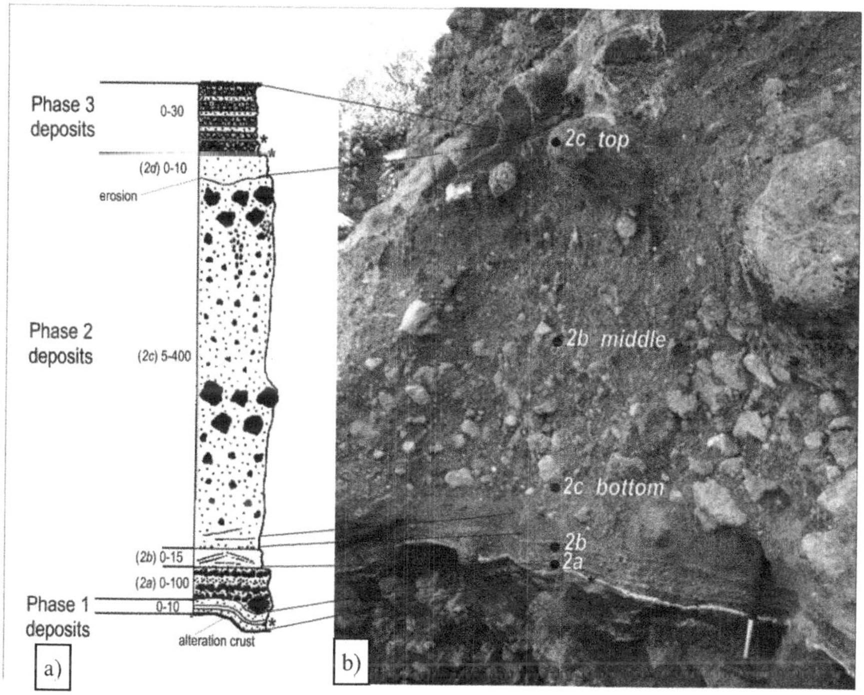

Figure 3.14. *a) Log (stratigraphic section) and b) photo of coarse-tail grading in the Breccia di Commenda deposits (Vulcano Island), modified from Gurioli et al. (2012)*

Equations to extract the velocity of the basal part of a dense or granular fluidized base PDC: These types of currents are those that mainly deposit mLT, bLT or dbLT lithofacies. In particular, the equations described below are most appropriate when mLT lithofacies are also characterized by coarse-tail gradings (with both juvenile and non-juvenile particles graded above a particular size threshold, but without the remaining matrix being graded, see Figure 3.14).

This type of PDC is currently very difficult to model and can be related to friction-based granular flows (Bursik et al. 2005) or Bingham-type pseudo-plastic flows (Kelfoun 2011). In the latter case, it is then necessary that a critical yield strength (τ_c) be overcome to initiate their motion.

Although there are different ways to calculate the yield strength, we report here what has been proposed for PDC units:

$$\tau_c = \frac{\Delta\rho_{mc} g r_{min}}{4} \quad \text{(Johnson 1970)} \qquad [3.11]$$

$$\tau_c = h_c \rho_c g \left(1 - \frac{h_{bl}}{h_c}\right) \sin\theta \quad \text{(Palladino and Valentine 1995)} \qquad [3.12]$$

where $\Delta\rho_{mc}$ is the density contrast between the smallest particle that has undergone grading and the surrounding matrix, r_{min} is the radius of the smallest particle that has undergone grading, h_c and ρ_c are the thickness and density of the current, respectively (see equation [3.14]), g is the gravitational acceleration, h_{bl} is the height of the basal part of the PDC (this should be set arbitrarily or based on the thickness of the deposit) and θ is the slope.

3.4.2.8. *Dynamic pressure*

Dynamic pressure (measured in Pa) is a measure of the lateral stress that a PDC exerts on natural features or structures such as buildings (see Figure 3.15). Dynamic pressure is important since it helps in quantifying the potential damage caused by a PDC, making it a key element in all physical vulnerability and risk studies (Zuccaro et al. 2008). Given the powerful destructive potential of PDCs, dynamic pressure cannot be measured during the eruption. We will therefore describe methods that focus on estimating the dynamic pressure of PDCs from field evidences, discarding calculations based on numerical models (Valentine and Wohletz 1989; Esposti Ongaro et al. 2002).

Figure 3.15. *Walls having been knocked down by the most energetic PDCs of the 79 CE eruption of Somma-Vesuvius volcano*

COMMENT ON FIGURE 3.15.– *Sites within the archeological excavations of Pompeii a) Villa dei Casti Amanti and b) Via Castricio. The white arrow indicates PDC direction (modified from Gurioli et al. (2007)).*

3.4.2.8.1. Post-eruptive methods to determine dynamic pressure

Dynamic pressure can be estimated using simple equations applied to field data, for example from fallen trees and poles, damage to buildings or by taking into account the entrainment of large lithic boulders in deposits.

– *Equations*: Dynamic pressure is defined, according to Valentine (1998), as:

$$P_{dyn} = \frac{1}{2}\rho_c u^2 \qquad [3.13]$$

where ρ_c is the density of the current and u is its velocity (see section 3.4.2.7.2). The density of the current can be defined by:

$$\rho_c = \rho_g(1 - C) + \rho_s C \qquad [3.14]$$

where ρ_g is the density of the gas phase (which can be considered equal to either that of air, pure gas, or a mixture of gas and ash), C is the particle concentration of the PDC (see section 3.4.2.6) and ρ_s is the density of the solid phase (see section 2.5.5). We recall that, as particle concentration varies within a current (and within a deposit), dynamic pressure varies as well within them.

– *Effects of PDCs on trees and poles*: This method, proposed by Clarke and Voight (2000), determines the dynamic pressure required to cause the observed damage to cylindrical elements (trees and poles). The authors consider the dynamic pressure to be uniform within the current or influenced by the stratification of a PDC. Since the results of the two calculations are very similar, we indicate here the procedure for the calculation of a uniform dynamic pressure, and we refer the interested reader to the cited article for the details applied to the case of a stratified dynamic pressure.

In this method, the dynamic pressure is linked to the bending moment (*M*, the bending exerted on an element by an external force), considering the damaged elements as cylinders:

$$P_{dyn} = \frac{M}{C_D r h^2} \qquad [3.15]$$

where r is the radius of the cylinder, h is the height above the breakpoint of the object and C_D is the drag coefficient (see section 2.5.4). The bending moment at the breaking point M_f can be calculated as follows:

$$M_f = \frac{I\sigma_f}{r} \qquad [3.16]$$

where σ_f is the ultimate failure strength and I is the area moment of inertia of the horizontal cross-section. This latter is a geometric property of a shape describing the distribution of points about an axis. It is a mechanical measure of the resistance of a body to bending moment. This moment of inertia is deduced from experimental data on similar materials (Beer and Johnston 1981). As a reference, Clarke and Voight (2000) used 50 MPa, the maximum breaking strength of Douglas fir. The area moment of inertia is calculated differently in the case where the cylinder is considered solid or hollow:

$$I = \frac{1}{4}\pi r^4 \quad \text{(solid cylinder)} \qquad [3.17]$$

$$I = \frac{1}{4}\pi ({r_o}^4 - {r_i}^4) \quad \text{(hollow cylinder)} \qquad [3.18]$$

where r_o and r_i are the outer and inner radii of the cylinder, respectively.

– *Damage to structures*: Baxter et al. (2005) determined dynamic pressure values taking into account the damage to structures on the island of Montserrat in the Lesser Antilles as a result of the 1997 Soufrière Hills volcano eruption. This analysis allowed for deriving a damage scale for the buildings, with each level characterized by an approximate range of dynamic pressure, derived from theoretical considerations and strength testing of building materials.

– Jenkins et al. (2013) conducted a similar but more focused analysis on several parameters of damaged building walls around Merapi volcano that allowed them to derive the following formula:

$$P_{dyn} = \left[\frac{\frac{2\mu+2\beta^2}{\alpha^2}}{3\beta-2\beta^2}\right] f_n \left(\frac{t}{w}\right)^2 \quad [3.19]$$

where μ is the ratio of vertical to horizontal modulus of rupture (2), β is a constant ($0 < \beta < 0.5$) giving the location of the point where the fracture line intersects the upper free edge of the wall (β is varied to minimize lateral pressure), α is the ratio of the height of the wall to its width, f_n is the modulus of rupture of the brickwork on the horizontal joint of the bed (0.15–0.45 N/mm^2) and t and w are the thickness and width of the wall, respectively.

– *Entrainment of large lithic blocks*: The work of Pittari et al. (2007) aimed at calculating the dynamic pressure required to move a series of large lithic boulders from the Abrigo ignimbrite deposits (Tenerife, Canary Islands). The basic concept is that a lithic boulder is entrained when the minimum applied force required to move the boulder (F) is equivalent to the drag force (D) exerted by a PDC:

$$F = \mu_s mg \cos\theta - mg \sin\theta \quad [3.20]$$

$$D = P_{dyn} A C_D \quad [3.21]$$

where μ_s is the static friction coefficient between the boulder and the substrate, m is the mass of the boulder, g is the gravitational acceleration, θ is the slope, A is the cross-sectional area of the boulder and C_D is the drag coefficient (see section 2.5.7). Imposing $F = D$ and solving for P_{dyn} leads to:

$$P_{dyn} = \frac{\mu_s mg \cos\theta - mg \sin\theta}{AC_D} \quad [3.22]$$

Pittari et al. (2007) experimentally determined that $\mu_s = 0.675$ for the studied boulders, whose shapes were approximately simple prisms or ellipsoids, using a theoretical value of $C_D = 0.8$.

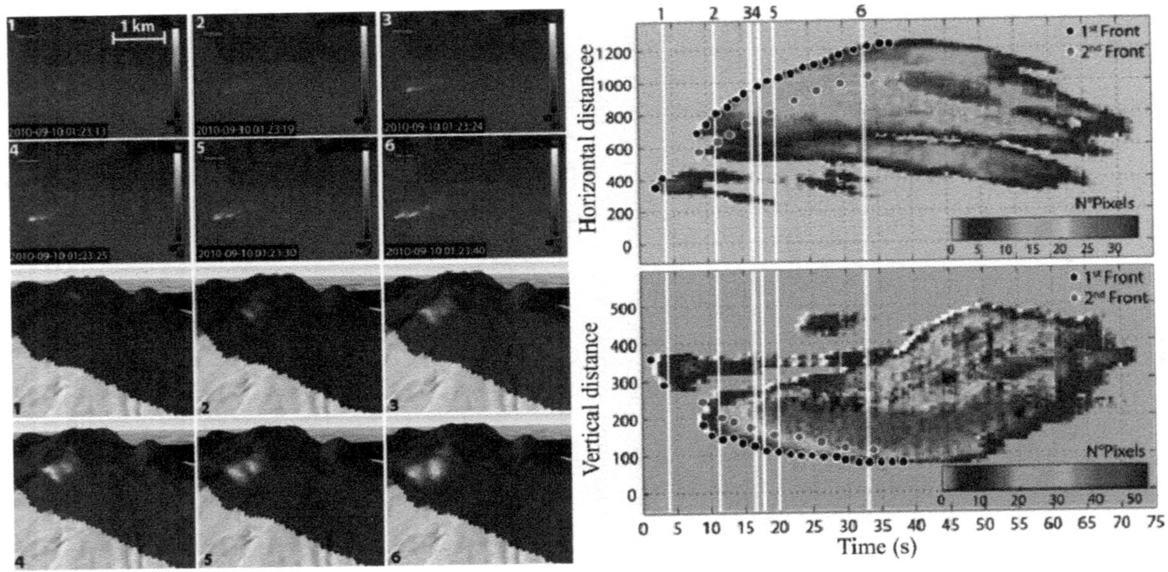

Figure 3.16. *a) Six screenshots associated with the September 10, 2010 PDC at Soufrière Hills volcano (Montserrat island, Lesser Antilles). PDCs were flowing along with the collapse scar produced by the February 11, 2010 partial dome collapse detected by the FLIR thermal camera and associated 3D views. b) Distribution of associated temperatures along (up) the horizontal (Hd) and (down) the vertical (Vd) thermal image axes, plotted against time. For a color version of this figure, see www.iste.co.uk/lenat/hazards3.zip*

COMMENT ON FIGURE 3.16.– *This analysis allows tracking of PDC thermal fronts (black dots) along vertical and horizontal directions, as well as calculating its propagation velocities. The white vertical lines refer to the thermal screenshots displayed on the left (modified from Delle Donne et al. (2014)).*

3.4.2.9. *Temperature*

Field evidences show that PDCs can be emplaced at any temperature (in °C or K) between magma and atmosphere, and that there is no systematic relationship between their temperature, mobility and volume. One of the main hazards associated with PDCs is their high temperature, which can inflict serious injury to the human body (Baxter et al. 1998) or generate secondary effects such as fire ignition (Cole et al. 2015).

3.4.2.9.1. Syn-eruptive methods to determine temperatures

PDC front temperature can be inferred, in syn-eruptive conditions, from thermal camera analysis (see Figure 3.16; Delle Donne et al. 2014), the principles of which are described in section 3.4.2.1. Similarly, infrared radar imagery from satellites (LANDSAT, MODIS or ASTER) can provide real-time measurements of PDCs (Blackett 2017).

3.4.2.9.2. Methods of determining temperatures after the eruption

Thermal effects on items: The simplest approach for obtaining the temperature of a PDC is to evaluate the effects of heat on materials (in the case where a PDC passes through a populated area). This is the case, for example, when partially or completely melted plastic objects are involved (Voight and Davis 2000; Charbonnier and Gertisser 2008) and for which their melting temperature is known. In particular, if charred wood pieces are present in a PDC deposit, the charcoal reflectance method of Scott and Glasspool (2005) can be used to obtain a minimum temperature value. The assumption of this method is that the level of charring of a log depends on the charring temperature, duration and burial mode to which the log is subjected (Pensa et al. 2015), which in turn affect the degree of reflectance of the log. This degree can be measured using optical microscopes in conjunction with known reflectance coefficients of other minerals.

Petrological evidences: The initial temperature T of a mixture resulting from a column or dome collapse can be obtained by the heat conservation equation between the erupted material at temperature T_0 and the ingested air with temperature T_a. If water is also involved, then it is necessary to introduce the specific heat of water vapor (Koyaguchi and Woods 1996). Assuming that the mixture between a PDC and the surrounding air is at atmospheric pressure, it follows that:

$$T = \frac{(1-\lambda)c_p T_0 + \lambda c_a T_a}{(1-\lambda)c_p + \lambda c_a} \qquad [3.23]$$

where λ is the mass (or volume) fraction of ingested air, c_p is the specific heat of the magma-volatile mixture involved during the eruption and c_a is the specific heat of air at ambient temperature. The mass (or volume) fraction of ingested air can be estimated by assuming an initial gas mass (or volume) fraction n_0 and calculating (based on petrological studies) the initial magma volatile content n_m:

$$\lambda = \frac{n_0 - n_m}{1 - n_m} \qquad [3.24]$$

Thermal remnant magnetization (TRM): If a hot PDC contains minerals with high iron content (hematite, magnetite), it is possible (under certain conditions) to obtain an estimate of the emplacement temperature (T_e) of a PDC. In general, when a hot iron-rich mineral cools in an external magnetic field (the Earth's magnetic field), it acquires a TRM, which retains the original orientation of the external magnetic field. TRM is a vector whose direction is parallel and whose intensity is directly proportional to the external magnetic field. TRM provides a method for quantitative determination of the emplacement temperature of particles in a volcanic deposit (Paterson et al. 2010). The method for estimating T_e of a PDC deposit is based on the following assumptions:

– the cold lithic blocks are incorporated and heated by the moving PDC, to be finally incorporated in the matrix of the deposit. They then cool down with the entire deposit. The process of re-heating and cooling to ambient temperature results in magnetic overprint;

– for each magnetic lithic block, there is a "blocking temperature" T_b (correlated to the size, type and shape of the grains) below which magnetization is effectively blocked and retained. T_b can be either the Curie temperature (580°C and 680°C for magnetite and hematite, respectively) or the ambient temperature, and it depends on the properties of each block (Cioni et al. 2004; Zanella et al. 2007; Benage et al. 2016);

– the ferromagnetic grains of a rock acquire a stable TRM at different temperature values, which depends on the mineralogy, size and shape of each grain;

– the TRM of grains with different T_b is mutually independent and add up to the total TRM of the rock.

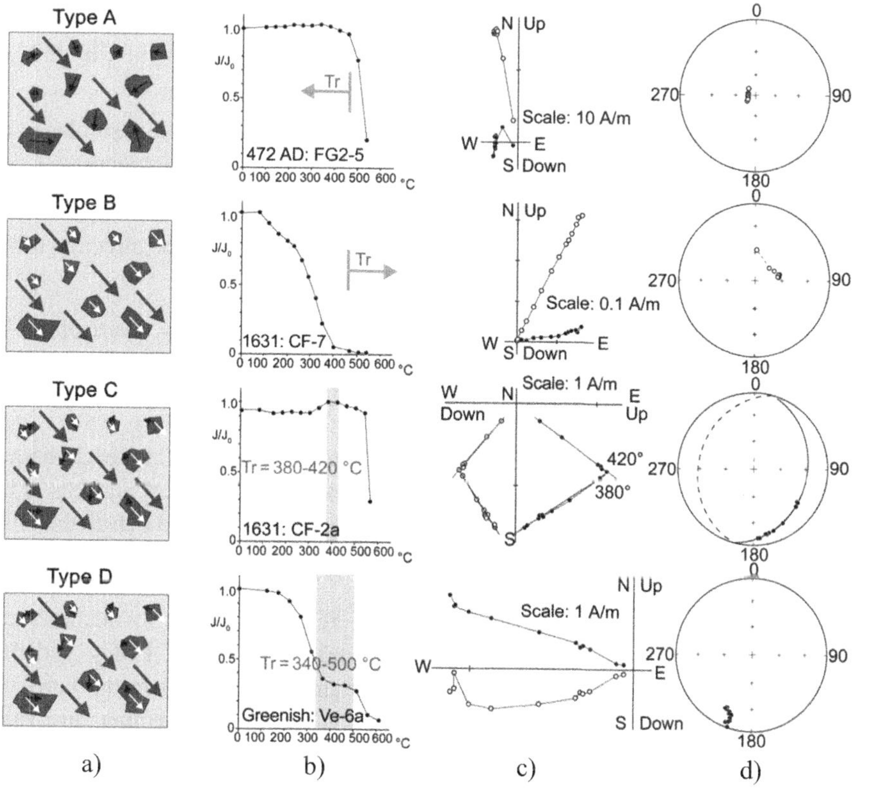

Figure 3.17. *Grain behavior during stepwise thermal demagnetization*

COMMENT ON FIGURE 3.17.– *Columns: a) grain type A, B, C, D (Cioni et al. 2004). Arrows: red = TRM of the fine-grained matrix, coincident with the ambient magnetic field at the time of eruption; blue = primary TRM acquired during grain formation; white = secondary TRM acquired during grain cooling in the deposit. b) Normalized intensity decay curve. Gray bar indicates the range of heating. c) Zijderveld (1967) diagram: solid/open point = declination/apparent inclination, numbers = T value (°C). d) Equal area projection: solid/open points = lower/upper hemisphere (modified from Zanella et al. (2015)).*

Magnetic overprinting depends on the relationship between the T_b spectrum of each lithic clast and the heating temperature T_r, which can be investigated by stepwise thermal demagnetization. We distinguish four types of grains on the basis of their TRM properties (see Figure 3.17).

– Type A grains have a single TRM component whose direction is scattered from grain to grain, and different from the paleo-magnetic direction at the time of eruption, which is recorded by the deposit matrix ($T_r <$ lowest T_b).

– Type B grains have a single TRM component whose direction is similar in all grains and close to the direction recorded by the deposit matrix ($T_r >$ highest T_b).

– Type C grains have two TRM components with different directions and T_b spectra. The direction of the high T_b component is occasional and the direction of the low T_b component is close to the paleo-magnetic direction.

– Type D grains have two TRM components with partially overlapping T_b spectra. They are similar to type C, but the two components are not clearly separated and, therefore, the T_r value can only be defined in the overlapping interval.

Given the time required for the individual constituents of a PDC to reach thermal equilibrium, it is more appropriate to interpret the results of the TRM analysis as the temperature reached by the entire deposit after its emplacement and before the onset of intensive cooling ($T_{dep} \neq T_e$) (Cioni et al. 2004).

Furthermore, the heating of fragments of a few millimeters in diameter can be considered synchronous with the transport and deposition of PDCs, and thus $T_r \approx T_{dep}$, whereas coarser grains need a longer time to heat: when the latter are heated, the whole deposit cools and equilibrium is reached at a temperature value lower than T_{dep}, implying that $T_r < T_{dep}$. Given these considerations, the best approach to assessing T_{dep} is to look for mutual consistency of experimental values of T_r at the scale of the sample site. Analysis of the plots (such as those in Figure 3.18) gives the range of T_r for each individual fragment, and the overlap of the ranges gives a reasonable estimate of T_{dep}.

To perform TRM measurements, several samples are taken manually in the field (after having indicated on a flat surface the direction and inclination of the sample) and/or by mechanical drilling in situ. A progressive thermal demagnetization (with an increase in temperature up to Curie temperature) is then performed in laboratory, which makes it possible to progressively isolate the most stable component of the magnetization (carried by the grains with the highest T_b). At each step, the intensity and direction of the magnetization vector of the sample are measured with a Spinner magnetometer.

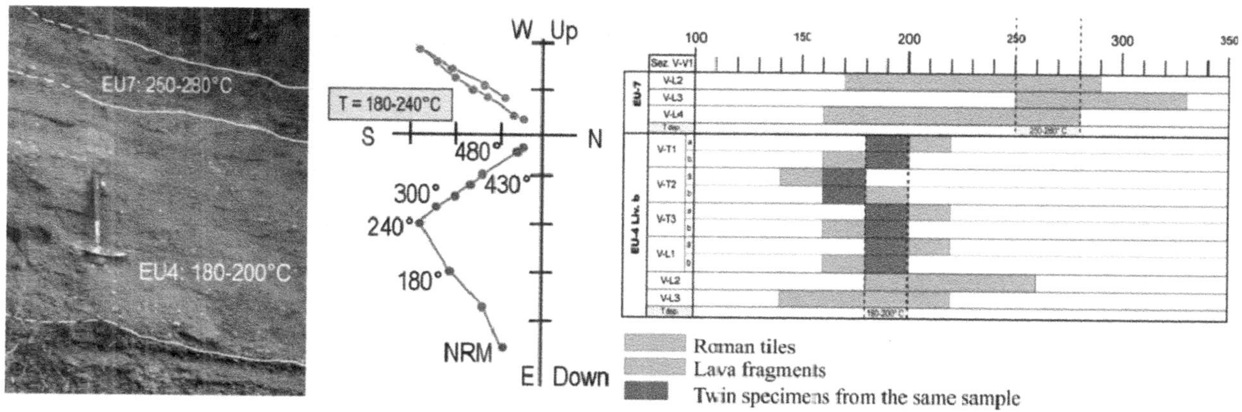

Figure 3.18. *Procedure for determining the T_{dep} for a PDC deposit. For a color version of this figure, see www.iste.co.uk/lenat/hazards3.zip*

COMMENT ON FIGURE 3.18.– *a) Two PDC units, in the archeological site of Pompeii. For each unit, several Roman tiles and lava fragments were collected and measured to obtain the* T_{dep} *of each fragment. b) An orthogonal Zijderveld projection diagram simultaneously displays the intensity and direction of the magnetization vector relative to each demagnetization step. The intersection of the two lines (a first component, passing through the origin, at high temperature, and a second component at low temperature) indicates the range of* T_{dep}. *c) The maximum overlap of each range of* T_{dep} *obtained from Roman tiles, lava fragments and twin samples from the same samples yields the final T of the entire deposit at this site.*

3.5. Perspectives

This chapter illustrates the data that can be collected from syn- and post-eruptive measurements on PDCs and their deposits, in order to extract useful information for hazard-related assessment.

Regarding syn-eruptive measurements, we are still at an early stage where new techniques are available to provide real-time measurements on PDC parameters. However, many technical difficulties remain for the application of these methods, which are mainly related to the destructive power and mobility of these currents.

In contrast, post-eruptive measurements on deposits have traditionally benefited from a longer research history. In addition, many advances have been made in our understanding of PDC dynamics in recent years, most of which have been achieved through numerical modeling (Roche et al. 2009) and large-scale analog experiments (Lube et al. 2015). Numerical modeling has also proven useful in the context of volcanic risk assessment, both for evaluating PDC invasion areas (Neri et al. 2015; Aravena et al. 2020) and quantifying some of the parameters described here, such as dynamic pressure or temperature (Esposti Ongaro et al. 2008). Despite these advances, field data analysis remains crucial because in many cases input parameters for numerical models and large-scale laboratory experiments depend on these data. It is therefore important to improve our knowledge of some key parameters that are still largely uncertain, such as current volumes or the TGSD of PDCs. For the former, the close link between PDCs and the underlying topography prevents the development of a generalized thickness decay rule with runout distance, like those developed for tephra fallout deposits. A similar problem also affects TGSD estimates: we are still far from a complete characterization of this parameter for PDC deposits, which has so far been restricted to small lobes or a few deposit units with specific characteristics. More field data are therefore needed to narrow this gap and provide new tools for assessing PDC-related hazards.

3.6. References

Andronico, D., Di Roberto, A., De Beni, E., Behncke, B., Bertagnini, A., Del Carlo, P., Pompilio, M. (2018). Pyroclastic density currents at Etna volcano, Italy: The 11 February 2014 case study. *J. Volcanol. Geotherm. Res.*, 357, 92–105.

Aravena, A., Cioni, R., Bevilacqua, A., de Michieli Vitturi, M., Esposti Ongaro, T., Neri, A. (2020). Tree-branching-based enhancement of kinetic energy models for reproducing channelization processes of pyroclastic density currents. *J. Geophys. Res. Solid Earth*, 125(7), e2019JB019271.

Bardintzeff, J.M. (1984). Merapi volcano (Java, Indonesia) and Merapi-type Nuée ardente. *Bull. Volcanol.*, 47, 433–446.

Baxter, P.J., Neri, A., Todesco, M. (1998). Physical modelling and human survival in pyroclastic flows. *Nat. Hazards*, 17, 163–176.

Baxter, P.J., Boyle, R., Cole, P.D., Neri, A., Spence, R., Zuccaro, G. (2005). The impacts of pyroclastic surges on buildings of the Soufriére Hills, volcano, Monsterrat. *Bull. Volcanol.*, 67(4), 292–313.

Benage, M.C., Dufek, J., Mothes, P.A. (2016). Quantifying entrainment in pyroclastic density currents from the Tungurahua eruption, Ecuador: Integrating field proxies with numerical simulations. *Geophys. Res. Lett.*, 43, 6932–6941.

Bernard, J. and Le Pennec, J.L. (2016). The milling factory: Componentry-dependent fragmentation and fines production in pyroclastic flows. *Geology*, 44, 907–910.

Blackett, M. (2017). An overview of infrared remote sensing of volcanic activity. *J. Imaging*, 3, 13.

Brand, B.D., Mackaman-Lofland, C., Pollock, N.M., Bendaña, S., Dawson, B., Wichgers, P. (2014). Dynamics of pyroclastic density currents: Conditions that promote substrate erosion and self-channelization – Mount St. Helens, Washington (USA). *J. Volcanol. Geotherm. Res.*, 276, 189–214.

Branney, M.J. and Kokelaar, P. (2002). Pyroclastic density currents and the sedimentation of ignimbrites. *Geol. Soc. Mem.*, 27, 143.

Brissette, F.P. and Lajoie, J. (1990). Depositional mechanics of turbulent nuées ardentes (surges) from their grain sizes. *Bull. Volcanol.*, 53, 60–66.

Brown, R.J. and Andrews, G.D.M. (2015) Deposits of pyroclastic density currents. In *The Encyclopedia of Volcanoes*, 2nd edition, Sigurdsson, H. (ed.). Academic Press, Amsterdam.

Bursik, M.I. and Woods, A.W. (1996). The dynamics and thermodynamics of large ash flows. *Bull. Volcanol.*, 58, 175–193.

Burt, M.L., Wadge, G., Curnow, R.N. (2001). An objective method for mapping hazardous flow deposits from the stratigraphic record of stratovolcanoes. A case example from Montagne Pelée. *Bull. Volcanol.*, 63, 98–111.

Bursik, M.I., Patra, A.K., Pitman, E.B., Nichita, C., Macias, J.L., Saucedo, R., Girina, O. (2005). Advances in studies of dense volcanic granular flows. *Reports on Progress in Physics*, 68, 271.

Campbell, C.S. (1990). Rapid granular flows. *Annu. Rev. Fluid Mech.*, 22, 57–90.

Cañón-Tapia, E. and Mendoza-Borunda, R. (2014). Magnetic petrofabric of igneous rocks: Lessons from pyroclastic density current deposits and obsidians. *J. Volcanol. Geotherm. Res.*, 289, 151–169.

Cas, R.A.F. and Wright, J.V. (1987). *Volcanic Successions Modern and Ancient: A Geological Approach to Processes, Products and Successions.* Allen and Unwin, St. Leonards.

Caulfield, J.T., Cronin, S.J., Turner, S.P., Cooper, L.B. (2011). Mafic Plinian volcanism and ignimbrite emplacement at Tofua volcano, Tonga. *Bull. Volcanol.*, 73, 1259–1277.

Charbonnier, S.J. and Gertisser, R. (2008). Field observations and surface characteristics of pristine block-and-ash flow deposits from the 2006 eruption of Merapi Volcano, Java, Indonesia. *J. Volcanol. Geotherm. Res.*, 177, 971–982.

Cioni, R., Marianelli, P., Sbrana, A. (1992). Dynamics of the A.D. 79 eruption: Stratigraphic, sedimentological and geochemical data on the successions from the Somma-Vesuvius southern and eastern sectors. *Acta Vulcanol.*, 2, 109–123.

Cioni, R., Gurioli, L., Lanza, R., Zanella, E. (2004). Temperatures of the A.D. 79 pyroclastic density current deposits (Vesuvius, Italy). *J. Geophys. Res. Solid Earth*, 109(B2), February.

Cioni, R., Tadini, A., Gurioli, L., Bertagnini, A., Mulas, M., Bevilacqua, A., Neri, A. (2020). Estimating eruptive parameters and related uncertainties for pyroclastic density currents deposits: Worked examples from Somma-Vesuvius (Italy). *Bull. Volcanol.*, 82.

Clarke, A.B. and Voight, B. (2000). Pyroclastic current dynamic pressure from aerodynamics of tree or pole blow-down. *J. Volcanol. Geotherm. Res.*, 100, 395–412.

Cole, P.D., Calder, E.S., Sparks, R.S.J., Clarke, A.B., Druitt, T.H., Young, S.R., Herd, R.A., Harford, C.L., Norton, G.E. (2002). Deposits from dome-collapse and fountain-collapse pyroclastic flows at Soufrière Hills Volcano, Montserrat. *Geological Society Memoirs*, 21, 231–262.

Cole, P.D., Neri, A., Baxter, P.J. (2015). Hazards from pyroclastic density currents. In *The Encyclopedia of Volcanoes*, 2nd edition, Sigurdsson, H. (ed.). Academic Press, Amsterdam. http://www.sciencedirect.com/science/article/pii/B9780123859389000547.

Costantini, L., Pioli, L., Bonadonna, C., Clavero, J., Longchamp, C. (2011). A Late Holocene explosive mafic eruption of Villarrica volcano, Southern Andes: The Chaimilla deposit. *J. Volcanol. Geotherm. Res.*, 200, 143–158.

Crandell, D.R. (1989). Gigantic debris avalanche of Pleistocene age from ancestral Mount Shasta volcano, California, and debris-avalanche hazard zonation. *US Geol. Surv. Bull.*, 1861, 1–32.

Cronin, S.J., Lube, G., Dayudi, D.S., Sumarti, S., Subrandiyo, S., Surono (2013). Insights into the October–November 2010 Gunung Merapi eruption (Central Java, Indonesia) from the stratigraphy, volume and characteristics of its pyroclastic deposits. *J. Volcanol. Geotherm. Res.*, 261, 244–259.

De La Cruz-Reyna, S. and Carrasco-Núñez, G. (2002). Probabilistic hazard analysis of Citlaltépetl (Pico de Orizaba) Volcano, eastern Mexican Volcanic Belt. *J. Volcanol. Geotherm. Res.*, 113, 307–318.

Delle Donne, D., Ripepe, M., De Angelis, S., Cole, P.D., Lacanna, G., Poggi, P., Stewart, R. (2014). Thermal, acoustic and seismic signals from pyroclastic density currents and Vulcanian explosions at Soufrière Hills Volcano, Montserrat. *Geol. Soc. London, Mem.*, 39, 169–178.

Dellino, P., Mele, D., Sulpizio, R., La Volpe, L., Braia, G. (2008). A method for the calculation of the impact parameters of dilute pyroclastic density currents based on deposit particle characteristics. *J. Geophys. Res. Solid Earth*, 113(B7), July.

Doronzo, D.M. (2012). Two new end members of pyroclastic density currents: Forced convection-dominated and inertia-dominated. *J. Volcanol. Geotherm. Res.*, 219–220, 87–91.

Druitt, T.H. (1992). Emplacement of the 18 May 1980 lateral blast deposit ENE of Mount St. Helens, Washington. *Bull. Volcanol.*, 54, 554–572.

Druitt, T.H. and Sparks, R.S.J. (1984). On the formation of calderas during ignimbrite eruptions. *Nature*, 310, 679–681.

Dufek, J. (2016). The fluid mechanics of pyroclastic density currents. *Annu. Rev. Fluid Mech.*, 48, 459–485.

Dufek, J. and Bergantz, G.W. (2007). Dynamics and deposits generated by the Kos Plateau Tuff eruption: Controls of basal particle loss on pyroclastic flow transport. *Geochemistry, Geophys. Geosystems*, 8(12), December.

Engwell, S.L., de' Michieli Vitturi, M., Esposti Ongaro, T., Neri, A. (2016). Insights into the formation and dynamics of coignimbrite plumes from one-dimensional models. *J. Geophys. Res. Solid Earth*, 121, 4211–4231.

Esposti Ongaro, T., Neri, A., Todesco, M., Macedonio, G. (2002). Pyroclastic flow hazard assessment at Vesuvius (Italy) by using numerical modelling II. Analysis of flow variables. *Bull. Volcanol.*, 64, 178–191.

Esposti Ongaro, T., Neri, A., Menconi, G., de'Michieli Vitturi, M., Marianelli, P., Cavazzoni, C., Erbacci, G., Baxter, P.J. (2008). Transient 3D numerical simulations of column collapse and pyroclastic density current scenarios at Vesuvius. *J. Volcanol. Geotherm. Res.*, 178, 378–396.

Eychenne, J., Le Pennec, J.L., Troncoso, L., Gouhier, M., Nedelec, J.M. (2012). Causes and consequences of bimodal grain-size distribution of tephra fall deposited during the August 2006 Tungurahua eruption (Ecuador). *Bull. Volcanol.*, 74, 187–205.

Fisher, R.V. (1990). Transport and deposition of a pyroclastic surge across an area of high relief: The 18 May 1980 eruption of Mount St. Helens, Washington. *Geol. Soc. Am. Bull.*, 102, 1038–1054.

Freundt, A. and Schmincke, H.U. (1986). Emplacement of small-volume pyroclastic flows at Laacher See (East-Eifel, Germany). *Bulletin of Volcanology*, 48, 39–59. https://doi.org/10.1007/BF01073512.

Gurioli, L., Cioni, R., Sbrana, A., Zanella, E. (2002). Transport and deposition of pyroclastic density currents over an inhabited area: The deposits of the AD 79 eruption of Vesuvius at Herculaneum, Italy. *Sedimentology*, 49, 929–953.

Gurioli, L., Houghton, B.F., Cashman, K.V., Cioni, R. (2005). Complex changes in eruption dynamics during the 79 AD eruption of Vesuvius. *Bull. Volcanol.*, 67, 144–159.

Gurioli, L., Zanella, E., Pareschi, M.T., Lanza, R. (2007). Influences of urban fabric on pyroclastic density currents at Pompeii (Italy) 1: Flow direction and deposition. *J. Geophys. Res. Solid Earth*, 112(B5), May.

Gurioli, L., Sulpizio, R., Cioni, R., Sbrana, A., Santacroce, R., Luperini, W., Andronico, D. (2010). Pyroclastic flow hazard assessment at Somma–Vesuvius based on the geological record. *Bull. Volcanol.*, 72, 1021–1038.

Hoblitt, R.P. (1985). Observations of the eruptions of July 22 and August 7, 1980, at Mount St. Helens, Washington (USA). Study document 1335, USGS, Reston.

Hughes, S.R. and Druitt, T.H. (1998). Particle fabric in a small, type-2 ignimbrite flow unit (Laacher See, Germany) and implications for emplacement dynamics. *Bull. Volcanol.*, 60, 125–136.

Isaia, R., D'Antonio, M., Dell'Erba, F., Di Vito, M., Orsi, G. (2004). The Astroni volcano: The only example of closely spaced eruptions in the same vent area during the recent history of the Campi Flegrei caldera (Italy). *J. Volcanol. Geotherm. Res.*, 133, 171–192.

Iverson, R.M. and Vallance, J.W. (2001). New views of granular mass flows. *Geology*, 29, 115–118.

Jenkins, S., Komorowski, J.C., Baxter, P.J., Spence, R., Picquout, A., Lavigne, F., Surono (2013). The Merapi 2010 eruption: An interdisciplinary impact assessment methodology for studying pyroclastic density current dynamics. *J. Volcanol. Geotherm. Res.*, 261, 316–329.

Karátson, D., Sztanó, O., Telbisz, T. (2002). Preferred clast orientation in volcaniclastic mass-flow deposits: Application of a new photo-statistical method. *J. Sediment. Res.*, 72, 823–835.

Kelfoun, K. (2011). Suitability of simple rheological laws for the numerical simulation of dense pyroclastic flows and long-runout volcanic avalanches. *J. Geophys. Res. Solid Earth*, 116(B8), August.

Kieffer, S.W. and Sturtevant, B. (1988). Erosional furrows formed during the lateral blast at Mount St. Helens, May 18, 1980. *J. Geophys. Res.*, 93, 14793–14816.

Komorowski, J.C., Jenkins, S., Baxter, P.J., Picquout, A., Lavigne, F., Charbonnier, S., Gertisser, R., Preece, K., Cholik, N., Budi-Santoso, A. et al. (2013). Paroxysmal dome explosion during the Merapi 2010 eruption: Processes and facies relationships of associated high-energy pyroclastic density currents. *J. Volcanol. Geotherm. Res.*, 261, 260–294.

Koyaguchi, T. and Woods, A.W. (1996). On the formation of eruption columns following explosive mixing of magma and surface-water. *J. Geophys. Res. B: Solid Earth*, 101, 5561–5574.

Kueppers, U., Putz, C., Spieler, O., Dingwell, D.B. (2012). Abrasion in pyroclastic density currents: Insights from tumbling experiments. *Phys. Chem. Earth*, 45–46, 33–39.

Lajoie, J., Boudon, G., Bourdier, J.L. (1989). Depositional mechanics of the 1902 pyroclastic nuéeardente deposits of Mt. Pelée, Martinique. *J. Volcanol. Geotherm. Res.*, 38, 131–142.

Lube, G., Cronin, S.J., Platz, T., Freundt, A., Procter, J.N., Henderson, C., Sheridan, M.F. (2007). Flow and deposition of pyroclastic granular flows: A type example from the 1975 Ngauruhoe eruption, New Zealand. *J. Volcanol. Geotherm. Res.*, 161, 165–186.

Lube, G., Cronin, S.J., Thouret, J.C., Surono, S. (2011). Kinematic characteristics of pyroclastic density currents at Merapi and controls on their avulsion from natural and engineered channels. *Bull. Geol. Soc. Am.*, 123, 1127–1140.

Lube, G., Breard, E.C.P., Cronin, S.J., Jones, J. (2015). Synthesizing large-scale pyroclastic flows: Experimental design, scaling, and first results from PELE. *J. Geophys. Res. Solid Earth*, 120, 1487–1502.

Middleton, G.V. and Southard, J.B. (1984). *Mechanics of Sediment Movement.* Society for Sedimentary Geology, Tulsa.

Neri, A., Bevilacqua, A., Esposti Ongaro, T., Isaia, R., Aspinall, W.P., Bisson, M., Flandoli, F., Baxter, P.J., Bertagnini, A., Iannuzzi, E. et al. (2015). Quantifying volcanic hazard at Campi Flegrei caldera (Italy) with uncertainty assessment 2: Pyroclastic density current invasion maps. *J. Geophys. Res. Solid Earth*, 120, 2330–2349.

Palladino, D.M. (2017). Simply pyroclastic currents. *Bull. Volcanol.*, 79, 53.

Palladino, D.M. and Valentine, G.A.V. (1995). Coarse-tail vertical and lateral grading in pyroclastic flow deposits of the Latera Volcanic Complex (Vulsini, central Italy): Origin and implications for flow dynamics. *Journal of Volcanology and Geothermal Research*, 69, 343–364. https://doi.org/10.1016/0377-0273(95)00036-4

Pallister, J.S., Schneider, D.J., Griswold, J.P., Keeler, R.H., Burton, W.C., Noyles, C., Newhall, C.G., Ratdomopurbo, A. (2013). Merapi 2010 eruption – Chronology and extrusion rates monitored with satellite radar and used in eruption forecasting. *J. Volcanol. Geotherm. Res.*, 261, 144–152.

Paterson, G.A., Roberts, A.P., MacNiocaill, C., Muxworthy, A.R., Gurioli, L., Viramonté, J.G., Navarro, C., Weider, S. (2010). Paleomagnetic determination of emplacement temperatures of pyroclastic deposits: An under-utilized tool. *Bull. Volcanol.*, 72, 309–330.

Pensa, A., Porreca, M., Corrado, S., Giordano, G., Cas, R. (2015). Calibrating the pTRM and charcoal reflectance (Ro%) methods to determine the emplacement temperature of ignimbrites: Fogo A sequence, São Miguel, Azores, Portugal, as a case study. *Bull. Volcanol.*, 77, 18.

Pittari, A., Cas, R.A.F., Monaghan, J.J., Martí, J. (2007). Instantaneous dynamic pressure effects on the behaviour of lithic boulders in pyroclastic flows: The Abrigo Ignimbrite, Tenerife, Canary Islands. *Bull. Volcanol.*, 69, 265–279.

Potter, D.B. and Oberthal, C.M. (1987). Vent sites and flow directions of the Otowi ash flows (lower Bandelier Tuff), New Mexico (USA). *Geol. Soc. Am. Bull.*, 98, 66–76.

Ripepe, M., Marchetti, E., Ulivieri, G. (2007). Infrasonic monitoring at Stromboli volcano during the 2003 effusive eruption: Insights on the explosive and degassing process of an open conduit system. *J. Geophys. Res. Solid Earth*, 112(B9), September.

Ripepe, M., De Angelis, S., Lacanna, G., Poggi, P., Williams, C., Marchetti, E., Donne, D.D., Ulivieri, G. (2009). Tracking pyroclastic flows at Soufrière Hills volcano. *Eos (Washington, DC)*, 90, 229–230.

Roche, O. (2015). Nature and velocity of pyroclastic density currents inferred from models of entrainment of substrate lithic clasts. *Earth Planet. Sci. Lett.*, 418, 115–125.

Roche, O., Phillips, J.C., Kelfoun, K. (2009). Pyroclastic density currents. In *Modeling Volcanic Processes: The Physics and Mathematics of Volcanism*, Fagents, S.A., Gregg, T.K.P., Lopes, R.M.C. (eds). Cambridge University Press, Cambridge.

Roche, O., Buesch, D.C., Valentine, G.A. (2016). Slow-moving and far-travelled dense pyroclastic flows during the Peach Spring super-eruption. *Nat. Commun.*, 7, 10890.

Rose, W.I., Pearson, T., Bonis, S. (1976). Nuée ardente eruption from the foot of a dacite lava flow, Santiaguito volcano, Guatemala. *Bull. Volcanol.*, 40, 23–38.

Scarpati, C., Sparice, D., Perrotta, A. (2014). A crystal concentration method for calculating ignimbrite volume from distal ash-fall deposits and a reappraisal of the magnitude of the Campanian Ignimbrite. *J. Volcanol. Geotherm. Res.*, 280, 67–75.

Scharff, L., Hort, M., Varley, N.R. (2019). First in-situ observation of a moving natural pyroclastic density current using Doppler radar. *Sci. Rep.*, 9, 1–9.

Scott, A.C. and Glasspool, I.J. (2005). Charcoal reflectance as a proxy for the emplacement temperature of pyroclastic flow deposits. *Geology*, 33, 589–592.

Self, S., Kircher, D.E., Wolff, J.A. (1988). The Ej Cajete Series, Valles caldera, New Mexico. *J. Geophys. Res.*, 93, 6113–6127.

Shea, T., Gurioli, L., Houghton, B.F. (2012). Transitions between fall phases and pyroclastic density currents during the AD 79 eruption at Vesuvius: Building a transient conduit model from the textural and volatile record. *Bull. Volcanol.*, 74, 2363–2381.

Sheridan, M.F. (1980). Pyroclastic block flow from the September, 1976, eruption of La Soufrière volcano, Guadeloupe. *Bull. Volcanol.*, 43, 397–402.

Silva Parejas, C., Druitt, T.H., Robin, C., Moreno, H., Naranjo, J.A. (2010). The Holocene Pucón eruption of Volcán Villarrica, Chile: Deposit architecture and eruption chronology. *Bull. Volcanol.*, 72, 677–692.

Sulpizio, R. and Dellino, P. (2008). Sedimentology, depositional mechanisms and pulsating behaviour of pyroclastic density currents. In *Developments in Volcanology*, volume 10, Gottsmann, J. and Martì, J. (eds). Elsevier, Amsterdam.

Sulpizio, R., Dellino, P., Doronzo, D.M., Sarocchi, D. (2014). Pyroclastic density currents: State of the art and perspectives. *J. Volcanol. Geotherm. Res.*, 283, 36–65.

Valentine, G.A.V. (1987). Stratified flow in pyroclastic surges. *Bull. Volcanol.*, 49, 616–630.

Valentine, G.A.V. (1998). Damage to structures by pyroclastic flows and surges, inferred from nuclear weapons effects. *Journal of Volcanology and Geothermal Research*, 87, 117–140. https://doi.org/10.1016/S0377-0273(98)00094-8.

Valentine, G.A.V. and Wohletz, K.H. (1989). Environmental hazards of pyroclastic flows determined by numerical models. *Geology*, 17, 641.

Voight, B. and Davis, M.J. (2000). Emplacement temperatures of the November 22, 1994 nuée ardente deposits, Merapi Volcano, Java. *J. Volcanol. Geotherm. Res.*, 100, 371–377.

White, J.D.L. and Houghton, B.F. (2006). Primary volcaniclastic rocks. *Geology*, 34, 677–680.

Wilson, C.J.N. (1985). The Taupo eruption, New Zealand II. The Taupo ignimbrite. *Philos. Trans. R. Soc. A: Math. Phys. Eng. Sci.*, 314, 229–310.

Wilson, C.J.N. and Houghton, B.F. (2000). Pyroclastic transport and deposition. In *Encyclopedia of Volcanoes*, Sigurdsson, H., Houghton, B.F., Rymer, H., Stix, J., McNutt, S. (eds). Academic Press, Cambridge.

Wilson, L., Sparks, R.S.J., Walker, G.P.L. (1980). Explosive volcanic eruptions IV. The control of magma properties and conduit geometry on eruption column behaviour. *Geophys. J. R. Astron. Soc.*, 63, 117–148.

Yamamoto, T., Takarada, S., Suto, S. (1993). Pyroclastic flows from the 1991 eruption of Unzen volcano, Japan. *Bull. Volcanol.*, 55, 166–175.

Zanella, E., Gurioli, L., Pareschi, M.T., Lanza, R. (2007). Influences of urban fabric on pyroclastic density currents at Pompeii (Italy) 2: Temperature of the deposits and hazard implications. *J. Geophys. Res. Solid Earth*, 112(B5), May.

Zanella, E., Sulpizio, R., Gurioli, L., Lanza, R. (2015). Temperatures of the pyroclastic density currents deposits emplaced in the last 22 kyr at Somma-Vesuvius (Italy). *Geol. Soc. Spec. Publ.*, 396, 13–33.

Zijderveld, J.D.A. (1967). The natural remanent magnetizations of the Exeter volcanic traps (Permian, Europe). *Tectonophysics*, 4, 121–153.

Zuccaro, G., Cacace, F., Spence, R.J.S., Baxter, P.J. (2008). Impact of explosive eruption scenarios at Vesuvius. *J. Volcanol. Geotherm. Res.*, 178, 416–445.

4

Gravity Monitoring of Volcanoes

Lydie GAILLER, Jean-François LÉNAT and Franck DONNADIEU
Laboratoire Magmas et Volcans, CNRS, IRD, OPGC,
Université Clermont Auvergne, Clermont-Ferrand, France

4.1. Introduction

The Earth's gravity field varies over time due to external and internal causes, but in minute proportions compared to its average value. Capturing these variations and identifying their causes therefore require precise measurement methods and complex analyses. The average value of the Earth's gravity is 981 Gal (the Gal is often kept as a unit in geophysics, with 1 Gal = 10^{-2} m.s^{-2}). The variations that we are going to focus on in monitoring will be very small, of the order of μGal (a few units or tens of μGal), or about 10^{-9} times the average gravity. We speak then of microgravimetry. The main external cause of gravity variation is the lunisolar tide which causes, as is well known, fluid movements (sea, water tables and atmosphere), and also a deformation of the solid Earth (up to several tens of centimeters). These effects can be corrected by models that are regularly refined or by continuous recordings concomitant with the monitoring measurements. The basic principle of microgravity monitoring is that changes in Earth's gravity can reflect deep mass transfer processes (Dzurisin 2006), often associated with surface deformation. However, as we will see, identifying and locating sources can be complex. Good knowledge of environmental signals and coupling gravity measurements with crustal deformation data is essential. Indeed, the combination of gravimetry and deformation allows access to the density (and therefore the probable nature) of the source of the observed variations. Deformations are only sensitive to

Hazards and Monitoring of Volcanic Activity 3,
coordinated by Jean-François LÉNAT.

variations in pressure or volume of the source, whereas gravimetry is sensitive to changes in mass. For a summary of the types of gravity variations and their measurement, reference can be made to the paper by Van Camp et al. (2017) as well as the review paper on gravimetric volcanic monitoring by Carbone et al. (2017). We will also describe the development of the technique of muon tomography of volcanoes, which allows us, using natural atmospheric muon fluxes, to characterize the 3D distribution of densities within edifices and thus potentially the variation of these distributions over time.

4.2. Instruments

Gravimetry measurements can be made either with relative gravimeters (see Figure 4.1) or with absolute gravimeters (see Figure 4.2) (simple or elaborate descriptions of these instruments are easily found on the Internet). As the name implies, relative gravimeters only measure the differences in vertical acceleration of gravity between stations. The gravity value can be accessed if the measurement network includes a station where it is known or measured with an absolute gravimeter. The measuring instruments have evolved with the technological progress. Relative gravimeters have been used for decades in the field. On the other hand, it is only recently that absolute gravimeters that can be transported in the field have been developed. It is easy to imagine that the development of portable and increasingly sophisticated instruments will accelerate in the future.

The resolution and accuracy of absolute gravimeters are better than those of relative spring gravimeters (different physical principles are generally used for these two types of instruments). But a distinction must be made between theoretical or laboratory-measured values and those obtained in the field. Greco et al. (2012) estimated that the uncertainty on data acquired at Etna with two absolute gravimeters (Micro-g LaCoste FG5 and an Italian prototype the IMGC-02) was 10 μGal or less, while laboratory values are on the order of 2–4. For a relative gravimeter (Scintrex CG-3 M), they give uncertainties up to 15 μGal in field measurements. Superconducting relative gravimeters have performances in sensitivity and drift very superior to those of spring relative gravimeters (an example is i-grav gravimeters installed on Etna in the framework of the Newton-g project).

Precision gravimeters have internal sources of error that can be minimized or corrected. For spring-based relative gravimeters, the main instrumental factors of error are (Seigel 1995) sensitivity to shock, vibration, internal temperature variations (if the power supply that controls thermal regulation is turned off), instrument leveling errors during measurements and calibration, which should be checked regularly.

Figure 4.1. *Field measurement with a relative gravimeter. The altitude at the time of the measurement is determined by GPS (whose antenna is visible). Photo: Jean-François Lénat*

Figure 4.2. *A portable A10 absolute gravity meter measuring on a geodetic benchmark (credit: Rob Carruth, USGS Arizona Water Science Center. Public domain)*

Network measurements are usually organized in loops in which at least one base station is regularly returned to. This procedure is very efficient to correct instrumental drifts or other variations like atmospheric pressure. Other strategies must be implemented in the case of fixed stations.

However, current technological advances (e.g. Newton-G project) will probably bring a revolution in gravity measurement, in particular around MEMS (micro-electro-mechanical systems) technologies.

4.3. Measurement procedures

Most of the gravity monitoring measurements concern ground measurements on benchmarks. As we are trying to detect low amplitude signals, special precautions are taken to ensure the best possible accuracy of measurements. This is called *microgravity monitoring*. We will see that in some cases where very large mass transfers (rock or liquid) are involved, less precise techniques can be implemented using ground or airborne measurements.

4.3.1. *Ground-level gravimetry*

Microgravity monitoring of a volcano can be performed either by periodic reiteration of an array or continuously with one or more permanent gravimeters. The two approaches can also be combined.

Typically, a periodic microgravity monitoring network on a volcano extends over an area of the order of 10 km, with reference stations that may be further away. Figure 4.3 shows the network of Etna. One of the particularities of this one is that the majority of the stations are measured using relative gravimeters and the other with absolute gravimeters. This configuration makes it possible to combine the speed of the measurements of the relative gravimeters with the precision of the absolute gravimeters. As shown in Figure 4.4, the network is divided into subnetworks, each associating relative measurements with an absolute measurement station. The subnetworks are then connected to each other. Stations far from the active areas of the volcano can be used as references to quantify variations. On a practical level, the station benchmarks must be materialized in a permanent way. These are often also geodetic benchmarks, which facilitate the concomitant and indispensable measurement (see below) of altitude variations between gravimetric reiterations.

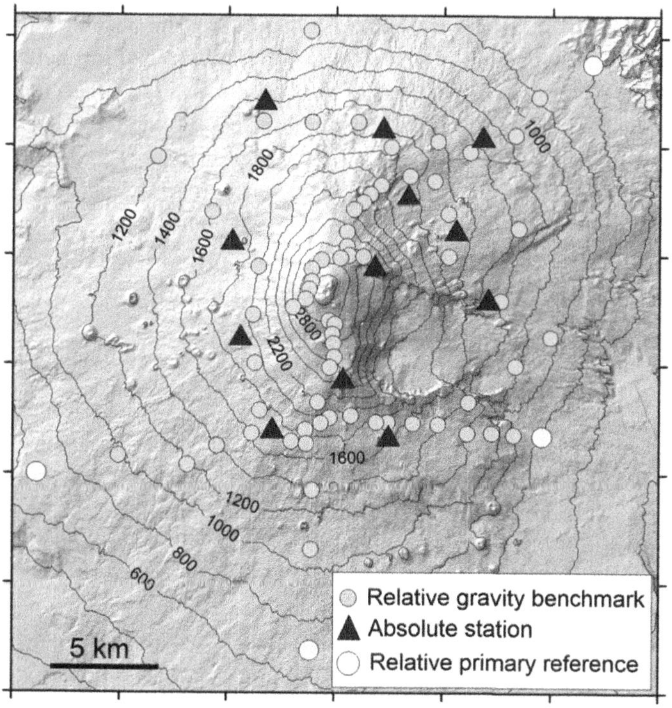

Figure 4.3. *Etna gravity monitoring network (from Greco et al. (2012)). For a color version of this figure, see www.iste.co.uk/lenat/hazards3.zip*

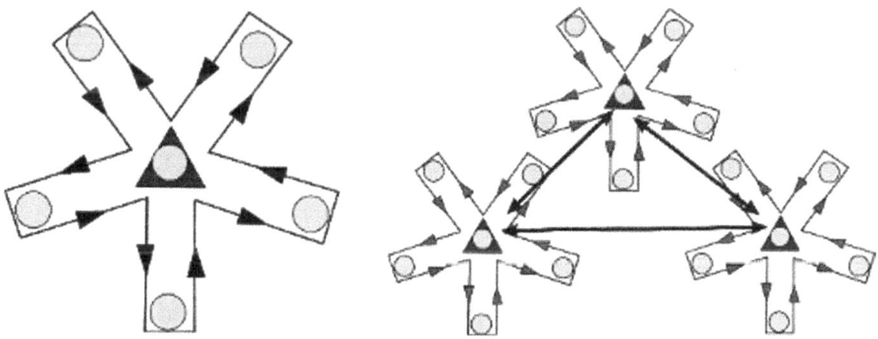

Figure 4.4. *Example of a microgravity monitoring measurement strategy with a mixed network of absolute (triangles) and relative (circles) gravimeters (from Greco et al. (2012))*

Continuous measurements have the advantage of giving access to the dynamics of variations (if variations are observed with a periodic repetition network, neither the date nor the dynamics of these variations are known). However, the installation and maintenance of continuous recording stations require relatively heavy resources (financial and logistical). Ideally, one needs a shelter, energy and, to improve the accuracy, a recording of the disturbing parameters (pressure, temperature, possible drift of the instrument's inclination). One of the best compromises is for one or more continuous stations to have access to the dynamics of the variations, and a repeat network to have access to the spatial distribution of the variations.

4.3.2. *On-board mobile gravimetry*

On-board mobile gravimetry cannot compete with microgravimetry because its accuracy is too low (of the order of a few mGal). However, for volcanic events involving very large masses (collapses, constructions, etc.), it can be an effective means of quantifying such disruptions. Moreover, these techniques allow the investigation of areas that are not easily accessible or at risk.

Many developments of gravity measurement from a moving vehicle (car, airplane, balloon, boat and, more recently, on-board autonomous underwater vehicles) have been made. A mobile gravimetry system commonly includes, in addition to one or more accelerometers, a device for measuring the acceleration of the vehicle with respect to an inertial reference frame and a device for determining the attitude of the vehicle and the measurement conditions (e.g. speed of the carrier, bathymetry and state of the swell for marine acquisition, flight altitude and state of turbulence for airborne acquisition). These complementary measurements allow the separation of the gravitational acceleration from the vehicle acceleration. Promising developments are underway around mobile underwater gravimetry systems (GraviMob of CNAM-ESGT-UBO), and the prospect of airborne measurements by drone is also the subject of numerous studies (principle of MEMS) (Middlemiss et al. 2016).

4.3.3. *Space gravimetry*

By focusing on very long wavelengths (>100 km), space-borne gravimetry may seem far removed from applications in volcanic monitoring. This is in fact the case at present, but the resolution and accuracy of these techniques are progressing and could provide useful information for volcanic monitoring, such as the monitoring of water, snow or ice masses, or very large deformations.

4.4. External phenomena generating temporal variations of gravity

4.4.1. *Seismic noise and wind*

Microseismic activity can be due to both natural causes (e.g. seismic activity of a volcano or the vibration of tree roots with the wind, coastal swell) and anthropogenic (urban areas, roads, etc.). The effects of these phenomena can reach several tens of μGal. But, as by nature they correspond more or less to white noise, they can be minimized statistically by accumulating and filtering measurements over a period of time.

4.4.2. *Atmospheric pressure*

A variation in atmospheric pressure (either meteorological or due to altitude variations between stations in a network) implies a variation in the mass of the air column above the measurement point. If atmospheric pressure is measured during gravity measurements, a correction can be applied (Merriam 1992). Otherwise, this correction will be taken into account with the drift between successive measurements at a base station. The variation of atmospheric pressure with altitude can, in general, be considered negligible (of the order of 0.05 μGal/ m).

4.4.3. *Land and ocean tides*

The attraction of the stars and, in the first place, of the Moon and the Sun (with an effect twice as great of the Moon) causes periodic variations of the gravity which can reach an amplitude of 0.3 mGal, with rates of variations reaching 30 μGal/hour. Gravimetric tides result from the combined effects of the displacement of water and air masses and the deformation of the Earth by gravitational effects, and the overloading (or discharge) of water masses that mechanically deform the Earth (Melchior and Ducarme 1989). These tides also cause the precession of the axis of rotation of the Earth, which also induces a small variation in gravity.

The knowledge of tidal terms is therefore crucial for many scientific applications: terrestrial and space geodesy, astronomy and, of course, microgravimetry. Global or regional models exist, allowing researchers to correct gravimetric and geodesic measurements for these effects. These models are sometimes implemented in the gravimeters, allowing a transparent correction during the measurements, or they can be applied in post-processing. The correction of the water mass overload effect is in general difficult (or less accurate) in coastal areas, as it depends on the bathymetry and shape of the coast. This can be improved with continuous recording near the coast, or with frequent returns to a coastal station.

4.4.4. *Hydrological reservoirs*

In areas with large seasonal differences in precipitation, there can be significant variations in the level of water tables, lakes and rivers and in the humidity of the vadose zone. For example, 10 cm of water remaining in the soil or in a water table close to the surface causes an increase in gravity of about 4 μGal. In an area where hydrogeological levels can vary to a significant extent for microgravity measurements, it is necessary either to estimate them or to repeat the measurements under the same seasonal conditions.

4.4.5. *Altitude variations and deformation*

If the external variations previously described have been corrected (or neglected), the gravimetric signal can be formulated as:

$$\Delta g_{res} = \Delta g_{obs} - \gamma . u_z - \Delta g_{def} \pm \Delta g_{\Delta topo} \quad [4.1]$$

with Δg_{obs} the difference in gravity measured between two dates, $\gamma . u_z$ the vertical gradient of gravity in free air multiplied by the vertical displacement of the station, Δg_{def} the Bouguer effect of deformation and $\Delta g_{\Delta topo}$ the variation in topography by construction (cone, flow, etc.) or destruction (collapse, slip, etc.). Δg_{res} is the residual anomaly associated with a subsurface mass change.

If the station has moved vertically, this must be taken into account because gravity decreases with distance from the center of the Earth (gradient in free air or GAL). The theoretical gradient (which corresponds to the general gradient on a global scale) is 0.3086 mGal/m. However, it has been observed that this gradient can be different locally (Rymer 1994). This is the case if a measurement is made near a topographic accident (e.g. a cliff) or in the presence of a significant static gravity anomaly. Deviations of up to 40% of the theoretical value are observed. It is therefore recommended to make local measurements of the gradient in the open air (e.g. by using a tripod to place the gravimeter at a height of about 1 m above the station).

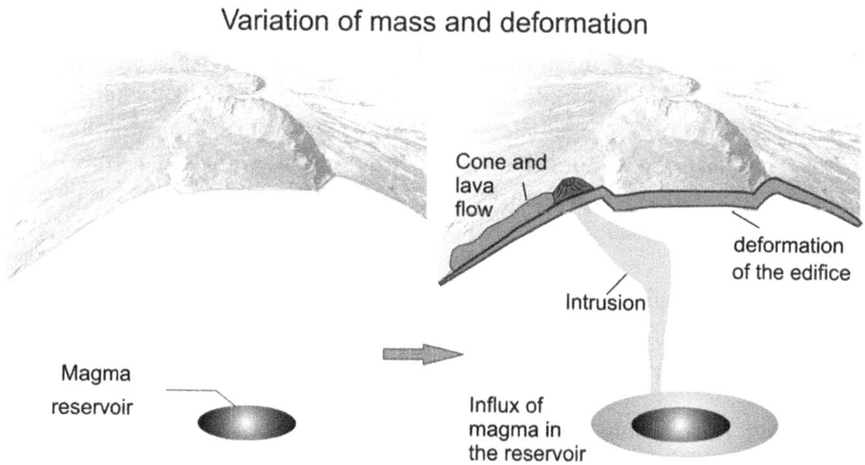

Figure 4.5. *Examples of positive mass changes in a structure*

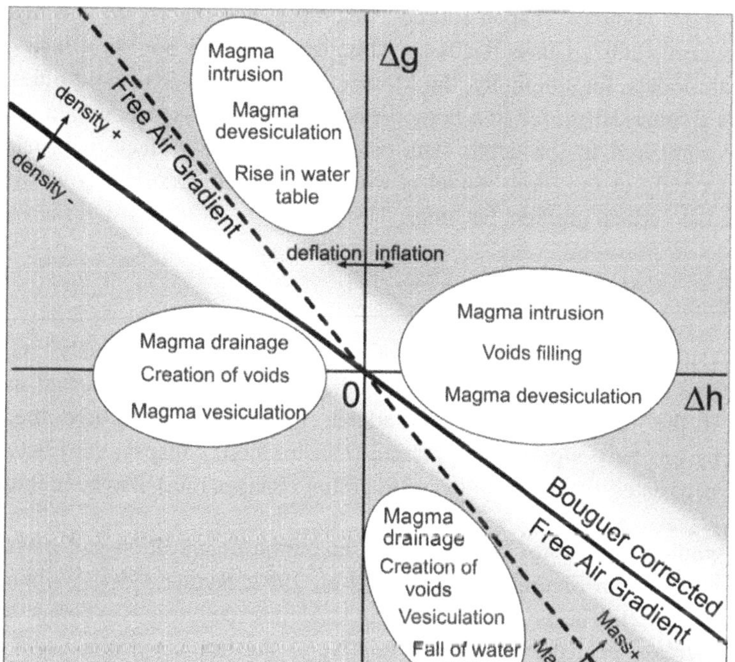

Figure 4.6. *Types of mass and density change with changes in gravity and altitude (from Williams-Jones and Rymer (2002))*

If the elevation has changed between two measurement dates, the station has nevertheless not moved in the open air (see Figure 4.5). For example, the ground has also risen in the case of inflation, and the effect of this additional volume of rock must be accounted for. The simplest way to account for this slice of ground is to identify it as an infinite horizontal layer, with thickness h equal to the vertical component of deformation and density ρ equal to that estimated for the terrain in the area. In this case, the total vertical gradient will be:

$$\frac{\Delta g}{\Delta h} = -GAL\ \mu Gal\ cm^{-1} + 2\pi G\rho \quad [4.2]$$

G being the universal gravitation constant.

However, in volcanology, if a horizontal layer can be assimilated to the setting of a sill as a source of deformation, it appears that other types are often more realistic. These can be models for which simple analytical solutions exist or more elaborate models that require numerical approaches. The former generally consider an isotropic elastic medium and a half-space. For the latter, numerical computation can consider more complex shaped sources, heterogeneous rheologies and topography (Currenti et al. 2007; Currenti 2014). Although the trend now is to use complex numerical models, for simplicity, the Mogi model (Mogi 1958) is still often used. It considers a spherical source in a homogeneous elastic half-space, with a very small diameter compared to the depth. The calculation of the surface deformation and gravity anomaly of the Mogi model is well explained in Dzurisin et al. (1980). In this case, the vertical gradient becomes:

$$\frac{\Delta g}{\Delta h} = -GAL\ \mu Gal\ cm^{-1} + \frac{4\pi G\rho}{3} \quad [4.3]$$

The residual anomaly is referred to as the "GAL corrected anomaly" (GALCA), if only the GAL correction was performed, or the "Bouguer corrected anomaly" (BCA), if the terrain gradient was added. In some cases where the surface topography has been significantly altered (construction, collapse, etc.) between the measurement dates, the contribution of the relief changes must also be included.

The gradients allow for the graphical separation of elevation and mass change effects, as shown in Figure 4.6 (Berrino et al. 1992; Rymer 1994; Williams-Jones and Rymer 2002). The position of the $\Delta g_{obs}/\Delta h$ pair determines whether there has been an increase or decrease in mass relative to the free-air gradient and whether there has been a change in density relative to the Bouguer-corrected gradient (relative to the density used to calculate the gradient).

Δg_{res} will be used for the calculation of the gravimetric sources.

4.5. Causes of gravimetric variations on volcanoes

At the scale of a volcanic edifice, the spatio-temporal variations in gravity, corrected for the effects described above, are generated by mass redistribution mechanisms at depth. They result from different phenomena mainly related to the dynamics of the hydrothermal system, the magmatic system at depth or its manifestations at the surface, or to volcano-tectonic events, often induced by the two previous ones.

4.5.1. *Evolution of hydrothermal systems*

Hydrothermal systems manifest themselves at the surface in the form of fluids and temperature anomalies (e.g. fumaroles, geysers, hot springs and acid lakes). At depth, the circulation of fluids and heat depends mainly on the porosity and permeability of the surrounding rocks because convective mechanisms are dominant. These systems are evolved by chemical processes (dissolution/ precipitation of minerals) and as a function of disturbances such as magmatic fluids and heat influx. For about a decade, work has been conducted to study and model the effect of these perturbations on deformation (Hurwitz et al. 2007; Hutnak et al. 2009; Ingebritsen et al. 2010) and associated gravity variations (Gottsmann et al. 2007; Todesco et al. 2010).

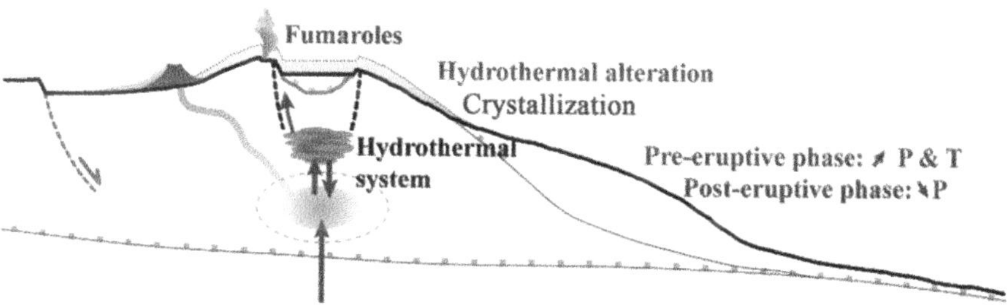

Figure 4.7. *Mass changes (>0 in red, <0 in blue) in response to the evolution of a hydrothermal system. P: pressure; T: temperature. For a color version of this figure, see www.iste.co.uk/lenat/hazards3.zip*

During an eruptive cycle, the entire hydrothermal system will evolve in response to changes in fluid pressure and temperature (see Figure 4.7):

– In the *pre-eruptive phase*, pressurization of the system (rise of magmatic fluids, increase of temperature) will tend to a negative variation of gravity.

– In the *post-eruptive phase*, pressure decreases (evacuation of interstitial fluids, opening of new fumarole vents at the surface, migration to more superficial levels) producing a positive variation of gravity. In parallel, the hydrothermal alteration process plays a major role in decreasing the permeability and the density of the environment. Conversely, the crystallization and the deposition of minerals generate a decrease in the porosity and thus an increase in the density.

4.5.2. *Evolution of magma systems*

Numerous studies have shown that microgravity variations are detectable before, during and after volcanic eruptions (Jachens and Eaton 1980) and can be indicators of magma transport to the surface, intrusion, storage, or, conversely, magma emptying or vesicularity change. Depending on the process(es) involved, the edifice undergoes an increase or decrease in mass (see Figure 4.8).

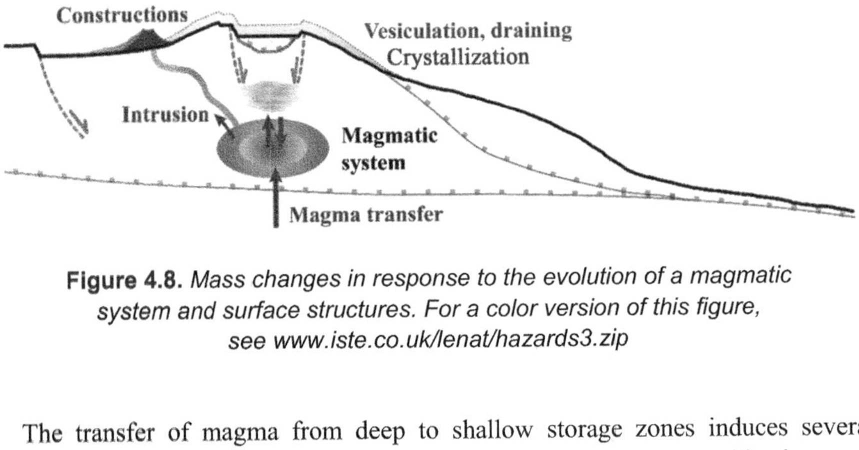

Figure 4.8. *Mass changes in response to the evolution of a magmatic system and surface structures. For a color version of this figure, see www.iste.co.uk/lenat/hazards3.zip*

The transfer of magma from deep to shallow storage zones induces several consequences: pressurization of the recharged chamber, crystallization and degassing (Tait et al. 1989). At the same time, intrusions can occur preferentially along pre-existing fractures, their filling generating an increase in mass.

Conversely, magma vesiculation and drainage phenomena will cause an overall decrease in mass within an edifice.

New volcanic constructions on the surface are also at the origin of the variations of gravity measured on the surface (e.g. cones and flows). Depending on the density and volumes involved, these structures modify the gravity measured at the surface.

4.5.3. *Volcano-tectonic phenomena*

Volcanic edifices are commonly the site of volcano-tectonic phenomena often linked to intrusive and hydrothermal activity, or to the presence of mechanical heterogeneities within the edifices. These phenomena are particularly expressed in the form of large instabilities such as flank slides or caldera collapses, whose dynamics can be very rapid. The instabilities can also be linked to a strong intrusive activity on the flanks of the edifices and generate block slides or debris avalanches involving volumes of a few hundred m^3 to several tens of km^3.

Resurgence, defined as the uplift of the floor of calderas after their collapse, is often directly related to inflation or pressurization of the magma system (Smith and Bailey 1968; Marsh 2014).

Such volcano-tectonic phenomena are at the origin of mass transfers (thus gravity variations) whose dynamics vary according to the process and time scales involved (see Figure 4.9).

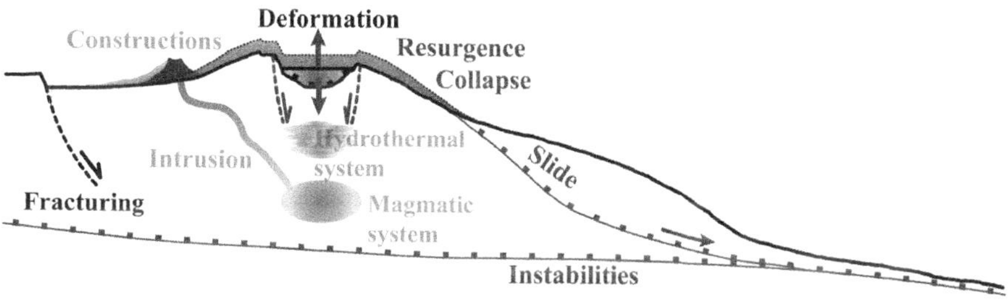

Figure 4.9. *Mass changes in response to volcano-tectonic events. For a color version of this figure, see www.iste.co.uk/lenat/hazards3.zip*

4.6. Case studies

4.6.1. *Monitoring and evolution of a magma reservoir at Kilauea*

Kilauea (Hawaii) is a shield volcano characterized in its summit area by a succession of effusive and explosive cycles over the past 2,200 years associated with caldera formation and filling (Swanson et al. 2012, 2014) (see Figure 4.10(a)). In 2008, a lava lake was emplaced in the caldera, within Halemaumau Crater. During the 2018 Kilauea major crisis, the lake was drained, until it was completely emptied, by downstream lava emissions (Neal et al. 2019). This event was accompanied by significant subsidence of the caldera (see Figure 4.10(b)). Monitoring the activity of this edifice is therefore a major issue and gravimetry has the capability to provide some important information:

– How do the structures inherited from the past and present evolution of the summit caldera influence the activity?

– How to characterize the structure and dynamics of a magmatic system (the lava lake)?

Gravity survey data imaged the geometry of the previous caldera (Poland et al. 2019) (see Figure 4.10(a)), the boundaries of which are weak zones prone for future collapse and for fluid (hydrothermal and magmatic) transfer. A microgravity monitoring of the lava lake has been carried out with two continuous gravity meters installed in May 2010 (see Figure 4.10) (Carbone and Poland 2012; Carbone et al. 2013). Coupled with measurements of the lake level, microgravimetry showed that the density of its upper part was very low (950 ± 300 kg m^{-3}), implying a very gas-rich and highly vesiculated surface magma. This explains the high explosiveness of the lake during slumps in its wall. In addition, gravity fluctuations were modeled by convection in a reservoir located at about 1 km depth.

4.6.2. *Follow-up of an intrusion and a collapse at Piton de la Fournaise*

To illustrate the potential of gravity monitoring methods, two examples of the Piton de la Fournaise crisis on Reunion Island are described below.

4.6.2.1. *Gravity variations related to the 1998 intrusion*

The March 1998 eruption, which occurred after more than 5 years of inactivity, was remarkable for its duration (196 days) and the volume of lava emitted (60 Mm^3) (see Figure 4.11(a)) (Staudacher et al. 1998). To better understand the mass redistribution phenomena within the edifice during the eruption, gravity variations were measured between November 1997 and March 1998 (see Figure 4.11(a)) on 29 stations of the Piton de la Fournaise reiteration network (Bonvalot et al. 2008).

Three other remote stations were used to check the calibration of the gravimeters between the two campaigns and two gravimeters were also installed continuously over the same period (see Figure 4.11(a)). A residual variation up to 100 μGal was observed (see Figure 4.11(b)). It can be explained by a model involving mass addition in the edifice distributed between dyke intrusion ($3.9–8.7 \times 10^9$ kg) and the feeding of a magmatic reservoir at sea level ($4.6–7.2 \times 10^{10}$ kg) (see Figure 4.11(c)).

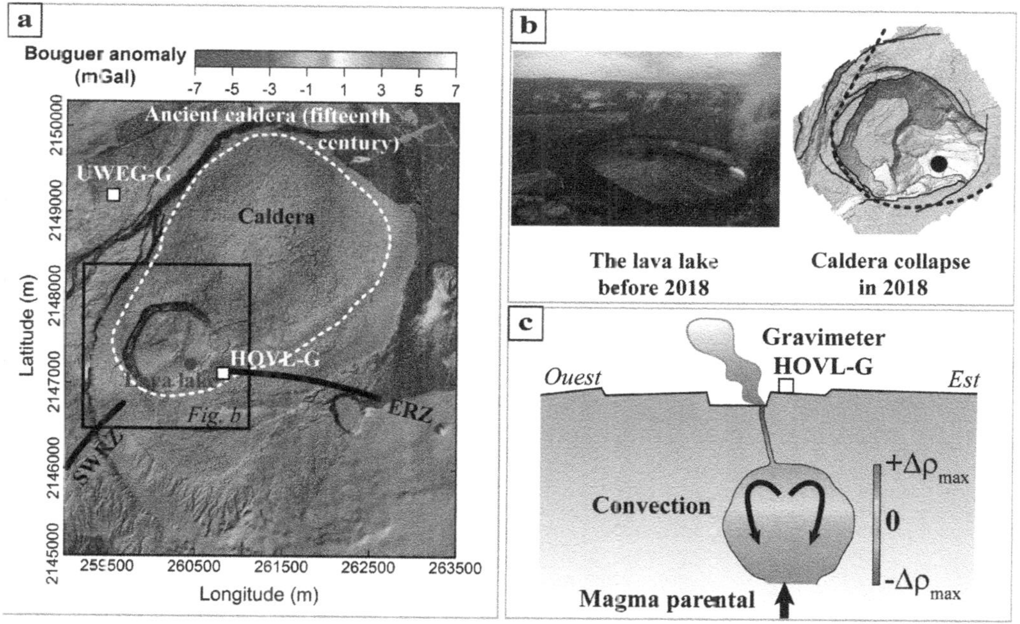

Figure 4.10. *a) Kilauea caldera prior to the 2018 crisis and associated Bouguer anomaly (Gailler and Kauahikaua 2018) that modeled the previous caldera boundaries (white dotted line). Location of continuous gravimeters (May 2010; Carbone and Poland 2012). b) The lava lake from 2008 to 2018 (photo: USGS, HVO) and consequence of the 2018 eruption on the summit zone (opentopo.org). c) Interpretive cross-section of the relationship between the surficial magma reservoir (continuous density gradient Δρ) and the lava lake from modeling of microgravity data (Carbone and Poland 2012). For a color version of this figure, see www.iste.co.uk/lenat/hazards3.zip*

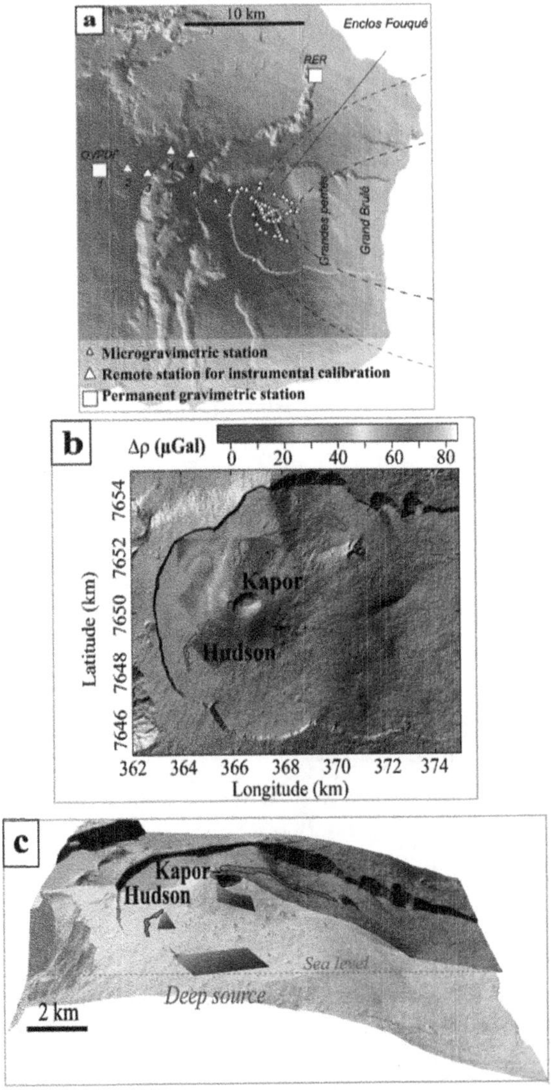

Figure 4.11. *a) Location of the microgravity reiteration network set up on Piton de la Fournaise between 1992 and 1995 (DTM: IGN). b) Eruptive centers of the 1998 eruption and gravity variation between November 1997 and March 1998 (source DTM: OVPF). c) Interpretation of the gravity variations observed after the 1998 eruption (Bonvalot et al. 2008). For a color version of this figure, see www.iste.co.uk/lenat/hazards3.zip*

4.6.2.2. *Gravity variations related to the collapse of the Dolomieu in 2007*

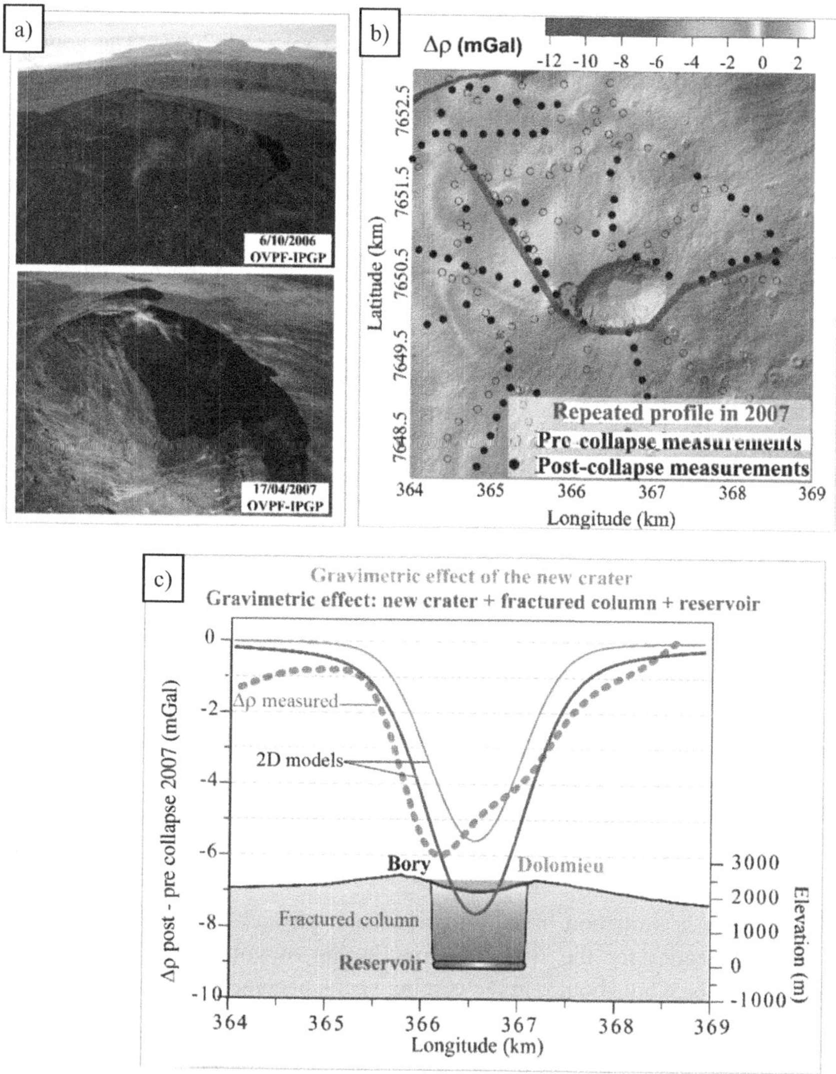

Figure 4.12. *a) The Dolomieu before and after its collapse in 2007 (OVPF photos). b) Gravity variations associated with the collapse of the Dolomieu in 2007. c) Direct 2D models and interpretations of the difference in gravity signal ante and post-collapse 2007 along the profile localized in (b) (Cailler et al. 2009). For a color version of this figure, see www.iste.co.uk/lenat/hazards3.zip*

At the top of the central cone of Piton de la Fournaise lie the Bory and Dolomieu craters (see Figure 4.12(a)). In April 2007, Dolomieu collapsed 320 to 340 m, with an estimated volume of about 100–120 $\times 10^6$ m^3 (Urai et al. 2007). The geological interpretation of the volcano-tectonic crisis suggests that the collapse is associated with the drainage of a superficial magma reservoir, by lateral intrusion and effusion of a volume of magma comparable to that of the collapse (Michon et al. 2007; Staudacher et al. 2009). Such an event involves mass transfers large enough to disrupt the gravity field (see Figure 4.12(a)).

This could be confirmed, with sufficient accuracy given the magnitude of the variations, by simple gravity survey measurements (Gailler et al. 2009). Gravity measurements were made after the collapse, in the area of previous surveys in the caldera and around the Dolomieu (see Figure 4.12(b)). Although not exactly co-located with previous measurements, these new data showed a clear decrease in gravity values in the summit area. 2D models of the difference between ante and post-collapse measurements (see Figure 4.12(c)) show that the observed mass deficit can be interpreted under two main assumptions (see Figure 4.12):

– the effect of the creation of the new crater alone (the collapsed rock volume is replaced by air);

– the combined effects of the creation of the crater, the fracturing of the rock column between the surface and the drained magma reservoir (decrease in the density of this column) and the drainage of the reservoir (mass departure).

4.7. Study and monitoring by muography

4.7.1. *Overview*

Muography is a method based on corpuscular physics that has been developed for several decades for the study and monitoring of volcanoes. It is still at the experimental stage but should, in the coming years, become a widely used technique in volcanology.

Atmospheric muons, observed in 1936 by Neddermeyer and Anderson (1937), are elementary particles produced by the interaction of cosmic radiation with atoms in the atmosphere. A muon has the electrical charge of an electron but a mass 207 times greater ("heavy electron"). It is an unstable particle with a lifetime of about 2 millionths of a second. During their short life, muons are able to travel tens of kilometers at a speed close to the speed of light. At sea level, the muon flux is about 170 muons per m^2/s (we are crossed by about 30 muons per second).

Because of their mass and energy, muons are able to penetrate water and rocks before being absorbed or disintegrated, at depths of less than 10 km. To give an order of magnitude, the muon flux is attenuated by a factor greater than 10 for 1 km of rock crossed. This attenuation is both a function of the thickness of the rock and the density of the medium. It is on this basis that the principle of muography was developed. By measuring the flux of muons passing through matter in one direction, one can obtain the average density of the column of matter passed through (integrated density), if one knows the topography (thus the length of the column of matter). By making observations of an object (e.g. a volcano) from different points of view, one can reconstruct the density distribution in 3D.

4.7.2. *Instruments*

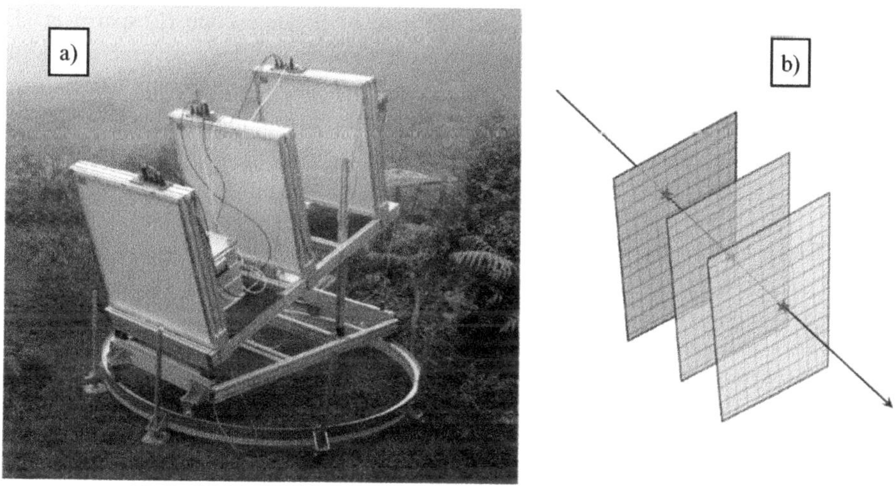

Figure 4.13. *a) A 3-plane telescope. Note the orientation devices (from Marteau et al. (2012)). b) Principle of reconstructing the trajectory of incident muons from the location of the impacts on the detector arrays*

Muon imaging in Earth sciences has benefited from techniques developed in corpuscular physics. The detectors use different technologies (Marteau et al. 2012): emulsions, scintillators and glass resistive plates. The detectors are used to make telescopes (see Figure 4.13). A muon telescope consists of several parallel planes. Each plane contains an array of detectors. The telescope can be oriented in azimuth and vertical. The aperture and resolution of the telescope are determined by the distance between the front and back planes, the size of the detectors and the size of the planes. Two planes are needed to access the orientation of the incident muons, but in order to refine the determination of this orientation, several planes are usually used. A resolution in the target of about 10 m can be reached.

4.7.3. *Measurements and data*

The processing of the measurements and their interpretation involve procedures that can be found in detail in specialized papers (Marteau et al. 2012; Ambrosino et al. 2015). The basic measurement is the target opacity, obtained by comparing the muon flux passing through the target (the volcano) with that coming from the atmosphere (open sky). The integrated density over the target traverse is obtained by calculation (Marteau et al. 2012).

The "exposure time" required to obtain a reliable image depends on both the thickness of the material traversed and the size of the telescope (Lesparre et al. 2010). With a detection area of 1 m^2 (a typical size), it will take several weeks of exposure for a structure of a few hundred meters thick and several months as soon as we are in the order of a kilometer. It will only take a few minutes in the case of small objects (e.g. detection of materials with a high atomic mass number in a load).

4.7.4. *Case studies*

A few emblematic works illustrate the use of muon imaging, whether for small objects (Borozdin et al. 2003; Jenneson 2004; Checchia 2016) or for the Pyramid of Chephren (Alvarez et al. 1970). In Earth science, applications are developing with subsurface measurements for radioactive storage sites and mining, but the range of targets is infinite. In volcanology, pioneering work has been done by Nagamine et al. (1995) and Tanaka et al. 2007.

While the method is becoming operational for the study of the internal structure of volcanoes, its use in monitoring is still in its early stages. Density variations within a volcano can be induced by a wide range of phenomena: magma intrusion, deformation, hydrothermal activity, water table variations, creation of fissures or craters and deposition or emission of products (see Figures 4.5, 4.7, 4.8 and 4.9). Muography has the ability to detect small variations in density and thus offers the opportunity to highlight subtle changes in the structure of volcanoes over time. Below is an example of structure determination and an example of structure variation.

4.7.4.1. *Imaging of the Soufrière of Guadeloupe*

The muon image shown in Figure 4.14 was obtained in approximately 5 weeks of exposure by Lesparre et al. (2012). The densest regions (RS2 and RS5) coincide with areas composed of massive rocks, while the less dense regions (RS1 and RS4) are interpreted as deeply hydrothermally altered areas within which known cavities exist. Good correlations are observed with models derived from gravity and electrical measurements.

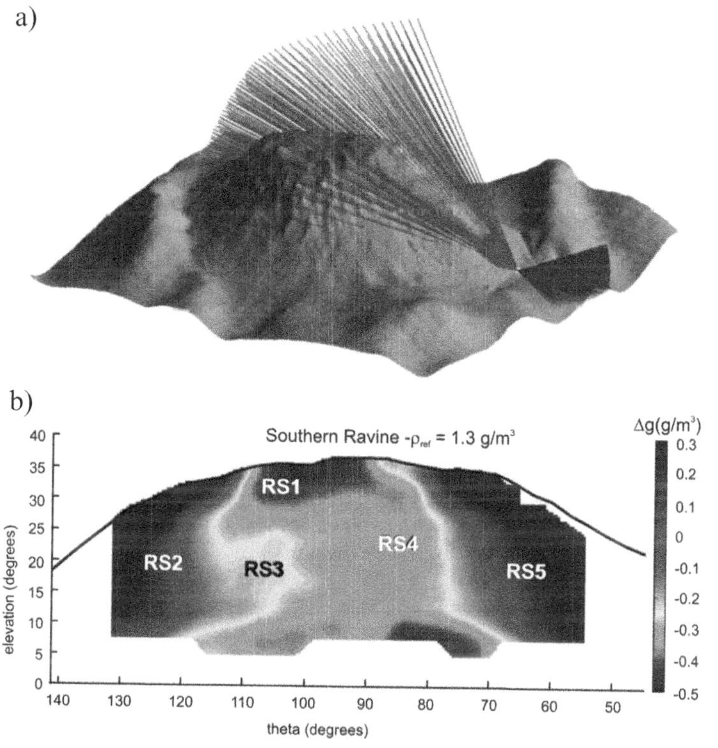

Figure 4.14. *Muon image of La Soufrière of Guadeloupe (from Lesparre et al. (2012)). a) A visualization of the muon trajectories reaching the telescope. b) Muon radiography of the dome of La Soufrière. Integrated density variations over the thickness of rock traversed are expressed relative to an average density of 1.3. The different zones are described in the text. For a color version of this figure, see www.iste.co.uk/lenat/hazards3.zip*

4.7.4.2. *Detection of a mass variation in the Asama volcano (Japan)*

During the eruption of Asama volcano on February 2, 2009, about 50,000 tons of tephra were ejected from the crater. Muography data had been acquired over this area since 2008 and continued to be acquired after the eruption (Tanaka et al. 2009). As shown in Figure 4.15, there is a subtle but significant variation between the pre- and post-eruption muon images. Interpretation of this variation leads to an estimate of the mass change in the crater of about 65,000 tons, a value comparable to the geological estimate.

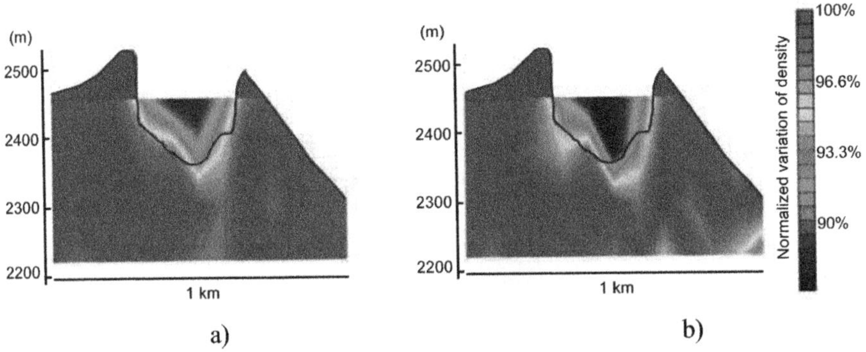

Figure 4.15. *Density variations (normalized to the highest density) integrated over the through-rock thickness of the Asama summit area before and after the February 2, 2009 eruption. a) Between January 6 and February 1 and b) between February 2 and 8. From Tanaka et al. (2009). For a color version of this figure, see www.iste.co.uk/lenat/hazards3.zip*

4.8. Perspectives

Microgravimetric monitoring thus contributes significantly to improve the monitoring of mass redistribution within structures, with the objective of discriminating their origin (superficial or deep, magmatic or hydrothermal). In parallel, quantifying the evolution of the density parameter over time is particularly interesting to follow the evolution of both magmatic and hydrothermal activity (e.g. evolution of density as a function of magmatic transfers and deep recharge) and to estimate the associated risks (water-magma interaction, major explosions). Many developments are underway to further improve and make more efficient the microgravity monitoring of structures. The miniaturization of sensors now opens the prospect of rapid surveys by drones and satellite monitoring is also the subject of continuous progress. Processing and modeling methods are also constantly evolving in order to provide 4D models, that is, spatio-temporal, more constrained and realistic (e.g. joint interpretation of different complementary datasets: gravimetry, seismic and deformation). Finally, muon imaging is in full development. Here again, the expected technical innovations and progress in data processing and interpretation suggest that this method will take a privileged place in the arsenal of methods for studying and monitoring volcanoes.

4.9. Acknowledgment

We thank Michel Diament for his constructive and accurate comments.

4.10. References

Alvarez, L.W., Anderson, J.A., Bedwei, F., El Burkhard, J., Fakhry, A., Girgis, A., Goneid, A., Hassan, F., Iverson, D., Lynch, G. et al. (1970). Search for hidden chambers in the pyramids. *Science*, 167(3919), 832–839 [Online]. Available at: https://www.science.org/doi/10.1126/science.167.3919.832 [Accessed May 24, 2022].

Ambrosino, F., Anastasio, A., Bross, A., Béné, S., Boivin, P., Bonechi, L., Cârloganu, C., Ciaranfi, R., Cimmino, L., Combaret, C. et al. (2015). Joint measurement of the atmospheric muon flux through the Puy de Dôme volcano with plastic scintillators and Resistive Plate Chambers detectors. *J. Geophys. Res. Solid Earth*, 120(11), 7290–7307.

Berrino, G., Rymer, H., Brown, G.C., Corrado, G. (1992). Gravity-height correlations for unrest at calderas. *J. Volcanol. Geotherm. Res.*, 53, 11–26 [Online]. Available at: http://www.sciencedirect.com/science/article/pii/037702739290071K.

Bonvalot, S., Remy, D., Deplus, C., Diament, M., Gabalda, G. (2008). Insights on the March 1998 eruption at Piton de la Fournaise volcano (La Réunion) from micro-gravity monitoring. *J. Geophys. Res. Solid Earth*, 113 [Online]. Available at: https://agupubs.onlinelibrary.wiley.com/doi/abs/10.1029/2007JB005084 [Accessed May 24, 2022].

Borozdin, K.N., Hogan, G.E., Morris, C., Priedhorsky, W.C., Saunders, A., Schultz, L.J., Teasdale, M.E. (2003). Radiographic imaging with cosmic-ray muons. *Nature*, 422, 277.

Carbone, D. and Poland, M.P. (2012). Gravity fluctuations induced by magma convection at Kīlauea Volcano, Hawai'i. *Geology*, 40, 803–806.

Carbone, D., Poland, M.P., Patrick, M.R., Orr, T.R. (2013). Continuous gravity measurements reveal a low-density lava lake at Kīlauea Volcano, Hawai'i. *Earth Planet. Sci. Lett.*, 376, 178–185 [Online]. Available at: http://www.sciencedirect.com/science/article/pii/S0012821X13003403 [Accessed May 24, 2022].

Carbone, D., Poland, M.P., Diament, M., Greco, F. (2017). The added value of time-variable microgravimetry to the understanding of how volcanoes work. *Earth-Science Rev.*, 169, 146–179 [Online]. Available at: http://www.sciencedirect.com/science/article/pii/S0012825216302598 [Accessed May 24, 2022].

Checchia, P. (2016). Review of possible applications of cosmic muon tomography. *J. Instrum.*, 11, C12072–C12072.

Currenti, G. (2014). Numerical evidence enabling reconciliation gravity and height changes in volcanic areas. *Geophys. J. Int.*, 197, 164–173.

Currenti, G., Del Negro, C., Ganci, G. (2007). Modelling of ground deformation and gravity fields using finite element method: An application to Etna volcano. *Geophys. J. Int.*, 169, 775–786.

Dzurisin, D. (2006). *Volcano Geodesy: New Geodetic Monitoring Techniques*, Springer-Praxis, Berlin.

Dzurisin, D., Anderson, L.A., Eaton, G.P., Koyanagi, R.Y., Lipman, P.W., Lockwood, J.P., Okamura, R.T., Puniwai, G., Sako, M., Yamashita, K. (1980). Geophysical observations of Kilauea volcano, Hawaii 2. Constraints on the magma supply during November 1975–September 1977. *J. Volcanol. Geotherm. Res.*, 7, 241–269.

Gailler, L. and Kauahikaua, J. (2018). Gravity signature of basaltic fill in Kīlauea caldera, Island of Hawai'i. In *Field Volcanology: A Tribute to the Distinguished Career of Don Swanson*, volume 538, Poland, M.P., Garcia, M.O., Camp, V.E., Grunder, A. (eds). The Geological Society of America, Reston [Online]. Available at: http://pubs.er.usgs.gov/publication/70200703 [Accessed May 24, 2022].

Gailler, L., Lénat, J.-F., Lambert, M., Levieux, G., Villeneuve, N., Froger, J.-L. (2009). Gravity structure of Piton de la Fournaise volcano and inferred mass transfer during the 2007 crisis. *J. Volcanol. Geotherm. Res.*, 184, 31–48.

Gottsmann, J., Carniel, R., Coppo, N., Wooller, L., Hautmann, S., Rymer, H. (2007). Oscillations in hydrothermal systems as a source of periodic unrest at caldera volcanoes: Multiparameter insights from Nisyros, Greece. *Geophys. Res. Lett.*, 34 [Online]. Available at: https://agupubs.onlinelibrary.wiley.com/doi/abs/10.1029/2007GL029594 [Accessed May 24, 2022].

Greco, F., Currenti, G., D'Agostino, G., Germak, A., Napoli, R., Pistorio, A., Del Negro, C. (2012). Combining relative and absolute gravity measurements to enhance volcano monitoring. *Bull. Volcanol.*, 74, 1745–1756.

Hurwitz, S., Christiansen, L.B., Hsieh, P.A. (2007). Hydrothermal fluid flow and deformation in large calderas: Inferences from numerical simulations. *J. Geophys. Res. Solid Earth*, 112 [Online]. Available at: http://dx.doi.org/10.1029/2006JB004689 [Accessed May 24, 2022].

Hutnak, M., Hurwitz, S., Ingebritsen, S.E., Hsieh, P.A. (2009). Numerical models of caldera deformation: Effects of multiphase and multicomponent hydrothermal fluid flow. *J. Geophys. Res. Solid Earth*, 114 [Online]. Available at: https://agupubs.onlinelibrary.wiley.com/doi/abs/10.1029/2008JB006151 [Accessed May 24, 2022].

Ingebritsen, S.E., Geiger, S., Hurwitz, S., Driesner, T. (2010). Numerical simulation of magmatic hydrothermal systems. *Rev. Geophys.*, 48 [Online]. Available at: https://agupubs.onlinelibrary.wiley.com/doi/abs/10.1029/2009RG000287 [Accessed May 24, 2022].

Jachens, R.C. and Eaton, G.P. (1980). Geophysical observations of Kilauea volcano, Hawaii 1. Temporal gravity variations related to the 29 November, 1975, M = 7.2 earthquake and associated summit collapse. *J. Volcanol. Geotherm. Res.*, 7, 225–240.

Jenneson, P.M. (2004). Large vessel imaging using cosmic-ray muons. *Nucl. Instruments Methods Phys. Res. Sect. A Accel. Spectrometers, Detect. Assoc. Equip.*, 525, 346–351 [Online]. Available at: http://www.sciencedirect.com/science/article/pii/S0168900204004607 [Accessed May 24, 2022].

Lesparre, N., Gibert, D., Marteau, J., Déclais, Y., Carbone, D., Galichet, E. (2010). Geophysical muon imaging: Feasibility and limits. *Geophys. J. Int.*, 183, 1348–1361.

Lesparre, N., Gibert, D., Marteau, J., Komorowski, J.-C., Nicollin, F., Coutant, O. (2012). Density muon radiography of La Soufrière of Guadeloupe volcano: Comparison with geological, electrical resistivity and gravity data. *Geophys. J. Int.*, 190, 1008–1019.

Marsh, B.D. (2014). On the mechanics of caldera resurgence [Online]. Available at: https://agupubs.onlinelibrary.wiley.com/doi/abs/10.1002/9781118782095.ch4 [Accessed May 24, 2022].

Marteau, J., Gibert, D., Lesparre, N., Nicollin, F., Noli, P., Giacoppo, F. (2012). Muons tomography applied to geosciences and volcanology. *Nucl. Instruments Methods Phys. Res. Sect. A Accel. Spectrometers, Detect. Assoc. Equip.*, 695, 23–28.

Melchior, P. and Ducarme, B. (1989). L'étude des phénomènes de marée gravimétrique. *Géodynamique*, 4, 3–14.

Merriam, J.B. (1992). Atmospheric pressure and gravity. *Geophys. J. Int.*, 109, 488–500 [Online]. Available at: https://onlinelibrary.wiley.com/doi/abs/10.1111/j.1365-246X.1992.tb00112.x [Accessed May 24, 2022].

Michon, L., Staudacher, T., Ferrazzini, V., Bachèlery, P., Marti, J. (2007). April 2007 collapse of Piton de la Fournaise: A new example of caldera formation. *Geophys. Res. Lett.*, 34 [Online]. Available at: http://dx.doi.org/10.1029/2007GL031248.

Middlemiss, R.P., Samarelli, A., Paul, D.J., Hough, J., Rowan, S., Hammond, G.D. (2016). Measurement of the Earth tides with a MEMS gravimeter. *Nature*, 531, 614–617.

Mogi, K. (1958). Relation of the eruptions of various volcanoes and the deformations of the ground surfaces around them. *Bull. Earthq. Res. Inst.*, 36, 99–134.

Nagamine, K., Iwasaki, M., Shimomura, K., Ishida, K. (1995). Method of probing inner-structure of geophysical substance with the horizontal cosmic-ray muons and possible application to volcanic eruption prediction. *Nucl. Instruments Methods Phys. Res. Sect. A Accel. Spectrometers, Detect. Assoc. Equip.*, 356, 585–595 [Online]. Available at: http://www.sciencedirect.com/science/article/pii/0168900294011699 [Accessed May 24, 2022].

Neal, C.A., Brantley, S.R., Antolik, L., Babb, J.L., Burgess, M., Calles, K., Cappos, M., Chang, J.C., Conway, S., Desmither, L. et al. (2019). The 2018 rift eruption and summit collapse of Kilauea Volcano. *Science*, 363(6425), 367–374.

Neddermeyer, S.H. and Anderson, C.D. (1937). Note on the nature of cosmic-ray particles. *Phys. Rev.*, 51, 884–886 [Online]. Available at: https://link.aps.org/doi/10.1103/PhysRev.51.884 [Accessed May 24, 2022].

Poland, M.P., Garcia, M.O., Camp, V.E., Grunder, A., Gailler, L., Kauahikaua, J. (2019). Gravity signature of basaltic fill in Kīlauea caldera, Island of Hawai'i [Online]. Available at: https://dx.doi.org/10.1130/2018.2538(13) [Accessed May 24, 2022].

Rymer, H. (1994). Microgravity change as a precursor to volcanic activity. *J. Volcanol. Geotherm. Res.*, 61, 311–328 [Online]. Available at: https://doi.org/10.1016/0377-0273(94)90011-6.

Rymer, H. (1996). Microgravity monitoring. In *Monitoring and Mitigation of Volcano Hazards*, Scarpa, R. and Tilling, R.I. (eds). Springer, Berlin.

Seigel, H.O. (1995). A guide to high precision land gravimeter surveys [Online]. Available at: https://scintrexltd.com/wp-content/uploads/2017/02/Guide-High-Precision-Land-Gravimeter-Surveys.pdf.

Smith, R.L. and Bailey, R.A. (1968). Resurgent cauldrons. *Mem. Geol. Soc. Am.*, 116, 613–662.

Staudacher, T., Bachèlery, P., Semet, M.P., Cheminée, J.-L. (1998). Report on Piton de la Fournaise (France). *Bulletin of the Global Volcanism Network*, 23, 2–9.

Staudacher, T., Ferrazzini, V., Peltier, A., Kowalski, P., Boissier, P., Catherine, P., Lauret, F., Massin, F. (2009). The April 2007 eruption and the Dolomieu crater collapse, two major events at Piton de la Fournaise (La Réunion Island, Indian Ocean). *J. Volcanol. Geotherm. Res.*, 184, 126–137 [Online]. Available at: http://www.sciencedirect.com/science/article/pii/S0377027308005945 [Accessed May 24, 2022].

Swanson, D.A., Rose, T.R., Fiske, R.S., McGeehin, J.P. (2012). Keanakāko'i Tephra produced by 300 years of explosive eruptions following collapse of Kīlauea's caldera in about 1500CE. *J. Volcanol. Geotherm. Res.*, 8–25, 215–216 [Online]. Available at: http://www.sciencedirect.com/science/article/pii/S0377027311003131 [Accessed May 24, 2022].

Swanson, D.A., Fiske, R.S., Rose, T.R., Mucek, A.E., Garcia, M.O., Mastin, L.G. (2014). Cycles of explosive and effusive eruptions at Kīlauea Volcano, Hawai'i. *Geology*, 42, 631–634.

Tait, S., Jaupart, C., Vergniolle, S. (1989). Pressure, gas content and eruption periodicity of a shallow, crystallising magma chamber. *Earth Planet. Sci. Lett.*, 92, 107–123 [Online]. Available at: http://www.sciencedirect.com/science/article/pii/0012821X89900253 [Accessed May 24, 2022].

Tanaka, H.K.M., Nakano, T., Takahashi, S., Yoshida, J., Ohshima, H., Maekawa, T., Watanabe, H., Niwa, K. (2007). Imaging the conduit size of the dome with cosmic-ray muons: The structure beneath Showa-Shinzan Lava Dome, Japan. *Geophys. Res. Lett.*, 34 [Online]. Available at: https://doi.org/10.1029/2007GL031389 [Accessed May 24, 2022].

Tanaka, H.K.M., Uchida, T., Tanaka, M., Takeo, M., Oikawa, J., Ohminato, T., Aoki, Y., Koyama, E., Tsuji, H. (2009). Detecting a mass change inside a volcano by cosmic-ray muon radiography (muography): First results from measurements at Asama volcano, Japan. *Geophys. Res. Lett.*, 36 [Online]. Available at: https://doi.org/10.1029/2009GL039448 [Accessed May 24, 2022].

Todesco, M., Rinaldi, A.P., Bonafede, M. (2010). Modeling of unrest signals in heterogeneous hydrothermal systems. *J. Geophys. Res. Solid Earth*, 115 [Online]. Available at: https://agupubs.onlinelibrary.wiley.com/doi/abs/10.1029/2010JB007474 [Accessed May 24, 2022].

Urai, M., Geshi, N., Staudacher, T. (2007). Size and volume evaluation of the caldera collapse on Piton de la Fournaise volcano during the April 2007 eruption using ASTER stereo imagery. *Geophys. Res. Lett.*, 34 [Online]. Available at: https://agupubs.onlinelibrary.wiley.com/doi/abs/10.1029/2007GL031551 [Accessed May 24, 2022].

Van Camp, M., de Viron, O., Watlet, A., Meurers, B., Francis, O., Caudron, C. (2017). Geophysics from terrestrial time-variable gravity measurements. *Rev. Geophys.*, 55, 938–992.

Williams-Jones, G. and Rymer, H. (2002). Detecting volcanic eruption precursors: A new method using gravity and deformation measurements. *J. Volcanol. Geotherm. Res.*, 113, 379–389 [Online]. Available at: http://www.sciencedirect.com/science/article/pii/S0377027301002724 [Accessed May 24, 2022].

5

Electrical and Electromagnetic Monitoring of Volcanoes

Lydie GAILLER, Jean-François LÉNAT and Philippe LABAZUY
Laboratoire Magmas et Volcans, CNRS, IRD, OPGC,
Université Clermont Auvergne, Clermont-Ferrand, France

5.1. Introduction

Electrical or electromagnetic (EM) methods give access to the distribution of the electrical resistivity of rocks at depth. Electrical resistivity is a parameter sensitive to the nature of the rocks, the composition of the fluids present and the temperature. Inside a volcano and in its roots, the resistivity of the rocks can vary in time under the effect of different phenomena: magma intrusion, fluid transfers, heating, variation of permeability and porosity in response to the formation of hydrothermal minerals or to stress variations. This internal dynamics can be understood with repeated measurements in time.

In addition to resistivity variations, electric and magnetic fields can be transiently disturbed by stress variations (see section 6.2.1) or fluid movements.

Electrical and EM monitoring methods have not yet reached an operational level comparable to that of seismic and deformation monitoring. Nevertheless, they constitute a promising field and, in certain cases, they can significantly enrich the knowledge of the internal dynamic phenomena of volcanic activity. A synthesis of these methods was made by Lénat (1995), and this chapter will introduce recent developments.

Hazards and Monitoring of Volcanic Activity 3,
coordinated by Jean-François LÉNAT.

5.2. Phenomena at the origin of resistivity and electric or EM field variations

5.2.1. *Reminder on the resistivity of rocks*

The electrical resistivity of a rock describes its ability to allow the movement of electrical charges. It is expressed in ohm-meter (Ω.m). We also use its inverse, the conductivity, expressed in Siemens per meter ($S.m^{-1}$). The physicochemical mechanisms that determine rock resistivity are described in geophysical treatises; one of the best sources is Glover (2015).

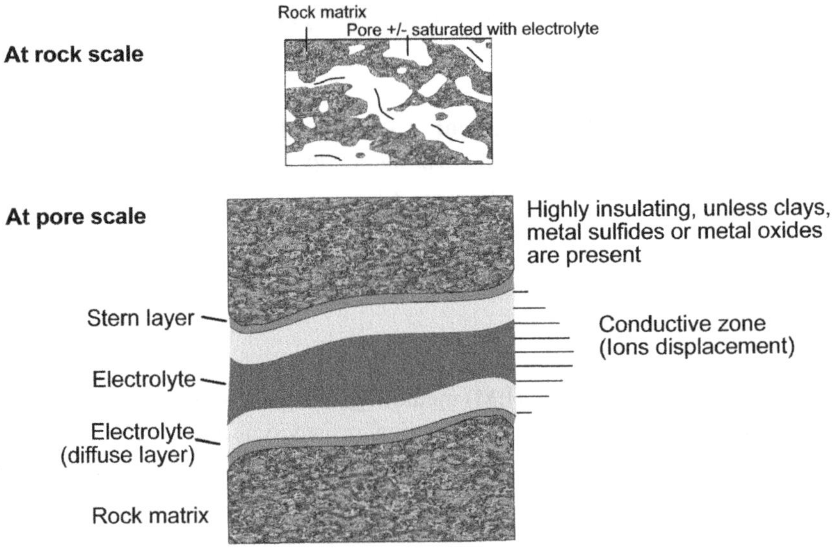

Figure 5.1. *Illustration of the electrolytic conduction of a rock*

In rocks, current conduction is primarily generated by the movement of ions from fluids (electrolytes) in the pores (*ionic conduction*) (see Figure 5.1). The matrix of a volcanic rock (minerals and glass) is, with few exceptions, highly insulating as compared to the fluids in the pores. The current transport efficiency depends on the resistivity of the electrolyte, the degree of electrolyte saturation and the density and connectivity of the pore network. The resistivity of the electrolyte depends on its ionic composition and temperature.

The presence of hydrated minerals disturbs this pattern. Indeed, clays and other hydrated minerals, through their structure, have a very large contact surface with the electrolyte of the pores and another component of conduction at the interface between the minerals and the electrolyte, called surface conduction, generally becomes predominant. This has the effect of strongly decreasing the resistivity of rocks containing these minerals. The nature and habitat of minerals formed in hydrothermal systems are generally described in publications related to geothermal energy (Browne 1978; Hersir and Árnason 2009).

Another exception is the presence of conductive or semi-conductive metallic minerals in the rock matrix. In this case, conduction is provided by electrons (*electronic conduction*). Common minerals of this type of conduction are sulfides, Fe and Ti oxides, and graphite.

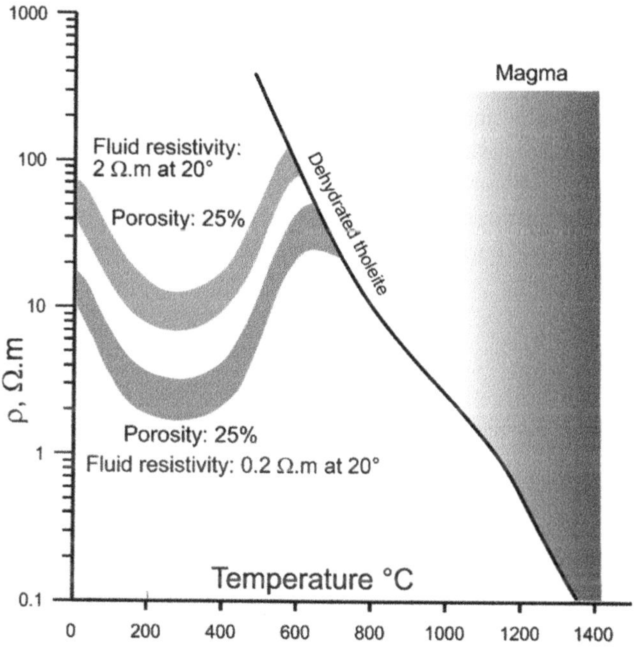

Figure 5.2. *Expected low-pressure resistivity for a fluid-saturated basalt as a function of porosity, fluid resistivity and temperature. Modified according to Kauahikaua et al. (1986)*

Magmatic liquids (magma chambers, intrusions, partial melting zones) are characterized by their low resistivity. Conduction is essentially ionic, charge transport being dominated by light alkalis (mainly Na), whose mobility is amplified by magma hydration (Gaillard 2004). Laboratory studies have highlighted the dependence of the electrical properties of magma liquids on different parameters, such as the composition of these liquids, pressure, temperature, their water content and oxygen fugacity (Gaillard and Marziano 2005; Ni et al. 2011; Guo et al. 2016). An application has been developed to calculate the electrical properties of magmatic liquids (Pommier and Le-Trong 2011).

For an ionic conduction, it is observed that the resistivity decreases with temperature until about 300°C before increasing again due to the increasing presence of steam (Rai and Manghnani 1977). Around 600°C, we reach the curve of a dehydrated basalt because the conduction of the rock matrix becomes superior to that of the electrolyte carried well beyond its critical temperature. The resistivity will decrease again until the fusion (see Figure 5.2).

5.2.2. *Causes of rock resistivity variation in the volcanic domain*

The volcanic edifice and its roots are dynamic domains where phenomena are likely to cause significant variations in the electrical resistivity of certain volumes of the infrastructure (see Figure 5.3). The objective of the monitoring will therefore be to measure and locate these variations and to interpret them in terms of volcanic dynamics, taking care to correct for meteoric influences on resistivity.

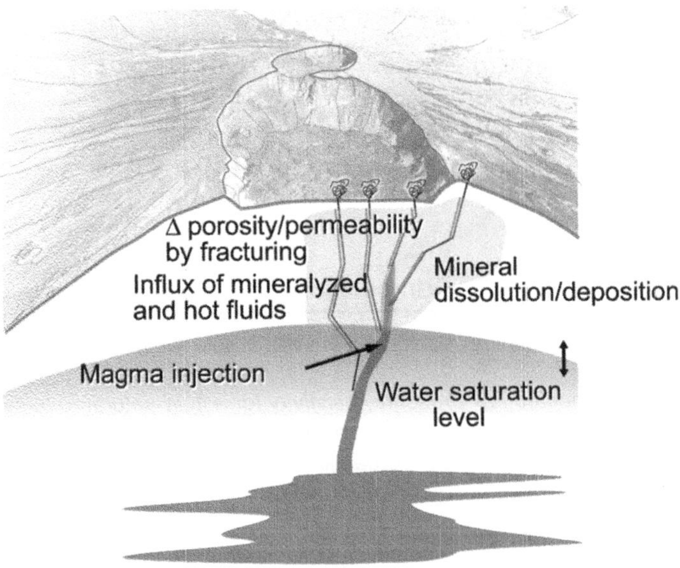

Figure 5.3. *Illustration of the different phenomena that can cause resistivity variations in volcanic structures*

The main sources of variation are as follows:

– Variations in the resistivity of the hydrothermal zone, either by dissolution of minerals and/or deposition of new mineral phases (generally conductive: clays, zeolites, hydrated minerals in general) or by the arrival of mineralized and/or hot fluids released by the magmatic system, or by the condensation of magmatic gases, or by variations in porosity/permeability due to clogging by mineral deposits or, conversely, to the opening up of fractures due to over-pressure or in connection with seismic activity.

– Variations in the level of fluid saturation of rocks due to fluctuations in aquifers or changes in hydrothermal circulation.

– Finally, the injection of magma will contribute to the modifications mentioned above, on the one hand, and will create regions with very low resistivity, on the other hand.

5.3. Natural and induced electric and EM fields

EM methods are based on the frequency or time relationships between various components of the magnetic field and/or those of the electric field. There are natural EM sources, the most important of which come from the interaction of the magnetosphere with solar wind particles or from atmospheric lightning. There may also be internal sources related to stress variations (see section 6.2.1) or originating from ocean movements and fluid circulations in rocks. Some monitoring methods use these natural sources, while others will use artificial fields created locally for measurement purposes or emitted by other activities (e.g. radio emissions).

5.4. Monitoring: instruments and procedures

5.4.1. *Direct current monitoring*

This type of monitoring is based on instruments and procedures used in geophysical surveying. The principle is to inject an electric current into the ground through two electrodes (injection pair classically designated AB) and to measure the induced potential difference with another pair of electrodes (designated MN) (see Figure 5.4). The geometrical relations between AB and MN allow the calculation of an apparent resistivity, at a given point in depth. By varying the geometry of the injection and the potential measurement, the subsurface can be investigated laterally and at depth. The set of measurements is used to calculate true resistivity distribution models. These methods are well described in geophysical treatises (Milsom and Eriksen 2011; Dubois et al. 2016).

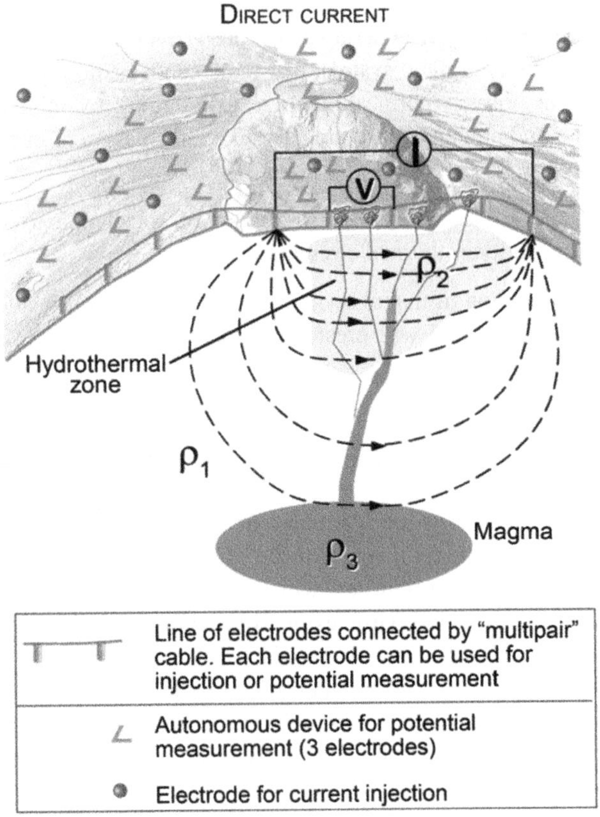

Figure 5.4. *Illustration of methods for measuring direct current resistivity. The distribution of the injected current lines (and thus of the potentials measured at the surface) depends on the resistivity of the rocks at depth and on the geometry of the device. For a color version of this figure, see www.iste.co.uk/lenat/hazards3.zip*

In order to show temporal variations in resistivity, the measurements must be repeated. This can be done with periodic measurement campaigns, but the tendency is to develop permanent devices that will be able to perform measurements at a higher frequency. Figure 5.4 illustrates two types of devices frequently used in prospecting. One consists of a line of electrodes along which the geometry of the injection and the potential measurement can be varied. In this case, we work in 2D. Another device consists of implanting independent equipment for the measurement of the potential difference (a semis of autonomous stations) and for the injection (one fixed electrode is generally placed outside the studied zone and another in or near this zone. This last electrode is moved in order to have a sampling of current lines favorable to the investigation of the whole area). This device is more adapted to a 3D investigation.

The monitoring networks are based on the principle of these prospecting devices. For technical, logistical and cost reasons, the number of injection and potential difference pairs can be reduced and adapted to the area to be monitored.

5.4.1.1. *Examples of direct current monitoring*

Few direct current (DC) experiments have been conducted (Yukutake et al. 1990; Zlotnicki et al. 2003) because this method involves placing electrical cable lines on the ground, which are very sensitive to lightning. The development of more compact systems, for example, the FullWaver system (Anon 2019), opens up new possibilities, especially in terms of 3D electrical tomography.

5.4.1.1.1. The example of the 1986 eruption of the Izu-Oshima volcano (Japan)

After the 1974 eruption of Izu-Oshima volcano, a resistivity monitoring network was installed in 1975 (see Figure 5.5(a)). It consisted of three injection electrodes located on one side of the summit crater and three potential measurement electrodes located on the other side. This device allowed several combinations of current injection and measurement of induced potential differences (A, B and C in Figures 5.5(a) and 5.5(b)). A series of measurements over a dozen years (see Figure 5.5(b)) has been published (Yukutake et al. 1990) and is the most comprehensive example of DC monitoring published to date. An interpretation of the signals preceding and accompanying the 1986 eruption has been proposed by Utada (2003) (see Figure 5.5(c)).

After the 1974 eruption, a drift in resistivities was observed (the resistivity values in Figure 5.5(b) are normalized to their value observed at the installation of the device: 84 Ω.m for A, 112 for B and 168 for C). Then they became stable again, with relatively small variations, some of which could be related to strong regional earthquakes. In early 1986 (stage 1 in Figure 5.5(b)), stronger variations were observed on pair C (the one with greater penetration), while pairs A and B showed a fairly similar increase. At the end of September (stage 2, see Figure 5.5(b)), sharp variations are observed on all pairs, with a greater amplitude on pair C. The eruption started on November 15 (stage 3, see Figure 5.5(b)).

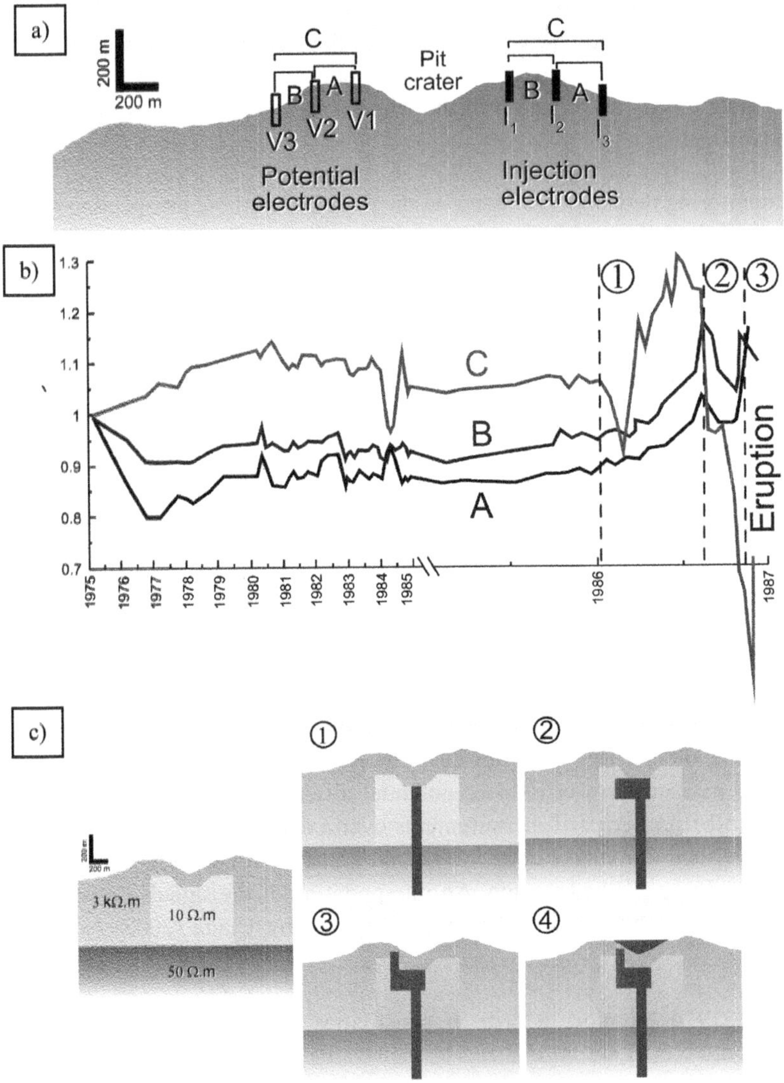

Figure 5.5. *Monitoring of the Izu-Oshima volcano with direct current*

COMMENT ON FIGURE 5.5.– *a) Measuring device. b) Time series of resistivity for different dipole-dipole geometries (resistivity values are normalized to their value at the beginning of the record). c) Model explaining variations by magma emplacement (in red, resistivity 0.5 Ω.m). Modified according to Utada (2003).*

Utada (2003) proposed a model, shown in Figure 5.5(c), explaining these variations by several stages of magma intrusion (resistivity 0.5 Ω.m) at relatively shallow depth. The initial resistivity distribution is based on previous electrical and magnetotelluric borehole investigations. Three stages of magma injection are defined, the fourth showing the post-eruption state. This model correctly explains the main variations in resistivity observed. However, it has limitations because the resistivity variations are only explained by the magma intrusion. It is easy to imagine that other phenomena have contributed to the variations. In particular, it is likely that the hydrothermal system was disturbed with an increase in permeability in response to the deformation of the edifice and the arrival of hot, mineralized magma gases.

Nevertheless, this experience has shown that volcanic crises can be preceded and accompanied by variations in resistivity at depth with significant amplitudes, well beyond the detection thresholds of measurement devices. This was confirmed more recently during the 2000 Miyakejima crisis (Zlotnicki et al. 2003).

5.4.2. *EM monitoring*

A wide variety of EM methods can be used in volcanic monitoring (see Figure 5.6). Some use natural signals. These are the magnetotelluric (MT) or audio-magnetotelluric (AMT) methods that use externally generated fields (Vozoff 1991). The spontaneous polarization (SP) method is based on natural signals related to fluid movements in the subsurface. Other methods use artificial signals (Fitterman 2015). These are generated by ground loops or induction lines or by antennas. Among these methods, a distinction is made between those that use a sinusoidal signal (FDEM, frequency domain electromagnetic) and those that use a stepped or pulsed signal (usually a succession of signals created by periodically injecting and cutting current in a loop or transmitter line; TDEM, time domain electromagnetic). With FDEM methods, we typically measure the amplitude and phase of the signal, while with TDEM methods, we measure the amplitude of the signal over time after the transmitter signal is cut off.

The general principle of these methods is based on a primary magnetic field that penetrates the soil where it induces an electric field that will itself generate a secondary magnetic field. The latter quantities will be influenced by the distribution of resistivity at depth. The depth of investigation will depend on the distribution of resistivities, the frequency of the primary field and the measuring device.

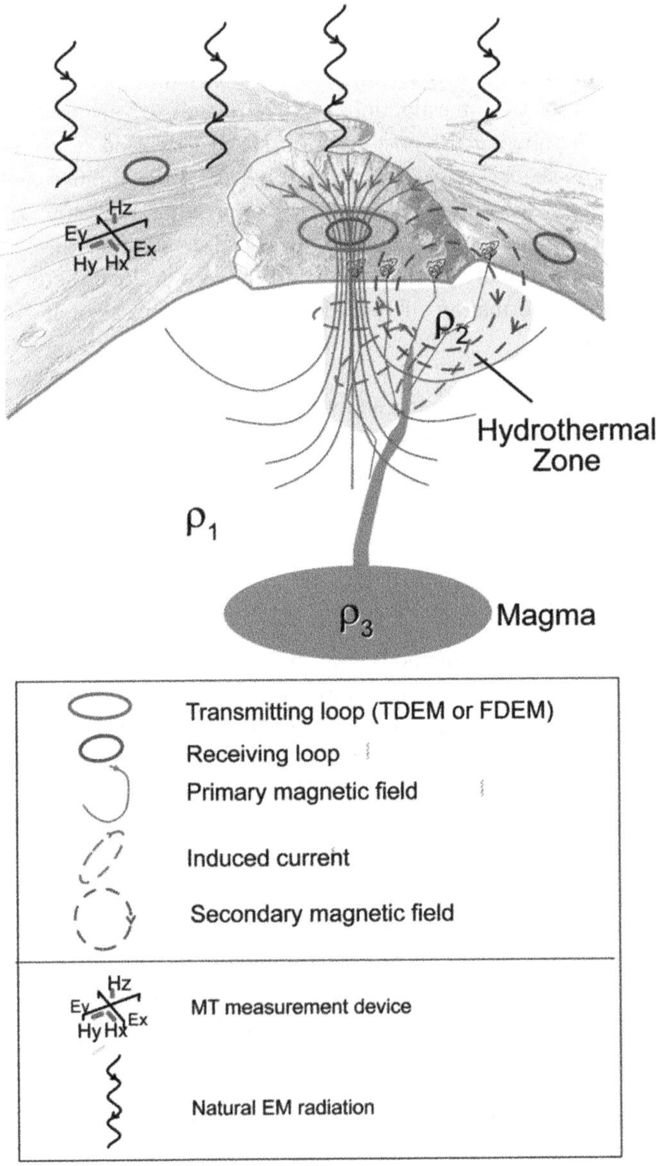

Figure 5.6. *Principles of depth resistivity measurement by active (artificially generated signals) or passive (natural signals) EM methods. For a color version of this figure, see www.iste.co.uk/lenat/hazards3.zip*

5.4.2.1. *Example of monitoring with a controlled EM source: the "ACTIVE" experiment on the Izu-Oshima volcano*

Since the DC monitoring experiment conducted between 1975 and 1987 (see above and Figure 5.5), it has been known that resistivity variations can be detected inside the Izu-Oshima volcano in response to internal activity. Despite this success, the DC monitoring experiment also showed that maintenance of a device involving long lines of surface cables was very difficult. Utada et al. (2007) experimented with an EM system to perform similar monitoring of the same area. This system was named ACTIVE (for Array of Controlled Transient-electromagnetics for Imaging a Volcanic Edifice).

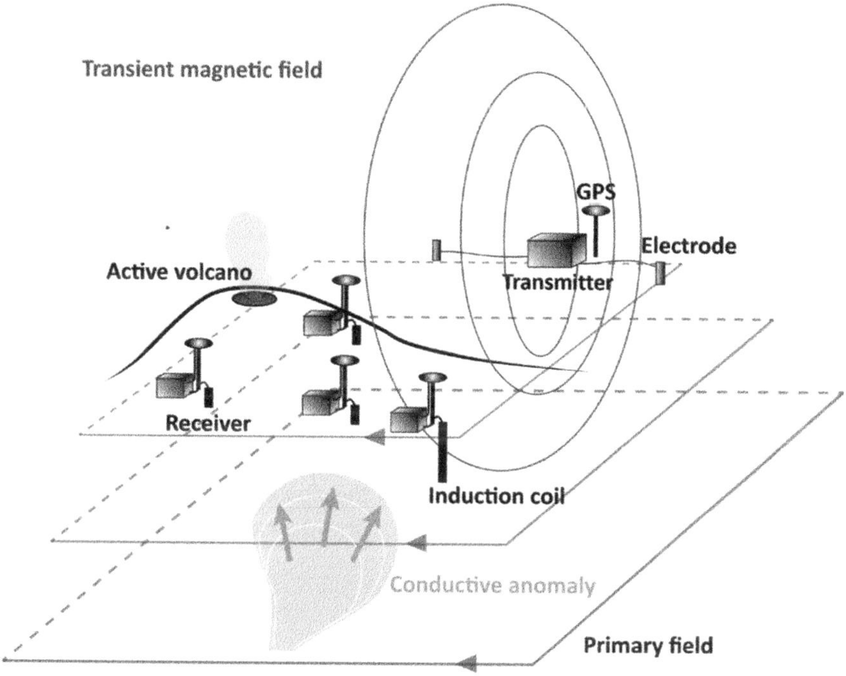

Figure 5.7. *Schematic of the ACTIVE network on Izu-Oshima volcano. From Utada et al. (2007)*

ACTIVE is similar to a system known as LOTEM (Long Offset Transient Electromagnetics) in geophysics (Strack et al. 1990). It consists of a 700 m long transmitter dipole and an array of five receivers that measure the vertical component of the magnetic field (see Figure 5.7). The whole system is synchronized by GPS clocks. Measurements are programmed at a certain periodicity. Over a period of

4 years (July 2002 to April 2005), the daily accuracy of the measurements has been estimated to be 1% or less. Variations (decreases) in resistivity of up to 7% have been attributed to two sources (see Figure 5.8). One is very superficial, in the area near the transmitter, and another at a depth of about 500 m below the crater. The interpretation of the data is based on 1D, 2D and 3D numerical modeling codes to calculate theoretical responses and model the evolution of the resistivity distribution at depth. The surface source does not seem to be related to a volcanic process. But the second one could be related to a temperature increase. This interpretation is consistent with other observations, such as the decrease in apparent resistivity observed by electrical probing and the increase in seismicity during this period. It is therefore a relatively accurate imaging method whose future developments will allow for reinforcing and improving the monitoring of volcanic activity.

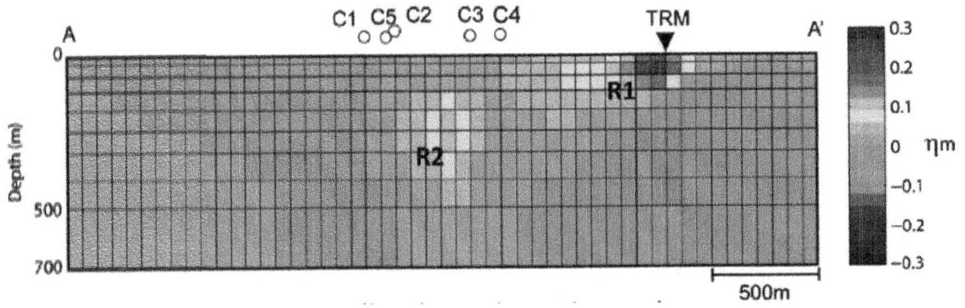

Figure 5.8. *Cross-section of a conductivity variation model during 4 years of measurements of the ACTIVE network. C1 to C5 show the location of the receivers, TRM the location of the transmitter, and R1 and R2 the two anomalies described in the text. Warm colors indicate an increase in conductivity (normalized in the diagram after Utada et al. (2007)). For a color version of this figure, see www.iste.co.uk/lenat/hazards3.zip*

5.4.2.1.1. Example of MT monitoring

Based on natural signals and offering a wide range of investigation depths, the MT method is a priori well suited for monitoring resistivity variations at depth. However, relatively few experiments have been published in volcanology (there are a few more on the monitoring of exploited geothermal sites). Of these, there are some on Sakurajima (Aizawa et al. 2011) and Piton de la Fournaise (Wawrzyniak et al. 2017). The latter is described below.

An important approach, valid for all electrical and EM monitoring methods, is to test a priori their sensitivity to variations that can be anticipated as a function of different phenomena. For example, one can consult the article by Carbonari et al. (2019) that studies, through models, the types of changes in the Solfatara hydrothermal system (Phlegrean Fields, Italy) detectable by MT monitoring.

5.4.2.1.2. Resistivity variations related to the March 9, 1998 eruption of Piton de la Fournaise (Wawrzyniak et al. 2017)

The March 9, 1998 eruption of Piton de la Fournaise was preceded by a migration of seismic activity to the surface from a depth of 5 km below sea level (Battaglia et al. 2005). The onset of this crisis preceded the eruption by 35 hours. Two fissures opened north of the central cone, followed by another on March 11 to the southwest (see Figure 5.9).

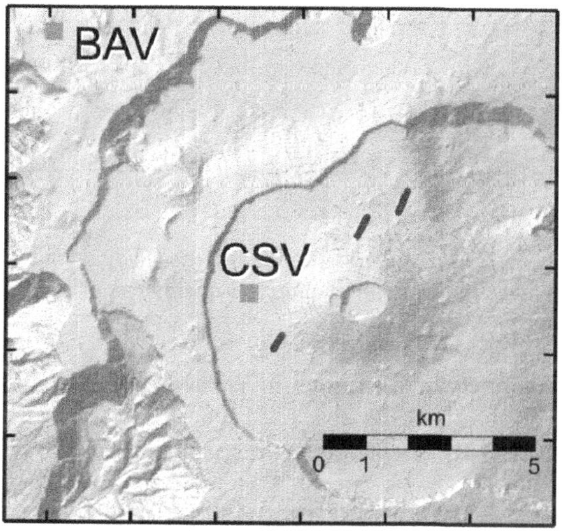

Figure 5.9. *Piton de la Fournaise EM network (1996–1999), in blue, and cracks from the March 1998 eruption, in black. For a color version of this figure, see www.iste.co.uk/lenat/hazards3.zip*

An EM monitoring network operated from 1996 to 1999 (Wawrzyniak et al. 2017). Two stations (BAV and CSV, see Figure 5.9) consisted of magnetometers (resolution 0.01 nT) recording the horizontal component of the magnetic field and electrodes for the electric field (resolution 5μV), for a measurement frequency of 20 s. The frequency band recorded was from 1.6 to 12.5 mHz (periods of about 625 to 80 s). This device allowed us to calculate an apparent resistivity for different frequencies.

Variations were observed before and especially during the eruption (see Figure 5.10) in bands corresponding to fairly long periods (80, 160, 320 and 640 s). Theoretically, signals at such periods are more sensitive to deep structures (tens of kilometers). The variations were best defined for periods of 80 and 120 s. The authors believe that the presence of magma with very low resistivity reduces the depth of investigation. Moreover, during the first phase of the crisis, it was observed that the ellipse of the phase tensor became elongated and oriented at N105°.

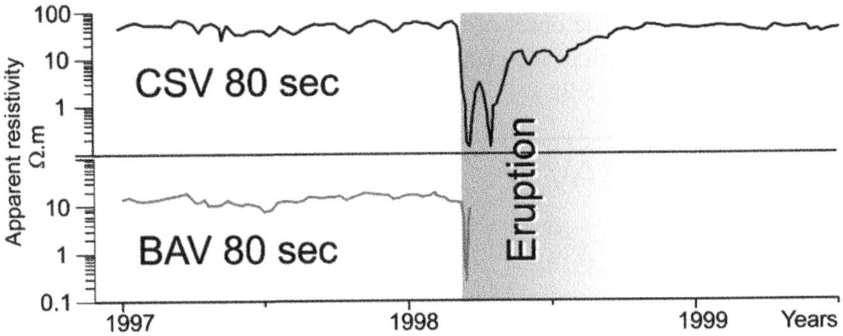

Figure 5.10. *Evolution of apparent resistivity during the 1998 crisis. Modified according to Wawrzyniak et al. (2017))*

The authors interpret the variations of apparent resistivity by the arrival of magma more conductive than the host rock and those of the ellipse of the phase tensor by the practically NS orientation of the magma intrusion.

5.4.2.2. *Example of FDEM monitoring*

A source-controlled EM monitoring system (0.01–8 Hz) was set up near the summit of Kilauea in 1979. Large and rapid variations were measured both before and after the August 1980 and January 1981 intrusive events, indicating the emplacement of shallow magmatic intrusions (Jackson et al. 1985).

5.4.2.3. *Example of VLF monitoring*

The VLF (Very Low Frequency) method is based on strong signals from military transmitters intended for communication with submarines. These emissions have a worldwide coverage in a frequency range of 15–30 kHz and penetrate the ground. At distances of a few hundred kilometers from a transmitter, wave propagation is guided between the ionosphere and the ground, and thus the magnetic field is horizontal and perpendicular to the transmitter. In uniform terrain, the components of these VLF signals remain constant, while in the presence of resistivity heterogeneities they are modified. In the presence of a conductive structure, a secondary magnetic field is induced and the combination with the primary field is no longer a horizontal field. The composite field angle (tilt angle) and the phase shift between the primary and secondary fields can therefore be measured in the form of the ellipticity (Paterson and Ronka 1971) (see Figure 5.11). The measuring instruments are very compact and lightweight, and therefore easy to implement in the field.

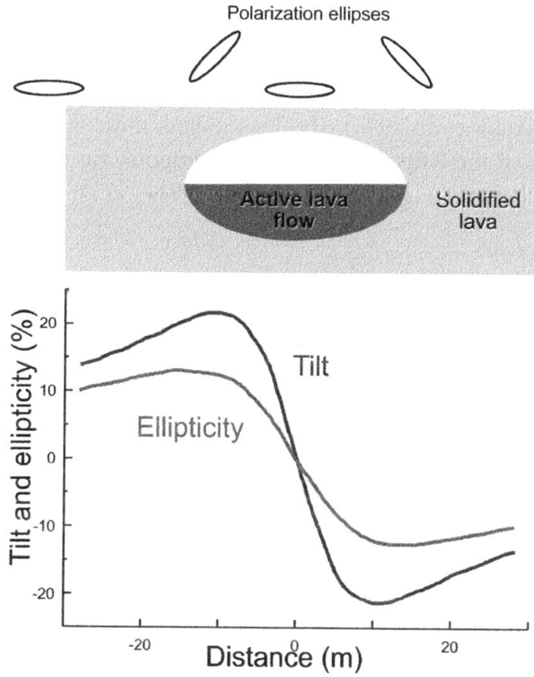

Figure 5.11. *VLF response (ellipticity and tilt – see text) above a lava flow circulating in a tube*

Due to the relatively high VLF frequencies, the depth of investigation is small (range 10–100 m). This method is particularly suitable for the detection of conductors in a resistant environment. The signal is maximum for structures elongated in the direction of the transmitter (the distribution of transmitters in the world generally allows us to approach this condition). Zablocki (1978a) described the main applications of the method in volcanology, both for static and for dynamic structures. One of the most notable applications in monitoring was the detection and characterization of active lava tubes at Kilauea (Jackson et al. 1987, 1988; Lénat 1995). Kauahikaua et al. (1996) formulated the principles for applying the method to lava tunnel monitoring (see Figure 5.11). VLF measurements above an active lava tube (thus a conductor in a resistive medium) are sensitive to the conductance (product of the cross-sectional area of the flow by its conductivity). It is therefore not necessary to know precisely the geometry of the flow because the conductance is modeled. For a plausible resistivity (based on experimental data), we obtain the cross-section of the flow. The uncertainties on the depth and conductivity will only affect the amplitude of the anomalies. If the detection of lava tubes is easy with VLF measurements on a lava field, the calculation of the flow rate requires the knowledge of the flow speed. The latter is usually determined by measurements between skylights, frequently present at the roof of the tubes, by various means of measurement (Kauahikaua et al. 1996). By this method, these authors were able to establish that the flux of the Kupaianaha (Kilauea) eruption varied from 25 000 m^3 to 54 000 m^3 per day, between April and November 1991.

5.4.2.4. *Examples of SP monitoring*

SP is a rather special method, offering the possibility of recording natural electrical currents in the frequency domain as well as in the static domain. This is the domain in which traditional SP is practiced. The natural telluric currents recorded in SP can result from magnetic induction as well as from currents generated by fluid movements (see Figure 5.1). Currents related to fluid movement result either from the application of electric fields, a pressure difference (electrokinetic or electrofiltration potentials), a thermal gradient (thermoelectric potentials) or a compositional gradient in the fluids (electrochemical potentials). Studies (Zablocki 1978b) tend to indicate that in volcanic systems electrokinetic (or electrofiltration) potentials dominate the SP signal. However, other mechanisms may contribute to the signal and it is generally difficult to distinguish between them. SP measurement requires the use of impolarizable electrodes (Petiau and Dupis 1980; Petiau 2000) to avoid electrode-to-soil contact potentials. The other constraint is to use a very high impedance voltmeter (much higher than the soil resistance between the two electrodes) in order not to short-circuit the natural signal. Examples of SP monitoring at Etna and Piton de la Fournaise are described by Lénat (1995). Here, we present the example of Miyakejima (Zlotnicki et al. 2003).

5.4.2.4.1. Temporal variations of SP anomalies of Miyakejima volcano during the 2000 crisis

Miyakejima volcano (Japan) is an active basaltic strato-volcano (eruptions in 1940, 1962, 1983 and 2000). The 2000 crisis was remarkable with a long intrusion to the northwest and the collapse of a small caldera of 1.6 km diameter and 450 m depth (Geshi et al. 2002). This volcano had been the subject of electrical and EM studies since 1995, with, in particular, SP monitoring with eight electrodes distributed along a 1 km E-W line passing through the summit (see Figure 5.12). This line was partly in line with a superficial hydrothermal system detected by AMT drilling.

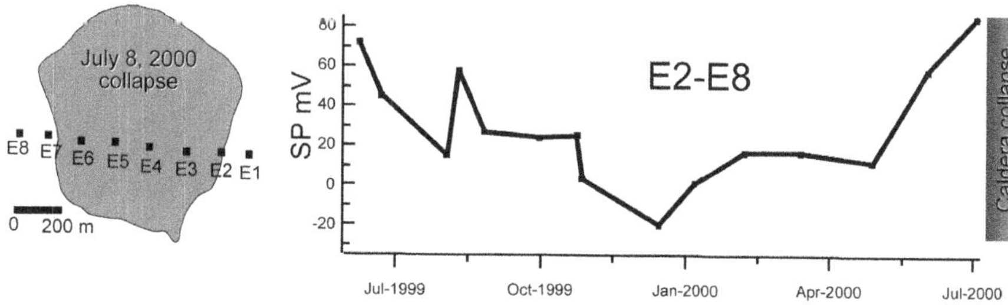

a) b)

Figure 5.12. *a) PS monitoring electrode array at Miyakejima before the 2000 crisis. b) Temporal variations in SP between E2 and E8. From Zlotnicki et al. (2003)*

The evolution since May 1999 of the potential difference between electrodes E2 and E8 is shown in Figure 5.12. Until December 1999, the decrease in potential difference is attributed by the authors to an increase in permeability in E2 allowing the infiltration of meteoric water (decrease in PS). Then, and particularly from May to June 2000, the hydrothermal circulation strengthens (increase in SP) until the beginning of the collapse of the caldera on July 8, which will be accompanied by phreatic explosions. The SP monitoring line is then destroyed. Many other geophysical and geological data were acquired during this crisis. The interpretation of the increase of the SP before the eruption in terms of an increase of the hydrothermal activity is quite coherent with the other observations.

5.5. Perspectives

Electrical and EM monitoring has often remained experimental or exploratory for volcano monitoring (but also in other geological contexts: geothermal, landslides, etc.). However, many published works prove that these methods provide original information with operational value. Spectacular instrumental developments and progress in the interpretation and modeling of the physico-chemical phenomena involved anticipate a more massive use of these techniques in volcanological monitoring.

5.6. Acknowledgment

Thanks to Jean Vandemeulebrouck for his meticulous proofreading and pertinent suggestions.

5.7. References

Aizawa, K., Kanda, W., Ogawa, Y., Iguchi, M., Yokoo, A., Yakiwara, H., Sugano, T. (2011). Temporal changes in electrical resistivity at Sakurajima volcano from continuous magnetotelluric observations. *J. Volcanol. Geotherm. Res.*, 199, 165–175 [Online]. Available at: http://www.sciencedirect.com/science/article/pii/S0377027310003446.

Battaglia, J., Ferrazzini, V., Staudacher, T., Aki, K., Cheminée, J.-L. (2005). Pre-eruptive migration of earthquakes at the Piton de la Fournaise volcano (Réunion Island). *Geophys. J. Int.*, 161, 549–558.

Browne, P.R.L. (1978). Hydrothermal alteration in active geothermal fields. *Annu. Rev. Earth Planet. Sci.*, 6, 229 [Online]. Available at: https://www.annualreviews.org/doi/10.1146/annurev.ea.06.050178.001305.

Carbonari, R., Di Maio, R., Piegari, E. (2019). Feasibility to use continuous magnetotelluric observations for monitoring hydrothermal activity. Numerical modeling applied to Campi Flegrei Volcanic System (Southern Italy). *Front. Earth Sci.*, 7, 262 [Online]. Available at: https://www.frontiersin.org/article/10.3389/feart.2019.00262.

Dubois, J., Diament, M., Cogné, J.-P., Mocquet, A. (2016). *Géophysique cours et exercices corrigés*. Dunod, Paris.

Fitterman, D.V. (2015). 11.10 – tools and techniques: Active-source electromagnetic methods [Online]. Available at: http://www.sciencedirect.com/science/article/pii/B9780444538024001937.

Gaillard, F. (2004). Laboratory measurements of electrical conductivity of hydrous and dry silicic melts under pressure. *Earth Planet. Sci. Lett.*, 218, 215–228 [Online]. Available at: http://www.sciencedirect.com/science/article/pii/S0012821X03006393.

Gaillard, F. and Marziano, G.I. (2005). Electrical conductivity of magma in the course of crystallization controlled by their residual liquid composition. *J. Geophys. Res. Solid Earth*, 110 [Online]. Available at: https://doi.org/10.1029/2004JB003282.

Geshi, N., Shimano, T., Chiba, T., Nakada, S. (2002). Caldera collapse during the 2000 eruption of Miyakejima Volcano, Japan. *Bull. Volcanol.*, 64, 55–68.

Glover, P.W.J. (2015). Geophysical properties of the near surface Earth: Electrical properties. In *Treatise on Geophysics*, 2nd edition, volume 11, Schubert, G.B.T. (ed.). Elsevier, Oxford [Online]. Available at: http://www.sciencedirect.com/science/article/pii/B9780444538024001895.

Guo, X., Chen, Q., Ni, H. (2016). Electrical conductivity of hydrous silicate melts and aqueous fluids: Measurement and applications. *Sci. China Earth Sci.*, 59, 889–900.

Hersir, G.P. and Árnason, K. (2009). Resistivity of rocks. *Short Course IV on Exploration for Geothermal Resources* [Online]. Available at: https://www.grocentre.is/gtp/moya/page/sc-10.

IRIS Instruments (2019). V-FullWaver, IRIS Instruments [Online]. Available at: http://www.iris-instruments.com/v-fullwaver.html.

Jackson, D.B., Kauahikaua, J., Zablocki, C.J. (1985). Resistivity monitoring of an active volcano using the controlled-source electromagnetic technique: Kilauea volcano, Hawaii. *J. Geophys. Res.*, 90, 12545–12555.

Jackson, D.B., Hort, M.K., Hon, K., Kauahikaua, J. (1987). Detection and mapping of active lava tubes using the VLF induction techniques, Kilauea Volcano, Hawaii. *Eos, Trans. Am. Geophys Union*, 68(44), 1543.

Jackson, D.B., Kauahikaua, J., Hon, K., Heliker, C.C. (1988). Rate and variation of magma supply to the active lava lake on the middle east rift zone of Kilauea volcano, Hawaii. *Geol. Soc. Am. (abstracts with program)*, 20, 397.

Kauahikaua, J.P., Mangan, M.T., Heliker, C.C., Mattox, T.N. (1996). A quantitative look at the demise of a basaltic vent; the death of Kupaianaha, Kilauea Volcano, Hawai'i. *Bull. Volcanol.*, 57, 641–648.

Lénat, J.-F. (1995). Geoelectrical methods in volcano monitoring. In *Monitoring Active Volcanoes: Strategies, Procedures and Techniques*, McGuire, B., Murray, J., Kilburn, C.R.J. (eds). UCL Press, London.

Milsom, J. and Eriksen, A. (2011). *Field Geophysics*, 4th edition. Wiley, Hoboken.

Ni, H., Keppler, H., Behrens, H. (2011). Electrical conductivity of hydrous basaltic melts: Implications for partial melting in the upper mantle. *Contrib. to Mineral. Petrol.*, 162, 637–650.

Paterson, N.R. and Ronka, V. (1971). Five years of surveying with the very low frequency – Electro magnetics method. *Geoexploration*, 9, 7–26 [Online]. Available at: http://www.sciencedirect.com/science/article/pii/0016714271900858.

Petiau, G. (2000). Second generation of lead-lead chloride electrodes for geophysical applications. *Pure Appl. Geophys.*, 157, 357–382.

Petiau, G. and Dupis, A. (1980). Noise, temperature coefficient, and long term stability of electrodes for telluric observations. *Geophys. Prospect.*, 28, 792–804.

Pommier, A. and Le-Trong, E. (2011). "SIGMELTS": A web portal for electrical conductivity calculations in geosciences. *Comput. Geosci.*, 37, 1450–1459.

Rai, C.S. and Manghnani, M.H. (1977). Electrical conductivity of basalts to 1550°C. In *Magma Genesis: Bulletin 96*, Dick, H.J.B. (ed.). Oregon Department of Geology and Mineral Industries, Portland.

Strack, K.-M., Lüschen, E., Kötz, A.W. (1990). Long-offset transient electromagnetic (LOTEM) depth soundings applied to crustal studies in the Black Forest and Swabian Alb, Federal Republic of Germany. *Geophysics*, 55, 834–842.

Utada, H. (2003). Interpretation of time changes in the apparent resistivity observed prior to the 1986 eruption of Izu–Oshima volcano, Japan. *J. Volcanol. Geotherm. Res.*, 126, 97–107 [Online]. Available at: http://www.sciencedirect.com/science/article/pii/S0377027303001197.

Utada, H., Takahashi, Y., Morita, Y., Koyama, T., Kagiyama, T. (2007). ACTIVE system for monitoring volcanic activity: A case study of the Izu-Oshima Volcano, Central Japan. *J. Volcanol. Geotherm. Res.*, 164, 217–243 [Online]. Available at: http://www.sciencedirect.com/science/article/pii/S0377027307001394.

Vozoff, K. (1991). The magnetotelluric method. *Electromagnetic Methods in Applied Geophysics*, Volume 2 [Online]. Available at: https://doi.org/10.1190/1.9781560802686.ch8.

Wawrzyniak, P., Zlotnicki, J., Sailhac, P., Marquis, G. (2017). Resistivity variations related to the large March 9, 1998 eruption at La Fournaise volcano inferred by continuous MT monitoring. *J. Volcanol. Geotherm. Res.*, 347, 185–206 [Online]. Available at: http://www.sciencedirect.com/science/article/pii/S0377027317303761.

Yukutake, T., Yoshino, T., Utada, H., Watanabe, H., Hamano, Y., Shimomura, T. (1990). Changes in the electrical resistivity of the central cone, Miharayama, of Oshima Volcano observed by a direct current method. *J. Geomagn. Geoelectr.*, 42, 151–168.

Zablocki, C.J. (1978a). Applications of the VLF induction method for studying some volcanic processes of Kilauea volcano, Hawaii. *J. Volcanol. Geotherm. Res.*, 3, 155–195 [Online]. Available at: http://www.sciencedirect.com/science/article/pii/0377027378900082.

Zablocki, C.J. (1978b). Streaming potentials resulting from the descent of meteoric water. A possible source mechanism for Kilauean self-potential anomalies. *Transactions Geothermal Resources Council*, 2(2), 747–748.

Zlotnicki, J. (2015). Dynamics of volcanic eruptions: Understanding electric signatures for activity monitoring. *Comptes rendus géoscience*, 347(3), 112–123.

Zlotnicki, J., Sasai, Y., Yvetot, P., Nishida, Y., Uyeshima, M., Fauquet, F., Utada, H., Takahashi, Y., Donnadieu, G. (2003). Resistivity and self-potential changes associated with volcanic activity: The July 8, 2000 Miyake-jima eruption (Japan). *Earth Planet. Sci. Lett.*, 205, 139–154 [Online]. Available at: http://www.sciencedirect.com/science/article/pii/S0012821X02010257.

6

Magnetic Monitoring of Volcanoes

Lydie GAILLER and Jean-François LÉNAT
Laboratoire Magmas et Volcans, CNRS, IRD, OPGC, Université Clermont Auvergne, Clermont-Ferrand, France

6.1. Introduction

The Earth's ambient magnetic field is the sum of the fields produced by several sources, both internal and external, stable or variable in time. Its amplitude varies at the Earth's surface between about 30,000 and 60,000 nT (nanoTesla), depending on the location. The most important source is the Earth's core (geodynamo). Its magnetic field varies slowly over time (secular variation, about 20 nT/year). The field created by the rocks of the lithosphere typically reaches a few nT to a few thousand nT. It is constant, except if certain geological phenomena that we will examine here disturb it. Finally, an external component is linked to solar activity. The particle fluxes emitted by the Sun interact with the Earth's magnetosphere and *ionosphere* causing, by various mechanisms, rapid variations (magnetic storms, up to several hundred nT), daily (diurnal variation whose amplitude is of the order of 20–30 nT) or at longer periods (27-day cycles, seasonal, 11 years, etc.). To measure variations in the magnetic field created by volcanic phenomena, generally of very low amplitude (a few nT or even less), it is necessary to be able to know the variations created by other internal or external phenomena. This implies the use of adapted equipment and strategies. If variations of volcanic origin are observed, it will then be necessary to identify the mechanisms responsible and to compare their theoretical effects with observations through models.

Hazards and Monitoring of Volcanic Activity 3,
coordinated by Jean-François LÉNAT. © ISTE Ltd 2022.

6.2. Phenomena generating magnetization and magnetic field variations

We will describe below the physical phenomena that can modify the magnetic field in a volcanic context. However, it is important to realize that several of these phenomena can occur simultaneously during volcanic eruptions (Johnston 2007). For example, a magma intrusion will generate both a thermal demagnetization phenomenon (*thermomagnetism*) and a stress variation (*piezomagnetism*). The heat and gases of the intrusion will disturb the fluid circulation by generating electrokinetic variations. All these phenomena can be associated with the recorded magnetic field variations.

6.2.1. *Piezomagnetism*

It is demonstrated in the laboratory that the magnetic properties (induced and remanent magnetization) of rocks vary according to the constraints imposed on them. For volcanic rocks, the main magnetizing minerals are titanomagnetites. The change in magnetization per unit volume as a function of stress can be expressed as (Johnston 1997):

$$\Delta I \approx K\sigma . I \qquad [6.1]$$

where I and ΔI are the magnetization and its variation due to a deviatoric stress σ and K is the stress sensitivity.

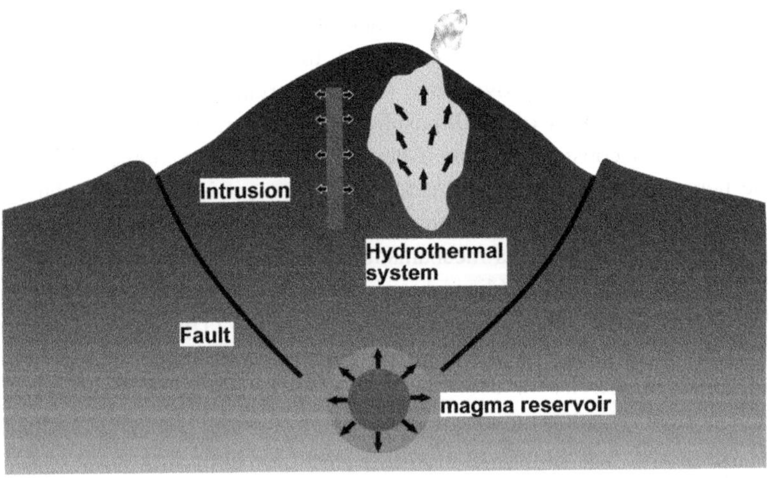

Figure 6.1. *Example of sources of stress variations in volcanoes*

The main sources of stress variation in volcanoes (see Figure 6.1) are related to magmatic activity (reservoirs, intrusions), hydrothermal activity and volcano-tectonic phenomena (fracturing, collapse).

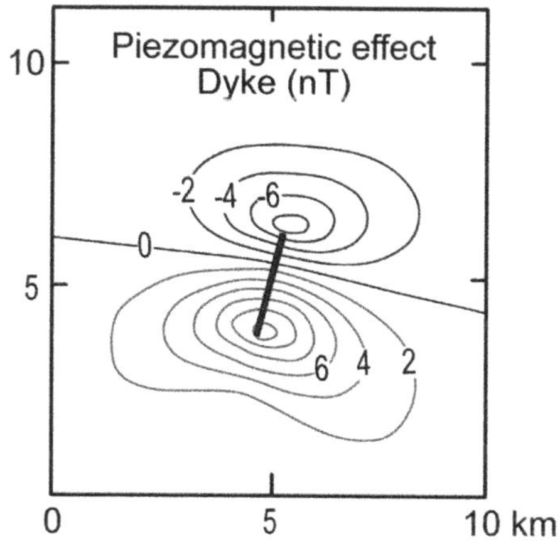

Figure 6.2. *Piezomagnetic effect for a pressure variation associated with the opening of a vertical dyke (opening: 3 m; depth: 1,000 m; height: 3 km). The inclination of the magnetic field and magnetization of the medium is about 50°. See Del Negro et al. (2004) for more details. In the case of a crack closure, the polarity of the anomaly would be reversed*

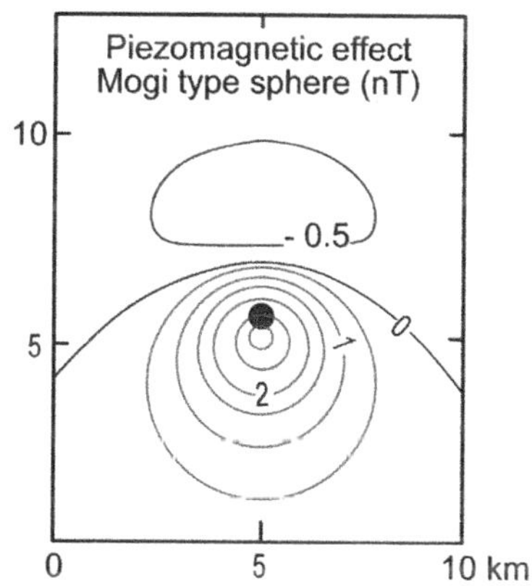

Figure 6.3. *Piezomagnetic effect for a pressure variation of +200 MPa of a sphere of 500 m diameter located at 3 km depth. The tilt of the magnetic field and magnetization of the medium is about 50°. See Currenti et al. (2009) for more details. In the case of a pressure drop of the sphere, the polarity of the anomaly would be reversed*

Several models have been developed to calculate the surface magnetic variations caused by the piezometric effect for different types of sources (sphere, dyke, fault, etc.) (see Figures 6.2 and 6.3) in order to compare them with observations. These models can be more or less sophisticated, ranging from a simple geometry source in a homogeneous magneto-elastic half-space (analytical formulation) to more heterogeneous models for the sources, medium and topography (finite element models) (Stacey 1962; Stacey et al. 1965; Stacey and Johnston 1972; Sasai 1991; Del Negro et al. 2004; Currenti et al. 2008, 2009).

6.2.2. *Thermomagnetism*

This magnetic property, studied by P. Curie during his thesis (Curie 1885), corresponds to the decrease of the magnetization (induced and remanent) of ferromagnetic bodies, until reaching the paramagnetic state (very weak magnetization in the presence of a field, which decreases rapidly with temperature) when the Curie temperature is reached. In the case of rocks, where the magnetization is carried by titanomagnetites poor in titanium, the Curie temperature is of the order of 580°C (see Figure 6.4).

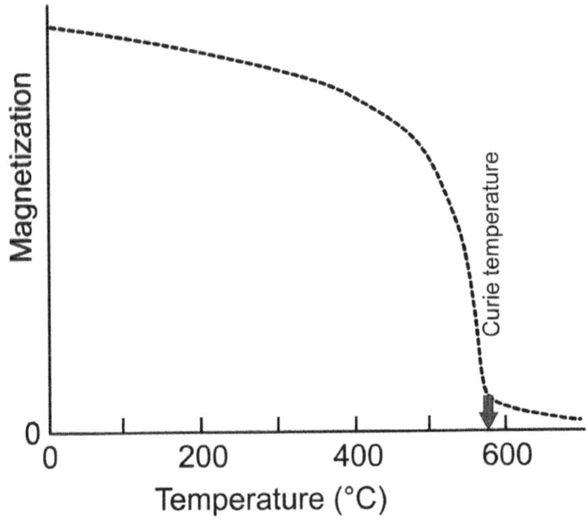

Figure 6.4. *Diagram illustrating the dependence of magnetization on temperature*

The heating, or cooling, of an area inside the volcano, by lowering or increasing its magnetization, will cause a variation of the magnetic field on the surface. This is called thermal *demagnetization* or magnetization of rocks.

6.2.3. *Electrokinetic effects*

During the circulation of fluids (e.g. hydrothermal) within a volcano, electric and magnetic fields are created (see Chapter 5 for an explanation of electrokinetic potentials) and may vary as a function of disturbances in fluid circulation (Mizutani et al. 1976; Fitterman 1979; Zlotnicki and Le Mouël 1990).

6.2.4. *Other mechanisms*

Alteration processes, natural or hydrothermal, can change the magnetization of the ground by transformations of magnetic minerals or by the deposition of new ones. These are slow processes. On the other hand, the processes of construction or destruction of magnetized volumes are faster. It can be the setting up of new lava flows, the construction of cinder cones or collapses (craters, calderas, flank slides).

6.3. Monitoring: instruments and procedures

Measurements can be made with magnetometers that measure the total field intensity (scalar magnetometers) or the components of the field (vector magnetometers). The scalar magnetometers most commonly used in geophysics are proton, Overhauser, cesium or potassium magnetometers. Vector magnetometers are mainly fluxgate. The operating principles of these instruments can be easily found in the literature and on the Internet. The accuracy required is less than 1 nT and is commonly achieved with current instruments.

Two approaches exist for monitoring measures:

– *periodic reiteration of measurements*;

– *quasi-continuous measurements* at fixed stations (see Figure 6.5).

We will see later, through the cases of Kilauea Iki and White Island, that the former is adapted to study slow variations of fairly large amplitude (tens of nT), while the latter allows the recording of rapid variations of small amplitude (a few nT or less). In both cases, retrieving variations of volcanic origin requires considering the difference in values between different stations (see Figure 6.5). This difference eliminates variations such as secular variation or variations of external origin (diurnal, etc.) that are assumed to be uniform for a relatively small area such as a volcano. A station far from the active area is usually used to calculate the differences. This operation is not, however, sufficient to cancel for each station the disturbances of non-volcanic origin, because there may be sources of disturbance based on local contrasts in electrical resistivity or magnetic susceptibility. It is estimated that these local disturbances may be as high as 5% or less of the signal (Del Negro et al. 2004). Various approaches have been developed to correct them (Del Negro et al. 2004).

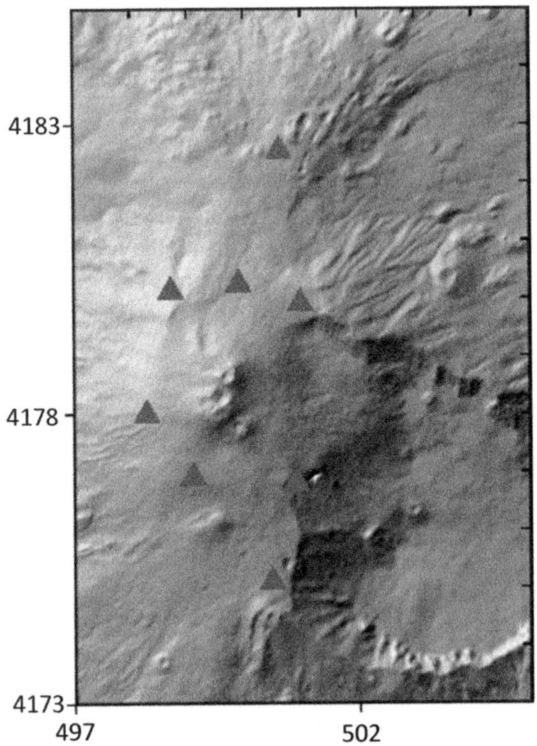

Figure 6.5. *Permanent magnetic monitoring network of Etna. A remote station not visible on this map serves as a reference station (according to INGV Catania). For a color version of this figure, see www.iste.co.uk/lenat/hazards3.zip*

6.4. Case studies

6.4.1. *Periodic monitoring*

6.4.1.1. *Kilauea Iki*

Kilauea Iki is a crater adjacent to the Kilauea caldera (see Figure 6.6). In 1959, an eruption on its rim discharged 38 million cubic meters of lava into this crater, creating a lava lake up to 135 m thick (Richter et al. 1970). This lava lake has progressively solidified over time from its rim, surface and floor. Its thermal and petrographic evolution has been studied by drilling (Helz and Thornber 1987) and by geophysical measurements (compiled by Hermance and Colp (1982)). The last drilling in 1988 showed the subsistence of a partially melted zone at a depth of 73 m over a thickness of about 27 m. The interest of these studies was great because this large volume of lava constituted an analog of an intrusion or a magmatic reservoir accessible to measurements.

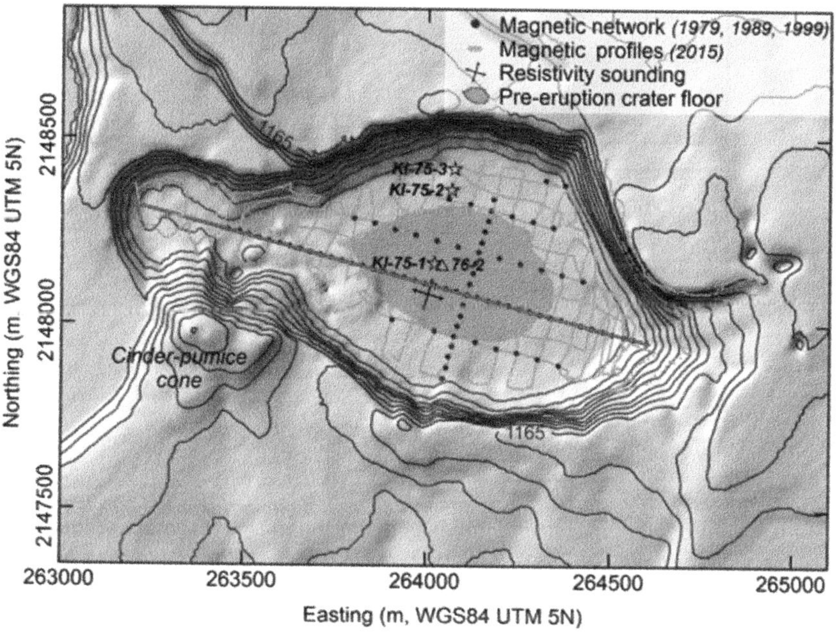

Figure 6.6. *Map of Kilauea Iki and location of magnetic measurements. From Gailler and Kauahikaua (2017). For a color version of this figure, see www.iste.co.uk/lenat/hazards3.zip*

Recently, Gailler and Kauahikaua (2017) published an interpretation of magnetic measurements made on the surface of Kilauea Iki from 1979 to 2015. Magnetic measurements detect the presence of non-magnetic areas, interpreted in this case as areas with temperatures above the Curie temperature. At Kilauea Iki, borehole measurements have determined that the Curie temperature is around 540°C (Zablocki and Tilling 1976). Gailler and Kauahikaua (2017) modeled the magnetic data using the borehole data to establish the structure of the 540°C isotherm over time (see Figures 6.6 and 6.7). Thermal calculations indicate that the lava lake is now at subsolidus temperatures, but the magnetic models demonstrate the presence of a significant lens above the Curie temperature in 2015. This example illustrates the ability of periodic magnetic surveys to detect high-temperature zones and monitor their evolution.

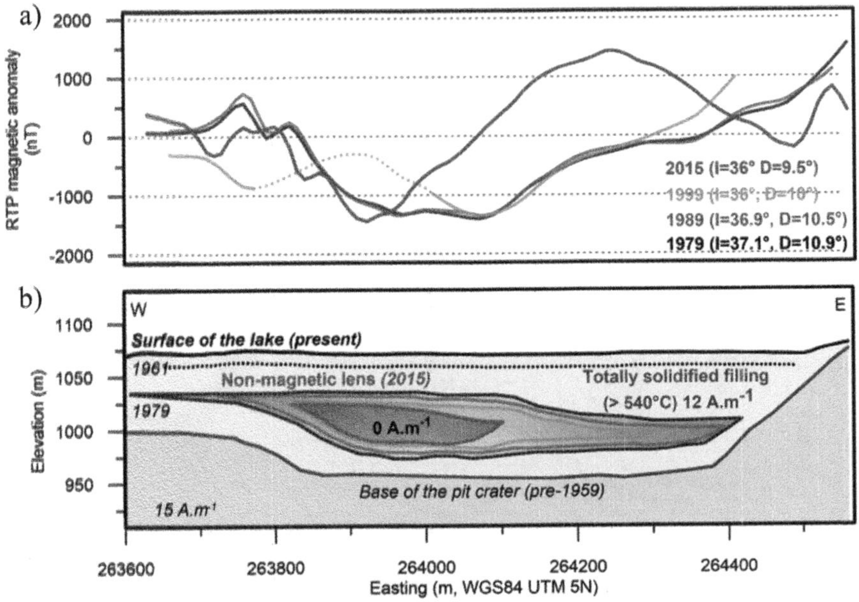

Figure 6.7. *a) Modeled magnetic anomalies along a nearly W-E profile at the surface of Kilauea Iki over time. b) Cross-section showing the evolution of the lens above the Curie temperature for the same epochs. From Gailler and Kauahikaua (2017). For a color version of this figure, see www.iste.co.uk/lenat/hazards3.zip*

6.4.1.2. *White Island*

White Island is a volcanic island about 2–2.5 km in diameter located about 50 km off the north coast of New Zealand. Its crater is the site of intense hydrothermal activity and low to moderate magnitude eruptions (Volcanic Explosivity Index (VEI): 1–3). As early as 1968, a network of magnetic measurements, materialized by benchmarks, was established (Hurst et al. 2004). Measurements have been made two to three times per year since then. In addition, a permanent magnetic network of three stations operated between 1993 and 1996.

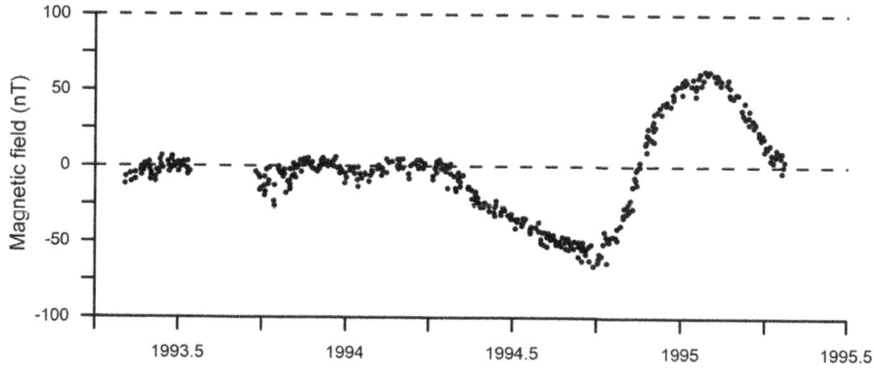

Figure 6.8. *Subcontinuous record of magnetic field variations at a station in White Iceland. From Hurst et al. (2004)*

Variations of several tens to hundreds of nT have been measured. The subcontinuous record for the years 1993–1996 (see Figure 6.8) establishes that the variations in magnetic field intensity are slow and progressive. Figure 6.9 shows the magnetic field variations measured from the periodic measurement network for a station located near a major fumarole area and its temperature. We note an inverse correlation between these two data. This can be easily interpreted in terms of thermal magnetization and demagnetization of the underlying rocks.

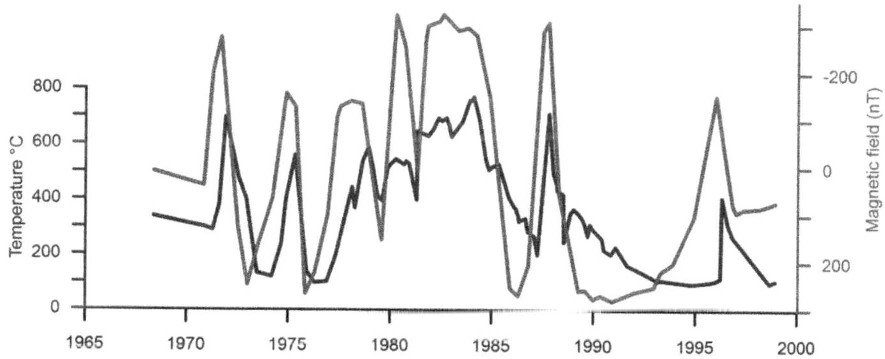

Figure 6.9. *In red: variations of the magnetic field intensity at a station near a fumarole zone (note the inverted scale). In blue, temperature of this fumarole zone. From Hurst et al. (2004). For a color version of this figure, see www.iste.co.uk/lenat/hazards3.zip*

6.4.1.3. *Mount St. Helens*

After the cataclysmic eruption of Mount St. Helens in May 1980, dome extrusion activity occurred episodically at some times and continuously at others. A network of periodic measurements was established in 1984. Analysis of magnetic field strength variations (Dzurisin et al. 1990) showed that the dome complex consisted of a core above the Curie temperature and a cooler, magnetized shell. The magnetic data allowed to model the gradual thickening rate of the shell and to estimate 18–36 years for the complete cooling time of the core.

6.4.2. *Continuous monitoring*

6.4.2.1. *Etna*

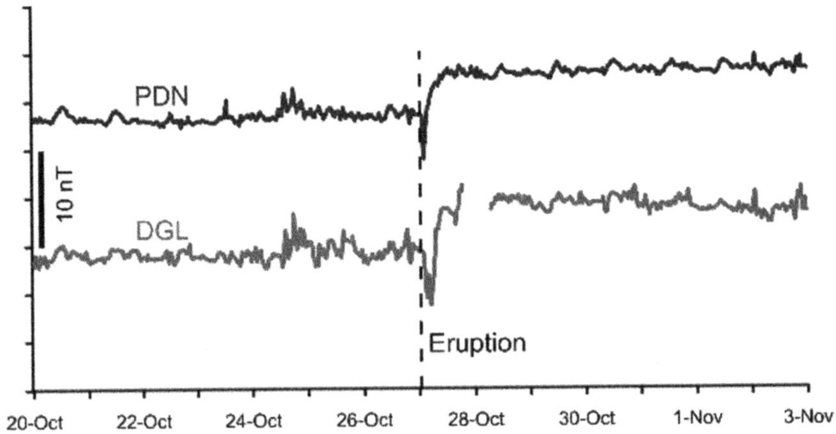

Figure 6.10. *Recording of magnetic field variations by stations north of the summit area of Etna during the October–November 2002 crisis*

The magnetic network of Mt. Etna is well developed (see Figure 6.5). On October 26, 2002, an eruption was preceded by 3 hours of intense seismic activity. Cracks have opened to the north and south of the summit area. To the north, cracks propagated on the NW flank, giving rise, as in the south, to lava flows. Magnetic stations south of the summit were damaged by falling tephra. On the other hand, in the north, the stations recorded signals concomitant with the activity (see Figure 6.10). There was first a decrease of 4–5 nT, then an increase of about 10 nT. Del Negro et al. (2004) interpreted these variations in terms of the piezomagnetic effect associated with the intrusion of a dyke. Initially, during the decline in signals, the dyke developed south of the stations. This phase is modeled by the dyke shown in Figure 6.3. The dyke then migrated to the N-NW of the stations, causing the reversal of the direction of the variation.

6.4.2.2. *Miyakejima*

The Miyakejima volcano (Japan) crisis of 2000 is famous for its spectacular volcanic phenomena: collapse of a small caldera, submarine eruption, formation of several tens of kilometers long lateral intrusion, strong explosive activity and emission of a considerable amount of gas (Nakada et al. 2005). Four main phases can be defined (see Figure 6.11). The first corresponds to an intrusion accompanied by seismic activity and deformation from June 26 to mid-August. This intrusion went offshore and resulted in a small submarine eruption on June 27. Then the subsidence of the summit (1.6 km in diameter and 450 m deep) occurred between July 8 and mid-August. Between August 10 and 29, violent explosions are generated in the collapsed area. Finally, a very strong degassing is observed. It lasted for months, gradually decreasing.

The volcano was monitored by, among other things, a permanent magnetometer network and magnetic variations of several tens of nT were observed (Sasai et al. 2002). Sasai et al. suggest that these variations are related to a set of volcanomagnetic effects:

– The main one is the disappearance of a large volume of magnetized rocks (formation of the caldera).

– Part of the signal may be due to thermal demagnetization of the rocks, as significant thermal anomalies have been recorded at the surface.

– The hydrothermal system having been very disturbed (geometrically by the collapse, thermally by the intrusion and chemically by the volcanic fluids), and magnetic signals of electrokinetic origin may contribute to the overall variations.

– Finally, the stress variations caused by the intrusion and collapse may have generated signals of piezomagnetic origin.

It is thus an exceptional case where all the volcanomagnetic mechanisms may have occurred.

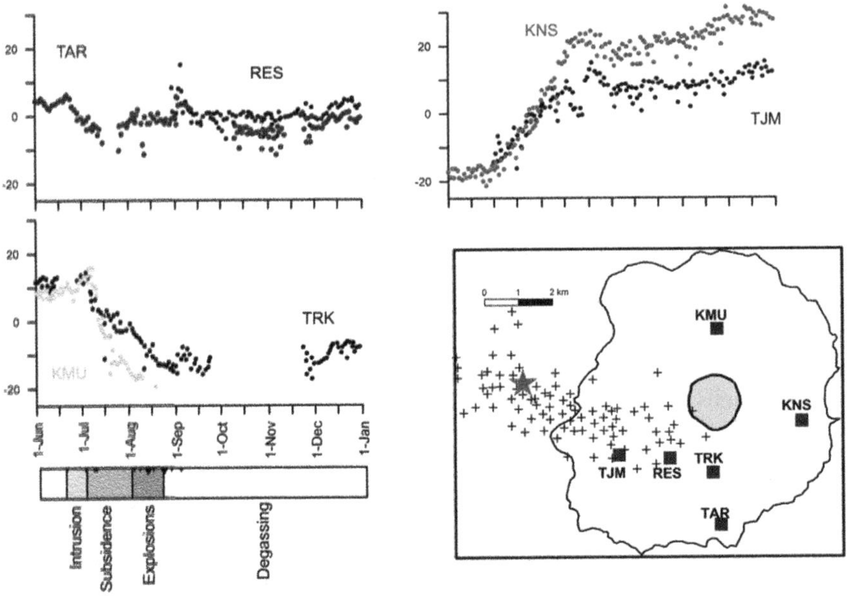

Figure 6.11. *Magnetic variations (in nT) during the 2000 Miyakejima volcano crisis. For a color version of this figure, see www.iste.co.uk/lenat/hazards3.zip*

COMMENT ON FIGURE 6.11.– *The temporal variations are shown on the graphs (after Nakada et al. (2005)). The cartouche at the bottom left shows the four main phases of the crisis. The map at the bottom right shows the location of the stations (squares), the epicenters of the intrusion-related earthquakes (crosses) and the location of the small submarine eruption (star).*

6.5. Perspectives

Magnetic monitoring has not yet acquired the standing of methods such as seismic or deformation monitoring. However, in this chapter, it was shown that magnetic signals could provide important constraints for the interpretation of volcanic crises. Magnetic monitoring is therefore a field called to a great development in the future. Several factors will contribute to this progress. Instrumental progress (sensitivity, miniaturization, association with displacement sensors and remote transmission technics) will benefit both periodic and permanent monitoring. For periodic monitoring, it is clear that the use of drones will facilitate the densification and frequency of surveys. Finally, satellite monitoring of electromagnetic fields is beginning to emerge (Zlotnicki et al. 2010). Of course, all of this must be accompanied by research on the magnetic properties of rocks, and the development of more sophisticated models and processing methods.

6.6. Acknowledgment

Thanks to Claire Bouligand for her meticulous proofreading of this chapter and for her pertinent suggestions.

6.7. References

Curie, M.P. (1885). *Propriétés magnétiques des corps à diverses températures*. Gauthier-Villars et fils, Paris.

Currenti, G., Del Negro, C., Sasai, Y. (2008). Time dependent piezomagnetic fields in viscoelastic medium. *Geophysical Journal International*, 172(2), 536–548.

Currenti, G., Negro, C., Del Di Stefano, A., Napoli, R. (2009). Numerical simulation of stress induced piezomagnetic fields at Etna volcano. *Geophysical Journal International*, 179(3), 1469–1476.

Del Negro, C., Currenti, G., Napoli, R., Vicari, A. (2004). Volcanomagnetic changes accompanying the onset of the 2002–2003 eruption of Mt. Etna (Italy). *Earth and Planetary Science Letters*, 229(1), 1–14.

Dzurisin, D., Denlinger, R.P., Rosenbaum, J.G. (1990). Cooling rate and thermal structure determined from progressive magnetization of the Dacite Dome at Mount St. Helens, Washington. *Journal of Geophysical Research: Solid Earth*, 95(B3), 2763–2780.

Fitterman, D.V. (1979). Theory of electrokinetic-magnetic anomalies in a faulted half-space. *Journal of Geophysical Research: Solid Earth*, 84(B11), 6031–6040.

Gailler, L. and Kauahikaua, J. (2017). Monitoring the cooling of the 1959 Kīlauea Iki lava lake using surface magnetic measurements. *Bulletin of Volcanology*, 79(6), 40.

Helz, R.T. and Thornber, C.R. (1987). Geothermometry of Kilauea Iki lava lake, Hawaii. *Bulletin of Volcanology*, 49(5), 651–668.

Hermance, J.F. and Colp, J.L. (1982). Kilauea Iki lava lake: Geophysical constraints on its present (1980) physical state. *Journal of Volcanology and Geothermal Research*, 13(1), 31–61.

Hurst, A.W., Rickerby, P.C., Scott, B.J., Hashimoto, T. (2004). Magnetic field changes on White Island, New Zealand, and the value of magnetic changes for eruption forecasting. *Journal of Volcanology and Geothermal Research*, 136(1/2), 53–70 [Online]. Available at: http://www.sciencedirect.com/science/article/pii/S0377027304001118 [Accessed May 24, 2022].

Johnston, M.J.S. (1997). Review of electric and magnetic fields accompanying seismic and volcanic activity. *Surveys in Geophysics*, 18(5), 441–476.

Johnston M.J.S. (2007). Volcano-electromagnetic effects. In *Encyclopedia of Geomagnetism and Paleomagnetism*, Gubbins, D. and Herrero-Bervera, E. (eds). Springer, Dordrecht.

Mizutani, H., Ishido, T., Yokokura, T., Ohnishi, S. (1976). Electrokinetic phenomena associated with earthquakes. *Geophysical Research Letters*, 3(7), 365–368.

Nakada, S., Nagai, M., Kaneko, T., Nozawa, A., Suzuki-Kamata, K. (2005). Chronology and products of the 2000 eruption of Miyakejima Volcano, Japan. *Bulletin of Volcanology*, 67(3), 205–218.

Richter, D.H., Eaton, J.P., Murata, K.J., Ault, W.U., Krivoy, H.L. (1970). Chronological narrative of the 1959-60 eruption of Kilauea Volcano, Hawaii. Professional Paper [Online]. Available at: https://doi.org/10.3133/pp537E.

Sasai, Y. (1991). Piezomagnetic field associated with the Mogi model revisited: Analytic solution for finite spherical source revisited: Analytic solution for finite spherical source. *Journal of Geomagnetism and Geoelectricity*, 43, 21–64.

Sasai, Y., Uyeshima, M., Zlotnicki, J., Utada, H., Kagiyama, T., Hashimoto, T., Takahashi, Y. (2002). Magnetic and electric field observations during the 2000 activity of Miyake-jima volcano, Central Japan [Online]. Available at: https://doi.org/10.1016/S0012-821X(02)00857-9.

Stacey, F.D. (1962). Theory of the magnetic susceptibility of stressed rock. *The Philosophical Magazine: A Journal of Theoretical Experimental and Applied Physics*, 7(76), 551–556.

Stacey, F.D. and Johnston, M.J.S. (1972). Theory of the piezomagnetic effect in titanomagnetite bearing rocks. *Pure and Applied Geophysics*, 97, 146–155.

Stacey, F.D., Barr, K.G., Robson, G.R. (1965). The volcanomagnetic effect. *Pure and Applied Geophysics*, 62, 96–104.

Zablocki, C.J. and Tilling, R.I. (1976). Field measurements of apparent Curie temperatures in a cooling basaltic lava lake, Kilauea Iki, Hawaii. *Geophys. Res. Lett.*, 3, 487–490 [Online]. Available at: https://doi.org/10.1029/GL003i008p00487.

Zlotnicki, J. and Le Mouël, J.L. (1990). Possible electrokinetic origin of large magnetic variations at La Fournaise volcano. *Nature*, 343, 633.

Zlotnicki, J., Li, F., Parrot, M. (2010). Signals recorded by DEMETER satellite over active volcanoes during the period 2004 August–2007 December. *Geophysical Journal International*, 183(3), 1332–1347.

List of Authors

Carole BERTHOD
Laboratoire Magmas et Volcans
CNRS, IRD, OPGC
Université Clermont Auvergne
Clermont-Ferrand
France

Sylvain CHARBONNIER
School of Geosciences
University of South Florida
Tampa
United States

Franck DONNADIEU
Laboratoire Magmas et Volcans
CNRS, IRD, OPGC
Université Clermont Auvergne
Clermont-Ferrand
France

Lydie GAILLER
Laboratoire Magmas et Volcans
CNRS, IRD, OPGC
Université Clermont Auvergne
Clermont-Ferrand
France

Lucia GURIOLI
Laboratoire Magmas et Volcans
CNRS, IRD, OPGC
Université Clermont Auvergne
Clermont-Ferrand
France

Claude JAUPART
Institut de physique du globe de Paris
Université de Paris
Académie des sciences
Paris
France

Philippe LABAZUY
Laboratoire Magmas et Volcans
CNRS, IRD, OPGC
Université Clermont Auvergne
Clermont-Ferrand
France

Jean-François LÉNAT
Laboratoire Magmas et Volcans
CNRS, IRD, OPGC
Université Clermont Auvergne
Clermont-Ferrand
France

Étienne MÉDARD
Laboratoire Magmas et Volcans
CNRS, IRD, OPGC
Université Clermont Auvergne
Clermont-Ferrand
France

Séverine MOUNE
Laboratoire Magmas et Volcans
CNRS, IRD, OPGC
Université Clermont Auvergne
Clermont-Ferrand
and
OVSG
Gourbeyre
Guadeloupe
and
Institut de physique du globe de Paris
Université de Paris
France

Émilie ROULLEAU
Laboratoire GéoSciences Réunion
Laboratoire Magmas et Volcans
CNRS, IRD, OPGC
Université Clermont Auvergne
Clermont-Ferrand
and
Université de La Réunion
Stratagem
Saint-Denis
France

Alessandro TADINI
Laboratoire Magmas et Volcans
CNRS, IRD, OPGC
Université Clermont Auvergne
Clermont-Ferrand
France

Simon THIVET
Laboratoire Magmas et Volcans
CNRS, IRD, OPGC
Université Clermont Auvergne
Clermont-Ferrand
France
and
Department of Earth and Environmental Sciences
Ludwig-Maximilians-Universitat
Munich
Germany

Jean-Claude THOURET
Laboratoire Magmas et Volcans
CNRS, IRD, OPGC
Université Clermont Auvergne
Clermont-Ferrand
France

Ivan VLASTELIC
Laboratoire Magmas et Volcans
CNRS, IRD, OPGC
Université Clermont Auvergne
Clermont-Ferrand
France

Index

R, S

T, V, W

Printed and bound by CPI Group (UK) Ltd, Croydon, CR0 4YY
13/09/2026
14960996-0003